THE BAMBOOS

Tree-grass—unique in the plant kingdom

Symbol of uprightness, chivalry and devotion

Inspiration of poets, artists and philosophers

Writing material of the ancients;

the stylus of contemporaries

Food, shelter and clothing of the people

Industrial substance of a thousand uses

Redeemer of waste places—protector of the soil

THE BAMBOOS

F. A. McClure

Smithsonian Institution Press
Washington and London

Reprinted by arrangement with Harvard University Press

Library of Congress Cataloging-in-Publication Data

McClure, Floyd Alonzo, 1897–1970.
The bamboos / F.A McClure.
p. cm.
Originally published: Boston : Harvard University, 1966.
Includes bibliographical references and index.
ISBN 1-56098-323-X
1. Bamboo. I. Title.
SB317.B2M29 1993
584′.93—dc20 93-32579

Printed in the United States of America
00 99 98 5 4 3

Book design by David Ford

This book was originally prepared under the auspices of
the Maria Moors Cabot Foundation.

My new Province is a land of bamboo groves;
Their shoots in spring fill the valleys and hills.
The mountain woodman cuts an armful of them
And brings them down to sell at the early market.
Things are cheap in proportion as they are common;
For two farthings I buy a whole bundle.
I put the shoots in a great earthen pot,
And steam them over the boiling rice.
Their purple nodules broken, they suggest an old brocade;
Their white skin gleams like new pearls.

—PO CHU-I, TRANSLATED BY WALEY (1929)

Contents

Foreword to the Paperback Edition

During his lifetime, McClure probably saw and studied more species of living bamboo than anyone before or since. He drew inspiration from bamboo in all its guises and manifold uses. He not only laid much of the groundwork for subsequent bamboo classification but also did pioneering work on morphology, propagation, and cultivation.

In recent years there has been a global surge of botanical and horticultural interest in bamboo. Groups of people within many nations have collaborated to explore and promote understanding and utilization of bamboo, and bamboo societies have proliferated worldwide. The American Bamboo Society, founded in 1979, supports research, conservation, the horticultural indroduction of new species, and provides information about bamboo through a newsletter and journal. There are also active bamboo societies in Europe, Japan, Asia, and Australia. Several international conferences have focused on bamboo and more are planned. Much of this interest has been inspired by the unique beauty, productivity, and usefulness of bamboo. As we become more aware of environmental degradation and depletion of resources, interest in bamboo will only increase.

A quarter century after publication, Floyd A. McClure's book, *The Bamboos*, remains a highly useful text and reference. The book is at once detailed and comprehensive, theoretical and practical. The contemporary student of bamboos will surely enjoy and benefit from McClure's life-long study and insight. This edition is a testament to his vision and influence.

GERALD BOL
President, American Bamboo Society

Introduction to the Paperback Edition

In the twenty-seven years since the original publication of F. A. McClure's landmark treatise, *The Bamboos*, many valuable contributions to our knowledge of these unusual and versatile grasses have been made by researchers and enthusiasts around the world, but this book endures as an excellent, comprehensive account of the woody bamboos. What makes this book a classic for anyone interested in these plants is that it represents the accumulated knowledge of a dedicated scientist and scholar who worked with bamboos for nearly four decades in both Asia and America. *The Bamboos* rests upon three underlying themes: one, the importance of understanding the biology of organisms; two, continuing evaluation of facts, methods, and perspectives; and three, integration of the basic and applied facets of scientific research. McClure's legacy, distilled in this book, consists of his insights into bamboos gleaned from many years of firsthand experience, his scientific integrity, and a humanistic perspective.

For those who are already bamboo enthusiasts, the availability of this popular book will be welcome. Those who are just beginning their odyssey into the world of bamboo will find a wealth of information on the structure of the bamboo plant, the propagation of bamboos, and a discussion of the most economically important species. Whatever one's level of knowledge, it is helpful to understand the changes in bamboo botany that have occurred since the publication of this book. Interest in all aspects of bamboo, including systematics (or taxonomy), ecology, economic uses and development, propagation, conservation, and cultural significance, has burgeoned in recent years. A synthesis of all the bamboo literature since 1966 would probably require another book, but I will attempt to summarize the subsequent major events and changes in the world of bamboo.

Perhaps the most startling development in bamboo taxonomy over the last quarter century or so is the change in our concept of what we consider to be a bamboo. McClure, Keng Yi-Li, Keng Pai-chieh, R. E. Holttum, and others focused entirely upon the "tree grasses" or woody bamboos, which with close to 1,000 species constitute the majority of bamboo diversity. But more recent studies by T. R. Soderstrom, C. E. Calderón, and others have shown that certain other groups of grasses, known as the herbaceous bamboos, belong to the same major evolutionary branch as the woody bamboos. According to Soderstrom (1985), before his death in 1970 McClure had come to agree that the herbaceous bamboos were in fact closely related to woody bamboos. Currently, studies employing molecular techniques are under way to examine in more detail the relationships among these bamboo groups. Are woody bamboos derived from herbaceous ones, or are herbaceous bamboos derived from woody ones? Are there any temperate herbaceous bamboos? Is rice a bamboo? These are all exciting questions that we hope to be able to answer in the near future.

Many new species of woody bamboos have been described, and along with the publication of several new genera, concepts of previously recognized genera such as *Arundinaria, Bambusa,* and *Arthrostylidium* have been modified. Controversy exists in China regarding the delimitation of *Arundinaria* and related genera, and this problem is currently being studied. In a posthumously published work, McClure (1973) resolved a number of taxonomic problems among the American woody bamboos by transferring many species from *Arundinaria* and *Arthrostylidium* to more recently described genera. New genera have been recognized, and *Swallenochloa* is no longer regarded as a genus separate from *Chusquea.* On the subject of economically important bamboos, it is worth noting that *Bambusa* and *Guadua* are generally regarded as distinct, and possibly not very closely related, genera. A monograph of *Guadua* is currently in preparation.

Many species names have changed as a result of these taxonomic decisions, but the names of the most commonly cultivated bamboos have remained relatively stable. One exception is the thorny bamboo of India, *Bambusa arundinecea* Retz., which is now correctly known as *Bambusa bambos* (L.) Voss. The most economically important American bamboo, *Guadua angustifolia* Kunth, retains this name, although when previously recognized as a member of *Bambusa,* its name was *Bambusa guadua* Humboldt & Bonpland.

The ecological importance of woody bamboos has been increasingly recognized over the last two decades. In contrast to other grasses, bamboos are typically associated with woody vegetation, and are common, even characteristic elements in tropical, subtropical, and temperate forests in many parts of the world. Because of their incredible capacity for rhizomatous growth, bamboos are often able to invade openings produced by cutting of the forests or other disturbances and can become extremely weedy, forming dense stands over large areas. Although much more research is needed, some interesting ecological studies on productivity, life history, and vegetation dynamics involving bamboos are available. The nature of interactions between bamboos and animals has been highlighted by the crisis in the panda populations in China caused in part by mass flowering and death of the bamboo species used by the pandas as their primary food source. Reports of other analogous interactions, involving rats, birds, frogs, monkeys, and beetles that spend at least part of their life cycles in close association with bamboo, are fascinating and merit further study.

Scientific technology in general has developed at a dizzying pace over the last three decades, giving us not only new instruments and methods for collecting information but also providing powerful computers with which to analyze data. This has had a direct effect on the development of practical techniques for bamboo propagation and crop improvement, as well as enhancement of research in bamboo systematics. Sophisticated techniques using tissue culture have allowed the propagation of a number of bamboo species from cells of vegetative buds or embryos when available. In turn, tissue culture has opened up possibilities for hormonal manipulation of bamboos to study their unusual flowering behavior, which may eventually permit hybridization and improvement of economically important bamboo species. Study of microscopic features of bamboo leaves and other plant parts using powerful electron microscopes now routinely complements light microscopy. One of the most remarkable advances in just the last few years is the ability to directly compare gene sequences among organisms in order to understand their evolutionary relationships. This ongoing research will provide the framework for further refinement of generic concepts and overall classification. Molecular biology may also help unlock some of the secrets of the cyclical, gregarious flowering for which the woody bamboos are so famous.

The past twenty-seven years have seen an increased interest in the use and development of bamboo and the conservation of bamboo resources, as well as a better understanding of the ecology and systematics of bamboos. A network of national and international bamboo societies has formed and plays a critical role in the exchange of information and plant material and in the conservation of bamboo species. Dr. McClure's devotion to bamboo, evident on every page of *The Bamboos*, unquestionably has inspired much of this activity, and I suspect that he would wholeheartedly applaud these developments.

What does the future hold for bamboo? Given the versatility and availability of bamboo in many developing nations, it will likely attain even greater importance as a renewable resource. Projects involving the utilization of bamboo are already under way in Costa Rica, Colombia, Ecuador, China, and India, among other countries. More species are being brought into cultivation, and the potential for their conservation is improving. Research into the ecology and systematics of bamboo will continue; the role of bamboo in forest and other ecosystems should be explored, new species remain to be discovered and described, and the evolutionary history of this unusual group of plants holds great intrinsic interest. The collaboration among bamboo researchers and enthusiasts on a global basis bodes well for the bamboos, as we come to understand more about these fascinating plants.

LYNN CLARK
Iowa State University

Literature Cited

McClure, F. A. 1973. Genera of bamboos native to the New World (Gramineae: Bambusoideae). Smithsonian Contr. Bot. 9: 1–148. [Ed. T. R. Soderstrom.]

Soderstrom, T. R. 1985. Bamboo systematics: Yesterday, today and tomorrow. J. Amer. Bamboo Soc. 6(1–4): 4–16.

Foreword

Born on a farm in Ohio, Floyd Alonzo McClure attended the Ohio State University at Columbus, with the intention of returning to the farm. But upon his graduation, in 1919, an unexpected offer of an instructorship in horticulture at Lingnan University in Canton, China, presented an irresistible opportunity for an adventurous young man interested in seeing the world and its peoples and plants. He stayed on at Lingnan, was made curator of the herbarium, and in 1931 became professor of botany there. In the meantime he had returned twice on furlough to the United States and had also led numerous field trips in the Chinese interior and in Indochina, in part as explorer for the United States Department of Agriculture. The effectiveness of his collecting was of course greatly helped by the fact that from his first years at Lingnan he had studied Cantonese and acquired a fluent command of it. It was on these field trips that McClure became interested in bamboos; he very early initiated a Bamboo Garden at Lingnan and established there many of the plants he had collected. This constant preoccupation with living plants, rather than preserved material, is characteristic of his interests, and it is that which has made the present book possible.

In 1932, McClure returned to the United States and Europe for a four-year period. He studied critical specimens in the bamboo collections at the United States National Herbarium in Washington, at the Royal Botanic Gardens, Kew, and at the British Museum; at Leiden he worked on other material. He obtained his doctorate at Columbus in 1935, and returned to China in 1936. At Lingnan he stimulated his colleagues in chemistry and engineering to make joint investigations of bamboo products, and he established a large experimental planting where 10,000 plants of bamboo could be studied both for the useful properties of the plants themselves and for their effects on the soil. He also studied and collected in Kwangsi and in the Hong Kong area. But in 1941 the war drove him back to the United States, and he has made the National Herbarium his headquarters ever since. From there he has visited almost every country in Latin America to study and collect bamboos, to direct programs of investigation, distribution, and propagation of bamboos there, work on bamboo propagation in Puerto Rico, and establish field plantings in Guatemala and five other Latin American countries. He also visited United States missions in the Far East, and introduced selected bamboos into various Pacific islands for trial. As consultant on bamboos to a United States paper manufacturer McClure has made field studies in Central America, Florida, and Texas and has collaborated in research on the pulping properties of selected species. In this he has been helped by his earlier studies of bamboo papers in China.

It was with a view to making McClure's lifetime study and experience with bamboos available to a wider circle that in 1955 the Maria Moors Cabot Foundation, through Professor Paul Mangelsdorf, invited him to try to set down at least a part of his accumulated knowledge. Surely no one can claim more extensive acquaintance at first hand with every aspect of this fascinating group of plants. We have been fortunate, too, in having the services of Elmer Smith, who has illustrated many botanical and zoological publications, and has given expression both to his botanical knowledge and his artistic feeling in preparing the very numerous illustrations. The book has been many years in preparation, and it contains not only a mass of information but much valuable insight. I have acted informally as editor and critic.

KENNETH V. THIMANN
Chairman, Executive Committee
Maria Moors Cabot Foundation

Preface

The presentation of this treatise in book form is the outcome of my response to an opportunity offered by the Maria Moors Cabot Foundation of Harvard University. The Foundation's Administrative Committee desired a summary of authentic information on selected aspects of the bamboos, to place this group of plants in perspective in relation to the planning of its research activities. I am deeply indebted to Dr. Paul C. Mangelsdorf, Chairman of the Committee, for the invitation to initiate this undertaking, and to his successor in this office, Dr. Kenneth V. Thimann, for sustained and sustaining interest and support.

My involvement in scientific studies of the bamboos dates from a period of service (1924 to 1927) under the United States Department of Agriculture, as Agricultural Explorer in China. I was instructed to secure living material of useful bamboos for trial in the Western Hemisphere. With facilities made available by the Lingnan University at Canton, China, budgetary aid from the Rockefeller Foundation through its "China Foundation," and personal subsidy generously provided by the late Helen Lyon Jones, it was possible eventually to assemble, principally from eight provinces of southeastern China, over 600 accessions of living bamboo plants for observation and experimental study. (Through the interest of Dr. R. E. Holttum, who recently visited Canton, it is now known here that this collection is still intact, and that it remains under the care of the erstwhile colleague who collaborated in establishing it.) The United States Department of Agriculture has accessioned 250 numbers of living plants of species selected from this collection during the 17 years from 1924 to 1940. In the pursuit of studies undertaken in that period I became greatly obligated to my Chinese colleagues for assistance of a most distinguished and effective quality.

Throughout the years 1933 to 1936 and 1941 to date, the Department of Botany of the Smithsonian Institution has extended beneficent hospitality to my herbarium studies on bamboo. The John Simon Guggenheim Foundation supported my taxonomic work on selected groups during the fiscal year 1942–43. During the fiscal year 1943–44, under the auspices of the Smithsonian Institution, I made a survey of useful bamboos of a special category available in the United States, Mexico, Honduras, Columbia, Venezuela, Brazil, and Puerto Rico for the Office of Scientific Research and Development of the U.S. National Research Council (McClure 1944). The post of Field Service Consultant on Bamboo with the Office of Foreign Agricultural Service of the U.S. Department of Agriculture (1944 to 1954) gave me the opportunity to study and collect bamboos in six countries of Central and South America, as well as in India, East Pakistan, and the islands of Java and Luzon. Ultimately I was able to establish living collections of elite economic species in Guatemala, El Salvador, Nicaragua, Costa Rica, Ecuador, and Peru. During this period it was my good fortune to have the use of research facilities, and the collaboration of members of the staff, at the Federal Experiment Station at Mayagüez, Puerto Rico, in carrying out exploratory studies designed to improve methods of bamboo propagation. At the suggestion of Donald D. Stevenson, the U.S. Department of Agriculture authorized me to serve, during this period, as its representative in the capacity of consultant on bamboo to Champion Papers Inc. This relation afforded an opportunity to make field studies in Jamaica and Trinidad, to design and supervise the establishment of an experimental bamboo plantation in Guatemala, and to participate in elaborate studies based on it. Since the termination, in 1956, of my latest period of service with the U.S. Department of Agriculture, financial support made available by the UNESCO Committee for Biological Studies in the Humid Tropics, the U.S. National Science Foundation, and the Maria Moors Cabot Foundation of Harvard University, respectively, has sustained my bamboo studies.

This account would not be complete without an acknowledgment of my indebtedness to innumerable persons of good will, in all walks of life, both here and in those parts of the world where my field studies have taken me. These encounters have left a lasting sense of gratitude for generous hospitality and for much practical help. I feel an obligation of a special nature to David Fairchild, Wilson Popenoe,

Robert Cunningham Miller, Conrad Chapman, Robert A. Young, Ross E. Moore, and Malcolm G. Lyon, each of whom has shared and actively supported a concern for the continuity, progress, and fruition of my work. The loyal encouragement of my efforts by Professor Adolph E. Waller, a favorite lecturer in Botany at the Ohio State University, has exerted sustaining influence of a rare quality.

By her tireless and efficient service in bringing up to date the assembling and processing of published papers on bamboo, and making many translations into English from a polyglot array of literature, beside carrying general secretarial responsibilities, Maude Kellerman Swingle lightened my burden and quickened the pace of my labors during the formative state of this book. Mamie Idella Herb garnered many references from the libraries of the Smithsonian Institution and the U.S. Department of Agriculture, helped process the literature, and transcribed with patient industry and gratifying precision, the numerous earlier drafts through which the manuscript has passed. Ruth Drury McClure has given constant support, enhancing the joys and devotedly sharing the trials of my pursuit of the bamboos through the years. With her help, the manuscript was read aloud and critically considered during each revision and the final touches were incorporated in the definitive draft.

William C. Kennard, erstwhile collaborator in the processing of data on bamboo propagation, has read with a critical eye that part of the manuscript which deals with this subject. For a painstaking and competent scrutiny of the whole original draft, and for valuable suggestions, I am grateful to R. E. Holttum, formerly Director of the Botanic Gardens at Singapore and author of a recently published taxonomic treatise on the bamboos of the Malay Peninsula. Among other reliable sources of information that have enabled me to extend beyond my personal experience the range of subjects covered in this book are several distinguished studies published by Willard M. Porterfield.

Permission has been graciously extended by the Philosophical Library, New York; Alfred Knopf, Inc., New York; The University Press, Cambridge; The Clarendon Press, Oxford; The Oxford University Press, London; Hiroshi Muroi; Yokendo, Ltd., Tokyo; the Ministry of Education, Government of India, and Encyclopedia Britannica, Inc., to reproduce textual material or illustrations, or both, from the copyrighted writings of Alfred North Whitehead,

Arthur Waley, Agnes Arber, C. R. Metcalfe, R. S. Troup, N. I. Vavilov, H. Muroi, Yoshio Takenouchi, P. N. Deogun, and myself, respectively. These courtesies are acknowledged individually at appropriate points in the book. The frontispiece, derived from a rubbing of a stone carving based on a painting by Hui Nien (known to connoisseurs of Oriental art for his skill in painting bamboos), was first used in this greatly reduced form by C. LeRoy Baldridge in the cover design of his book, *Turn to the East,* privately published and copyright by him in 1925.

Excepting those credited to other sources, the photographs and the original sketches for the line drawings are my own. With his well-known artistic and technical skill, Elmer W. Smith has tidied my pencil sketches, rendered them in ink, and reconditioned illustrations adopted from published sources. Several of the original drawings were made by him specially for this book.

As a last word, I wish to thank the staff of the Harvard University Press, and Joseph D. Elder, the Science Editor, in particular, for distinguished contributions to the degree of excellence that has been achieved in the production of this volume.

F. A. M.

Smithsonian Institution
Washington, D.C.
November 28, 1964

Introduction

The bamboos have age-old connections with fishing, papermaking, landscape gardening, handicrafts, medicine, art, and even poetry. Having long been creatively exploited at the handicraft level, various kinds of bamboo minister in a comprehensive manner to the material needs of native peoples of both the old World and the New. In some of the heavily populated regions of the tropics certain bamboos supply the one suitable material that is sufficiently cheap and plentiful to meet the vast need for economical housing (McClure 1953). Beyond this, bamboo culms provide the raw material for hundreds of implements of daily necessity in the home and in the pursuit of a livelihood, as well as musical instruments and countless sorts of toys for children. The young shoots of certain species are eaten as a vegetable; the living plants of others are cultivated in decorative or protective hedges about homes and villages. Oriental scholars, artists, and epicures have celebrated the importance of bamboo and praised its admirable qualities in paintings, rhymed couplets, and lyrical tributes (Po Chu-i, translated by Waley 1929; Li 1942; McClure 1937, 1956*b*, 1961*a*).

The suitability of bamboo tools and bamboo fibers for making paper was demonstrated centuries ago by Chinese artisans (McClure 1928; Sung 1929). As a result of recent fiber-dimension studies and improvements in mill techniques, fine papers of many varieties and adaptations can now be made from the pulp of certain bamboos. High-grade bamboo pulps can be used in the pure state for coated or uncoated book and magazine papers. The pulp of certain bamboos excels in the field of soft facial tissues and for thin, India-type papers where opacity is at a premium. Fiber dimensions with a high length-to-diameter ratio give many bamboo pulps a special versatility.

As a result of modern technological advances and applied

research, new uses for bamboo products have been developed and new significance found in old uses. Tabashir, precipitated within the culm internodes of many tropical bamboos, consists largely of amorphous silica in microscopically fine state. It has been shown to have, both in the purified form and in combination with several elements and compounds, excellent properties as a catalyst for certain chemical reactions (Netherlands Patent No. 53,471). The most successful of Edison's early electric lamps had for its light-giving element a carbonized filament of bamboo. (McClure 1948). Bamboo fibers were still used in special-purpose carbon-filament lamps as recently as the year 1908 (Bryan 1926). Japanese scientists have demonstrated that charcoal prepared from certain bamboos has properties that render it superior to that derived from conventional sources for use in electric batteries (Miyake and Sugiura 1950, 1951), and in India bamboo charcoal is used for pharmaceutical purposes. Oriental jewelers use the silica-charged charcoal from bamboos in preference to that from other sources. From the white powder abundantly produced on the outer surface of young culms of a Chinese bamboo, Chang (1938) isolated, among other substances, a crystalline compound related to the female sex hormone. Piatti (1947) reported the successful preparation of liquid diesel fuel from bamboo culms by distillation. In Japan, Kato (1911) extracted from bamboo shoots the enzymes nuclease and deaminase, besides an unnamed enzyme that dissolves fibrin, and another of emulsin type, capable of hydrolyzing salicin. Komatsu and Sasaoka (1927) isolated glucuronic acid and L-xylose, in a crystalline state, from the juice of bamboo shoots. Yoshida and Ikejiri (1950) found an aqueous extract of bamboo shoots superior to conventional media for the culture of certain pathogenic bacteria (*Shigella* and *Brucella*). This extract, when added to tuberculin, increased the intensity of responses to the skin test.

The use of dried mature bamboo leaves for deodorizing fish oils has been patented in Japan (Japanese Patent No. 175,685). In both the Eastern and Western Hemispheres, the foliage of many bamboos has long been used as a major or a supplementary source of forage (McClure 1958). The improvement of the management of grazing range in the coastal plain of North Carolina, where a native bamboo is the principal source of browse for beef cattle, is the object of a project in which three state agencies and three agencies of the U.S. Department of Agriculture have collaborated for nearly two decades with very fruitful results (Shepherd 1952). However, young shoots of many tropical species of bamboo contain lethal concentrations and amounts of cyanogens (Valt'er *et al.* 1910, 1911). The digestive processes of the herbivora destroy the poison, but in India cattle sometimes die when they are allowed to eat too freely of the toxic young shoots. Boiling readily drives off the volatile cyanide, so there is no risk involved in eating cooked shoots.

Bamboos are set off from other grasses by the predominance of certain "bambusoid" structural characters, many of which are considered to be "primitive." The most easily recognizable vegetative features that distinguish the bamboos are the prominent development of a rhizome system, the woodiness and strong branching of the culms, the presence of petioles on the leaf blades, and the difference in form between the sheaths clothing young culm shoots and those borne on the leafy twigs. To these may be added such floral characters as well-developed lodicules, in most species three in number, and a style consisting typically of a single column, bearing one, two, or three (rarely more) stigmas (see Figs. 48 and 53). No single character is diagnostic, however, and the boundary between the bamboos and the other grasses is not sharp (see Parodi 1961).

Of very uneven geographical distribution, bamboos appear more or less prominently in the natural vegetation of many parts of the tropical, subtropical, and mild-temperate regions of the world, from sea level to the snow line. Bamboos are found in the greatest abundance and variety on the southern and southeastern borders of Asia, from India through China and Japan to Korea. In the vast Eurasian continent north, west, and northwest of Tibet and China, however, no native bamboo has been found. The island of Madagascar is particularly rich in endemic genera and species, having more known kinds than all of Africa. Three endemic species have been described from Australia. In the Western Hemisphere, the known natural distribution of bamboos extends from 39°25′ N in eastern United States (personal observation) to 45°23′30″ S in Chile (Dusén, 1903–1906:18ff) and even to 47° S in Argentina (Parodi 1945:64).

The natural distribution of bamboo in the world has been greatly modified by human intervention. For example, many natural stands have been more or less completely destroyed in clearing land for agriculture. The canebrakes of the United States have been greatly reduced in number and extent. *Guadua aculeata,* a giant species once abundant locally in several countries of Central America, has been almost completely eliminated in some areas. On the other hand, *Bambusa vulgaris,* a pantropic species of unknown origin, has been naturalized in large areas on the island of Jamaica in the wake of a sort of migratory agriculture in which stakes freshly cut from living culms of this bamboo, and stuck in the ground to support yam vines, take root and produce thriving groves. *Sinobambusa tootsik,* a Chinese bamboo, once highly prized as a garden ornamental in Honolulu, has shown that it can become a troublesome weed there, for it escaped from cultivation and now dominates many acres of once pure native vegetation. In Africa, Europe, the British Isles, and the United States, introduced bamboos have found an important place as ornamentals and as a source of fishing poles, garden stakes, and fencing, as well as material for handicrafts and interior trim. Many little-known exotic species of promise are still confined to introduction gardens, where they await basic studies and the prolonged critical research necessary to test their potential usefulness in the local economy.

Relatively little of a basic nature is known about any of the bamboos. Among the disciplines that have neglected this group more or less completely are those dealing with anatomy, cytology, genetics, plant breeding, biochemistry, morphology, morphogenesis, ecology, and the physiological phenomena related to vegetative growth, flowering, and fruiting. Even the commonly practiced methods of propagation are still restricted to the traditional procedures of the plantsman's art, and current applications of the newer knowledge of hormones and the techniques of controlling vegetative growth and sexual reproduction have not touched the problems involved in the practical management and exploitation of plants of this group. Furthermore, the taxonomic treatment of the bamboos, which is basic to all other angles of study, is still in a very retarded state. An amplified and modernized description of this important tribe of the grasses is presented on pp. 282ff. Revised descriptions of the tenable genera are being prepared, with special emphasis on the redefinition of their respective boundaries, and the systematic use of fundamental characters hitherto generally neglected.

One objective of this treatise is to make a start at removing the bamboos from the shadow of insufficient knowledge, and to bring them into fresh perspective in the light of personal experience and of published observations judged to be authentic and pertinent to the aspects touched on herein. The fragmentary nature of present knowledge about members of this generally neglected group of plants renders impossible, as yet, the synthesis of an adequate picture of either their essential character or their relation to members of other tribes of the Gramineae,[1] and to extragramineous groups such as the Arecaceae, Juncaceae, Liliaceae, Flagellariaceae, and Restionaceae (see Stebbins 1956). A thoroughly satisfactory taxonomic exploration of the relationships of the bamboos, and the exploitation of their technological potentialities, await basic studies by specialists in diverse disciplines.

The three principal parts of this account are designed to give the reader a concept of (1) what to look for in studying a bamboo plant, (2) what is known about bamboo propagation (with suggestions for improving conventional procedures), and (3) distinguishing characters, and known technical properties, of several bamboos that are outstanding in the importance they have been given as economic plants. The appendices touch on additional aspects of the bamboos, and provide supplementary reference material.

Graduate students and specialists in the several disciplines concerned with original research in the plant sciences will find numerous invitations to undertake studies designed to illuminate areas where recorded knowledge of the bamboos is deficient. An effective exploration of the physiological foundations of gross morphology and anatomy should improve our understanding of the evolving diversity of form that is at once the major concern and, often, the dispair of the taxonomist.

Perspectives

Although the following description of the bamboo plant is presented primarily from the point of view of gross morphology, the

[1]Traditional perspectives upon the morphological, physiological, and genetic barriers that supposedly prevail between major divisions of the grass family cannot fail to be modified by facts brought to light through an incredible feat accomplished by scientists working independently in India and Formosa, respectively. I refer to the apparently successful hybridization between a bamboo (subfamily Bambusoideae) as "male" parent and a sugarcane (subfamily Panicoideae) as "female" parent, first reported by Venkatraman (1937). Morphological and anatomical studies (Kutty Amma and Ekembaram 1940; Loh *et al.* 1950) and cytological studies (Janaki Ammal 1938) appear to have established the hybrid nature of the resulting progenies. However, since the matter seems still to hover in the realm of controversy, further attempts at such crosses and thorough studies of the progeny should be carried out before final conclusions are formed with respect to the significance of the published evidence.

physiological and biochemical setting of morphogenesis has been constantly in mind during its preparation.[2]

The external form of the homologous parts of a bamboo plant varies with their respective positions on the several segmented axes, and with the increase in size of the whole plant. These differences in outward form reflect underlying physiological processes that may operate through biochemical gradients as well as by discrete steps (see Schaffner 1927; Porterfield 1933; Watson 1943; Thompson 1944; Thimann 1954; Prat 1954; Wardlaw 1956, 1959; Heslop-Harrison 1959; Zimmerman 1961). Possibly it is for this reason that some of the outward manifestations of gross morphology in the bamboo plant are not easily amenable to the rigid application of a hierarchy of technical terms or reasoned concepts. On the other hand, the fact that some structures do not yield readily to a regime of sharply defined categories is, perhaps in part at least, an indictment of our insufficient knowledge. Again, the fault may lie partly in the point of view from which we make our observations, or in the nature of the assumptions on which we base our interpretations.

The best picture of any given bamboo that can be pieced together by the taxonomist, on the basis of the morphology of conventional herbarium specimens alone, is sadly inadequate. The increasing diversity and the improving integration of recorded botanical data have resulted in a swelling chorus of appeals for a broader approach to the study of taxonomy as well as other aspects of biology (see Vavilov 1940; Sprague 1944:235; Bush and Lowell 1953; Metcalfe 1954:440; Prat 1960:60 *et passim;* Kety 1960). For a third of a century, I have had a growing conviction that a really successful conquest of the bamboos requires an interdisciplinary approach. Ideally, this would involve the collaboration of a group of specialists representing such points of view as anatomy, biochemistry, histology, physiology, morphology, genetics, ecology, taxonomy, and utilization, to name the most obviously pertinent disciplines. Instead of working independently, as in the past, on study materials drawn from diverse sources—materials all too often of uncertain identity—they should work together (at least in conscious collaboration) on documented material from a common source. In this way, all of the data recorded for a given

[2]Takhtajan (1959:129) expresses the opinion that "the evolutionary botany of the future will be erected on the basis of a synthesis of morphology and physiology." Macleod and Cobley (1961:viii) aver that "the cornerstones of botanical philosophy are still the triumvirate of morphology, taxonomy and physiology."

botanical entity could be integrated, with confidence in their common pertinence. As the publication and integration of such data progressed, the recognition, delineation, and differentiation of critical species could be given greater depth of focus, the classification of the bamboos could gradually be made more consistent and more workable (functional), and the approach to their economic exploitation could be fruitfully broadened.

The hitherto generally neglected branching habit as expressed in the rhizome, the culm, and the inflorescence is herein given special attention. Also emphasized are the nature and role of the prophyllum, which is very important to an understanding of branching habit, particularly in the indeterminate inflorescence. Takenouchi's excellent illustrations (1931*a*; 1932) of the forms assumed by the prophyllum in the vegetative axes of the common Japanese species are a pioneer contribution to the elucidation of this category of bamboo structures.

Published dissertations on the bamboos often leave basic concepts, observations, and generalizations inadequately documented. A special effort has been made here to avoid this fault by citing illustrative species and, according to the requirements of the case, a particular specimen or collection.

In my effort to achieve an improved understanding and portrayal of the bamboo plant, I have sought fresh ways of looking at familiar things, and have challenged established concepts and usages that cast a shadow on certain areas where more light is needed.[3] Critical attention directed at (1) the elimination of ambiguities and (2) the disciplined communication of my own observations has led to the adoption of some new terminology and to the modification of some of the old. Terms in these two categories are highlighted by means of asterisks and daggers, respectively, in the Glossary.

Attention is directed especially to the taxonomic scope of the subject matter of this work. It is essentially restricted to the plants universally recognized as bamboos (the tribe Bambuseae), although occasional reference is made to features of members of the related tribe, Streptochaeteae. Entirely omitted from the treatment are the tribes Olyreae, Phareae, and Parianeae, recently allocated to the subfamily Bambusoideae by Parodi (1961).

[3]Koestler (1964) touches obliquely upon the history and uses of this device in his chapter entitled "The Evolution of Ideas." The taxonomist of reflective temperament will discover in Koestler's work much discourse that is instructive as well as diverting.

Part I The Bamboo Plant

1 Vegetative Phase: The Maturing Plant

As most commonly encountered, a bamboo plant is somewhere on its way through the usually long, purely vegetative phase of development (or maturing) that intervenes between the seedling stage (p. 122) and the reproductive phase (p. 82). The component parts of the plant as they appear in the more familiar vegetative phase will be described first. The different forms assumed by the various kinds of structure as they appear in either of the two more rarely observed stages will then be more easily understood. Four major categories of vegetative structures may be recognized: segmented axes, sheathing foliar organs, buds, and roots.

Segmented axes

The basic frame of the bamboo plant consists of a ramifying system of segmented vegetative axes (Fig. 1). These axes may be differentiated as rhizomes, culms, and culm branches. There is no central trunk or main axis. Each component axis, whether rhizome, culm, or branch, consists of a series of nodes and internodes. It is clothed (at least during the period of its active growth) with enveloping sheaths that face alternate sides of the axis at the successive nodes. With certain minor exceptions, each sheath subtends a bud or a branch complement. Adventitious roots are initiated at the nodes of segmented axes—in certain parts of the plant to be detailed later (p. 79).

The term "segmented" refers to the regular alternation of nodes and internodes that characterizes the axes of the rhizome system, the culms, and the several orders of culm branches. This segmentation is seen as an expression of physiological periodicity. The mechanism of the origin of this type of structure poses intriguing, and as yet unsolved, problems relating to morphogenesis. These

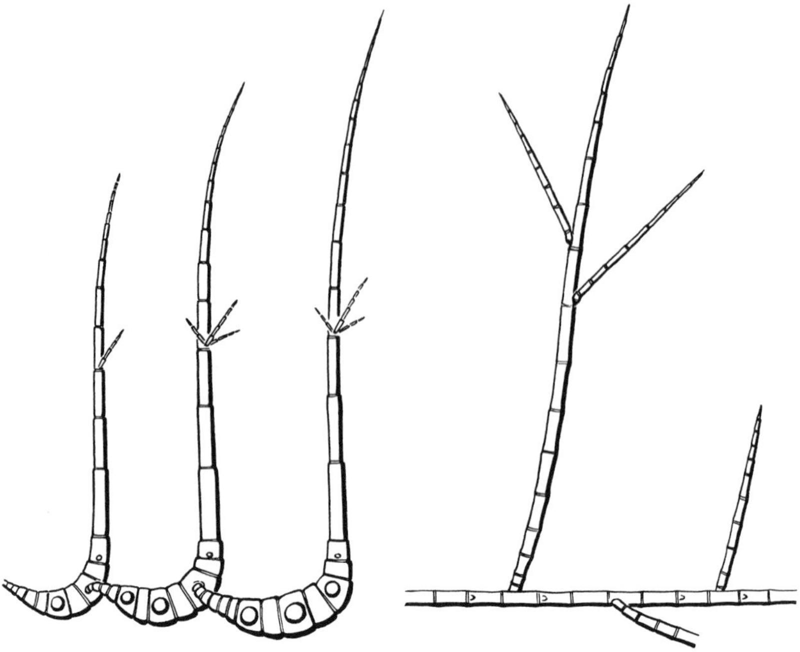

Fig. 1. The basic frame of a bamboo plant consists of ramifying systems of segmented axes. The characteristics of the system of branching shown by the rhizomes, and of that shown by the culms, have distinctive value for the recognition of the taxon to which a given plant belongs.

problems are ably discussed by Porterfield (1933). They require further investigation by means of the tools and techniques of anatomy, biochemistry, and physiology.

The nodes of the segmented axes are important centers of morphogenetic activity. In a growing point, it is at the loci later recognizable as nodes that sheath primordia are initiated and segmentation first comes into focus (see pp. 61 and 123). Only at the nodes do roots and branches emerge. The final growth in length of each segmented axis takes place only near the nodes. Variations in the form and other features of the nodes and internodes (particularly those of the culm) afford useful characters for the differentiation of taxonomic entities.

The neck

Each segmented axis is a branch of another segmented axis. The basal portion of each segmented vegetative axis of every category consists of a series of nonfistulose, usually very short, segments clothed with persistent, relatively small scalelike sheaths. The nodes here are always without buds, and usually lack roots and root primordia. In all axes of the plant, this basal part of each axis is typically more slender than the part distal to it. Convention has already established the use of the term "neck" (Arber 1934: 67) for the constricted form it always assumes in the underground parts of the plant (see text, p. 13 and Figs. 2, 3, and 6). The superficial resemblance of this transitional structure to a neck is in some bamboos almost completely lost elsewhere in the plant. In the branches of the culm in all species of some genera (*Arundinaria* and *Phyllostachys,* for example) it is very short and is not appreciably constricted (Fig. 27).

As compared with other segmented axes, the "neck" appears to be less specialized, notably in its internal structure. Superficially it appears, in its most common form, as a transitional structure of no special importance. However, the diversity of its manifestations of form and behavior, especially in certain of the underground axes, marks the neck as a focus of particular interest. The basis of this assertion is made plain in the following pages (see especially pp. 17 and 128).

The rhizome system

Rhizome axes are typically subterranean, and the rhizome system constitutes the structural foundation of the plant (Figs. 1–4). Since this part of the plant is wholly or almost entirely out of sight and is not easily accessible, it is usually ignored by collectors, generally neglected by taxonomists, and practically unknown to contemporary anatomists.[1] Consequently, the rhizome system is one of the least well understood parts of the bamboo plant. This neglect has unfortunate consequences, because the rhizome system performs important functions in the life of the plant. Moreover, an understanding of the form of the rhizome system is prerequisite to an understanding of the clump habit in any bamboo (Figs. 16, *1* and 17). By the term "clump habit" I refer to the spatial relation of the culms that make up the visible part of an individual bamboo

[1]Shibata (1900, Figs. 1–14) illustrated some anatomical features of the rhizome of *Arundinaria hindsii* (*sensu bot. jap., non* Munro), *Phyllostachys pubescens* (as *Ph. mitis*), *Pseudosasa* (as *Arundinaria*) *japonica, Sasa* (as *Bambusa*) *nipponica* and *Sasa* (as *Bambusa*) *palmata.* McClure (1963) illustrated anatomical features of the rhizome that differentiate *Arundinaria tecta* from *A. gigantea.*

plant, whether the plant is caespitose or diffuse in its spread in space.

The rhizome system assumes, in plants of different species and genera, a number of more or less sharply distinct forms and habits

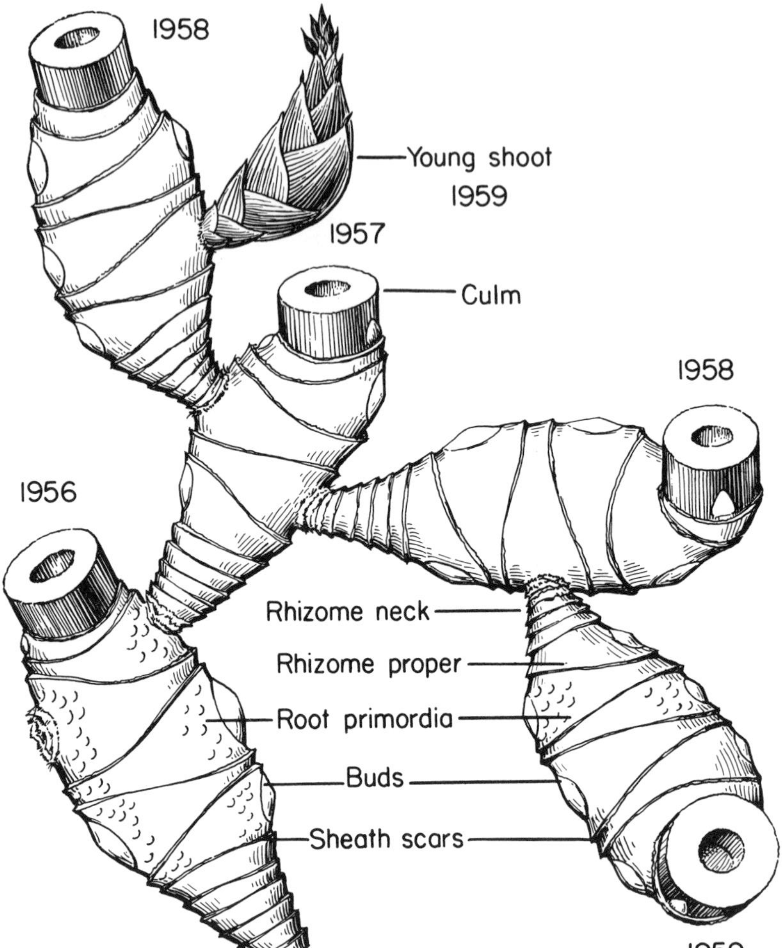

Fig. 2. Pachymorph rhizome system of *Bambusa tuldoides,* as seen from above, showing individual rhizomes produced in successive years, and the base of the culm invariably terminal to each. The sheaths have been removed (note sheath scars) to reveal lateral buds, root primordia, and other details. Roots emerge on the lateral and lower surfaces of the rhizome proper in an all-over pattern, while as a rule the upper surface is less completely covered; in this area the root primordia may remain dormant indefinitely.

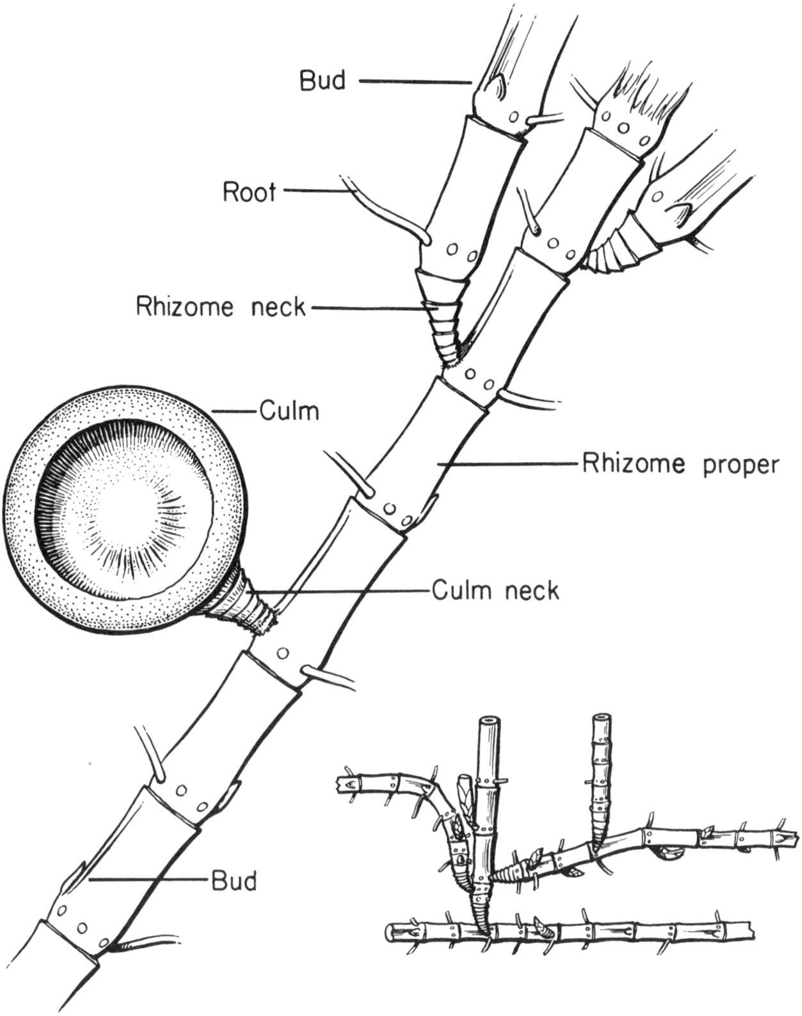

Fig. 3. Leptomorph rhizome system of *Phyllostachys bambusoides,* as seen from above, showing the base of a culm (in cross section) and two new rhizomes, all produced from lateral buds. Root primordia emerge in a single verticil at each node of the rhizome proper, equally on all sides, and develop promptly. Those that have been cut off are shown as small circles. Roots (not shown) are also borne at the several subterranean nodes of the obconical base of the culm, immediately above (distal to) the culm neck. The smaller, unlabeled figure (from Takenouchi 1931*a*:Pl. 1, Fig. 4) is a lateral view based on the same species. Note the abrupt change in direction of the axis on the left, from vertical to horizontal.

Fig. 4. An early illustration of the type of bamboo rhizome herein named leptomorph, exemplified by *Phyllostachys viridis*. The terminal bud of the rhizome in this bamboo and all other species of the genus *Phyllostachys* that have been studied is normally diageotropic. The rhizome is pictured here as turning upward at length (*left*) to form a culm (see metamorph II axes, p. 30). The manifestation of negative geotropism by the terminal bud of the rhizome is facultative, not obligate, in bamboos of this genus. The tillering of culms shown at the right is also facultative, and rare, in species of *Phyllostachys*. It can appear in bamboos with leptomorph rhizomes only where there are buds at the underground nodes of the culm. From A. and C. Rivière 1879:Fig. 3 (as *Ph. mitis*). See p. 51.

of growth (Fig. 10). These diverse manifestations afford challenging openings for fruitful studies from the points of view of physiology and anatomy as well as gross morphology. They may be used for the recognition and description and, to a limited extent, the classification of taxonomic entities.

Arber (1934:67) refers to the individual axes (branches) of the bamboo rhizome system as "rhizome segments." For the rhizome in *Bambusa vulgaris* (see Fig. 5) this term appears at first sight to be quite suitable. However, when the clearly segmented nature of the individual rhizome axis is considered, Arber's use of the term is seen to be ambiguous. Its application to the individual branches of the slender, wide-ranging type of rhizome (see Fig. 3) is even more clearly inappropriate. It seems best, therefore, to apply the term "rhizome" to each individual branch or axis of the rhizome system. Each rhizome develops from a bud—most commonly from a bud on another rhizome. Under certain conditions a rhizome may emerge from a culm, starting as a bud either at a subterranean node or, more rarely, at an aerial node of the culm (see p. 245 and Fig. 97, *4–6*).

The individual rhizome is a segmented axis consisting typically of two parts: the rhizome proper and the rhizome neck (Figs. 2 and 3). The neck is basal to the rhizome proper, and is the part that develops first. In its most common form, the rhizome neck is relatively short and obconical in shape but it may be greatly elongated and obterete (Fig. 13). Its basal diameter is smaller than that of either of the axes it joins together. In bamboos with pachymorph rhizomes (p. 24) the rhizome neck carries the rhizome primordium that is apical to it downward to a depth that is characteristic for each species, the age of the plant, and the nature of the substrate. This behavior of the neck often assumes a dra-

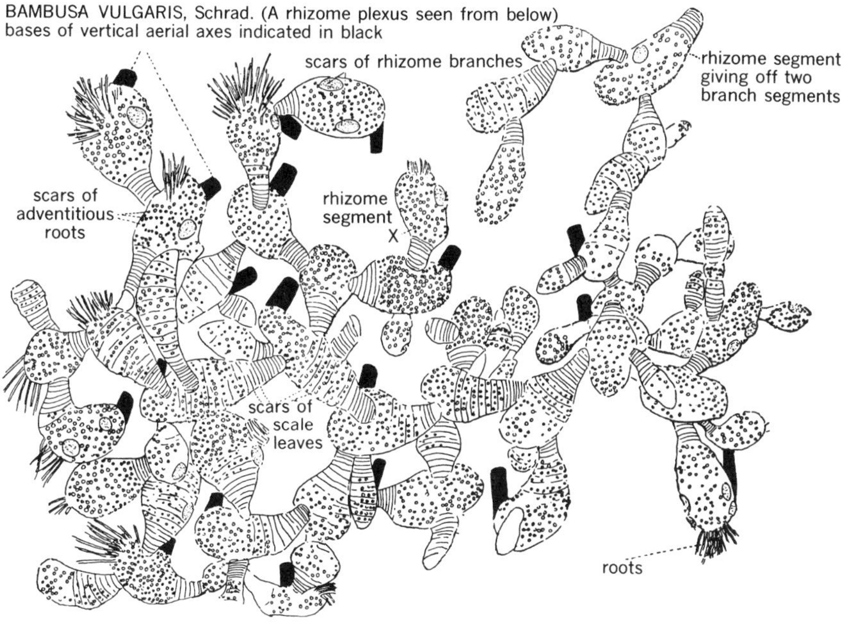

Fig. 5. This slightly stylized sketch by Agnes Arber (1934:Fig. 23) gives only an inkling of the complicated mass of interwoven axes that the pachymorph rhizome system of a bamboo plant becomes in time. Portion of a specimen (No. 205) of *Bambusa vulgaris* preserved in the Museum at the Royal Gardens, Kew, seen from below; greatly reduced. See also Fig. 2 and pp. 24f.

matic aspect in young seedlings; see p. 128 and Fig. 60,*i* (*Dendrocalamus strictus*), also Fig. 61 (*Melocanna baccifera*). Again, in large plants of *Guadua angustifolia* the rhizome necks show, in their "start-and-stop" growth and in their physical orientation, a special adaptation related to the support of the culms (Fig. 6).

Arber (1934:67) apparently was the first to use the term "neck" for this part of the rhizome, attributing its detailed study to Shibata. As a matter of fact, the structure described by Shibata (1900:443, Figs. 13, 14) was the neck ("*Stiel*") of the young culm ("*Schössling*") and not that of the rhizome. Shibata says in a footnote: "This '*Stiel*' furnishes the sole path for the movement of materials between the rhizome and the culm that develops from the shoot." This statement is incomprehensible to the reader who is unaware that Shibata was referring only to bamboos in which

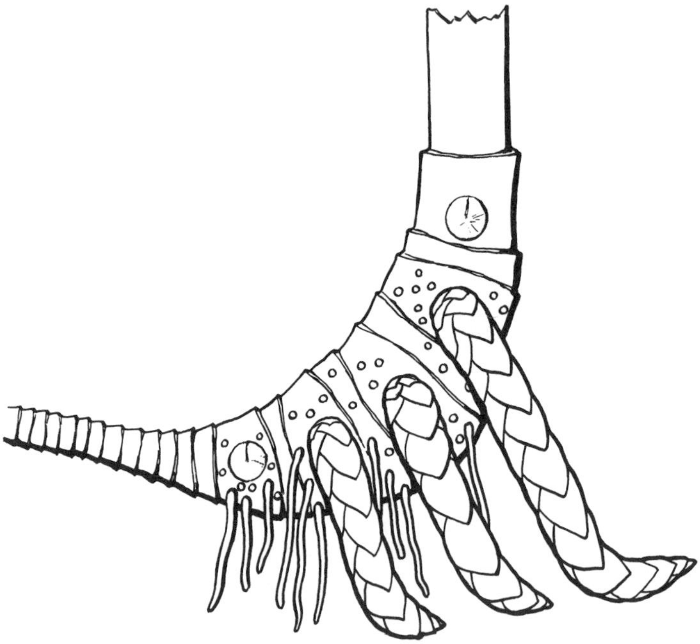

Fig. 6. In large plants of *Guadua angustifolia,* the lateral buds of the pachymorph rhizome push prematurely and make a limited growth to form only the neck portion of the next order of branches, thus providing a collective "foot" to support the heavy culm that is produced at the same time from the terminal bud of the mother rhizome. In this way, the tough, elongated necks of as yet incomplete rhizomes form an indispensable support for the 100-ft culm.

the culms develop from the lateral buds of long, slender rhizomes, commonly referred to in earlier literature as *traçant,* monopodial, or indeterminate, and hereinafter referred to as leptomorph (Figs. 3, 4, and 8, *B*).

The rhizome proper is characterized by its typically subterranean position, by the presence of roots or root primordia and prophyllate buds (always solitary) at all or most of its nodes, and by the relatively simple and uniform character of the sheaths that clothe it (Fig. 34). Each diageotropic rhizome axis normally orients itself so that the plane of distichy in the insertion of sheaths and buds or branches lies at right angles to the force of gravity.

Two basic forms of rhizome

The rhizome manifests its character in two divergent forms, each with important variations. Differences in the gross morphology, combined with differences in growth habit, give rise to characteristics by which a number of more or less clearly distinct forms of rhizome system and clump habit may be recognized (see p. 32 and Fig. 10).

It appears that in their classic work, *Les Bambous* (1879:312–322), the Rivières were the first to publish (in a Western language, at any rate) a clear distinction between the two basic forms assumed by the bamboo rhizome. These were taken as characteristic of two distinct groups of bamboos (see p. 209) described as:

"1. Bamboos of autumnal growth, and caespitose clump habit;
"2. Bamboos of spring growth, and a generally spreading [*traçant*] clump habit" (p. 312).

An unidentified species of *Gigantochloa*—under the name *Bambusa macroculmis* Rivière and Rivière (1879:Fig. 2)—was chosen as exemplifying the *first* group. Its rhizome is described as developing in a horizontal direction for a short distance, after which it turns upward to form a culm (as in Figs. 2 and 7).

Phyllostachys viridis (under the name *Ph. mitis*) was chosen as exemplifying the second group. Its rhizome is described as ranging widely in a horizontal direction, with more or less widely spaced culms arising from lateral buds at the nodes (as in Figs. 3 and 17).

The next contribution to the subject is that of Houzeau de Lehaie (1906–1908:150), who called the rhizome of bamboos of the *first* group a "caulo-bulbe," and illustrated the rhizome of *Bambusa vulgaris* (under the name *B. thouarsii*) as an example (Fig. 7). Houzeau characterized the rhizome of bamboos of the *second* group as "long and slender, with indefinite subterranean development." As an example, he illustrated the rhizome of an unspecified member of the genus *Phyllostachys.* This author concluded his discussion with the challenging statement that "the essential difference between caespitose bamboos and running [*traçant*] bamboos

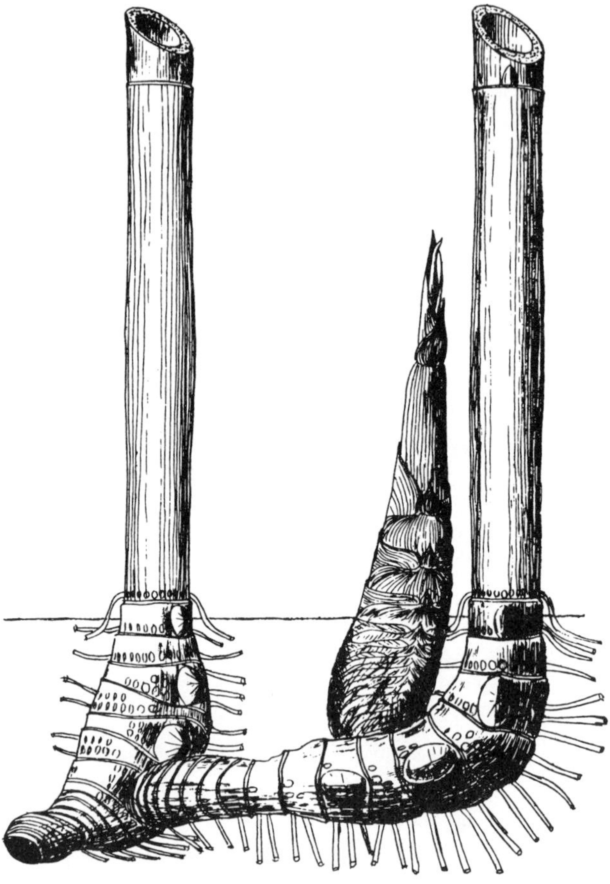

Fig. 7. *Bambusa vulgaris* shows the type of rhizome herein named pachymorph. Under the synonymous name, *Bambusa thouarsii,* this species was chosen by Houzeau de Lehaie (1906:Pl. III) to illustrate the development of a typical "caespitose" bamboo. See Fig. 5.

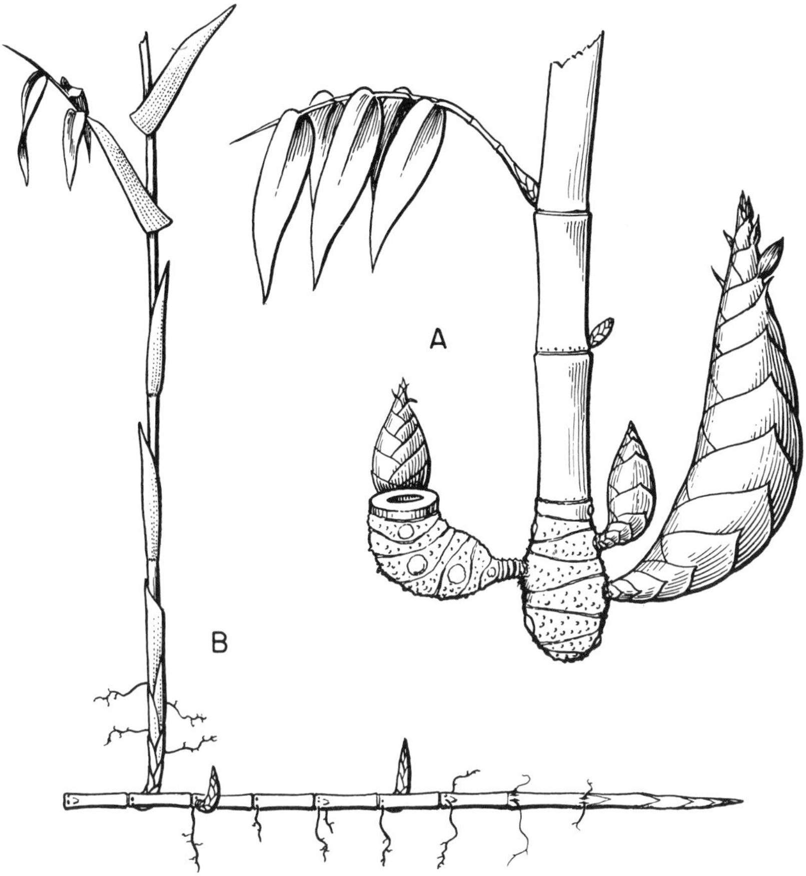

Fig. 8. Types of rhizome designated by McClure (1925:Figs. 1, 2) as (*A*) sympodial (*Bambusa beecheyana*); (*B*) monopodial (*Arundinaria amabilis*). These terms were later, and perhaps more appropriately, applied by Japanese botanists (notably Nakai) to the manner of origin of culms associated, respectively, with the two types of rhizome.

is neither generic nor specific, but is of a physiological nature." The recent discovery of the occurrence (in *Chusquea fendleri*) of both types of rhizome in the same plant (see Fig. 10, *11*) appears to confirm this assertion. However, Houzeau's observation loses its intended force when we reflect that all morphological features recognized as "characters" at either the specific or the generic level are basically physiological in origin.

McClure (1925) first proposed the terms monopodial and sympodial to characterize the branching habit of the rhizome in the two main groups of bamboos called by the Rivières "traçant" and "caespitose," respectively. Takenouchi (1926:44) adopted McClure's terminology in listing three "kinds" of rhizomes, and the genera of Japanese bamboos that exemplify them. However, four distinct forms of growth (Fig. 9) were actually illustrated. Takenouchi later (1931*a*:Fig. 1, *1–4*) again illustrated four dif-

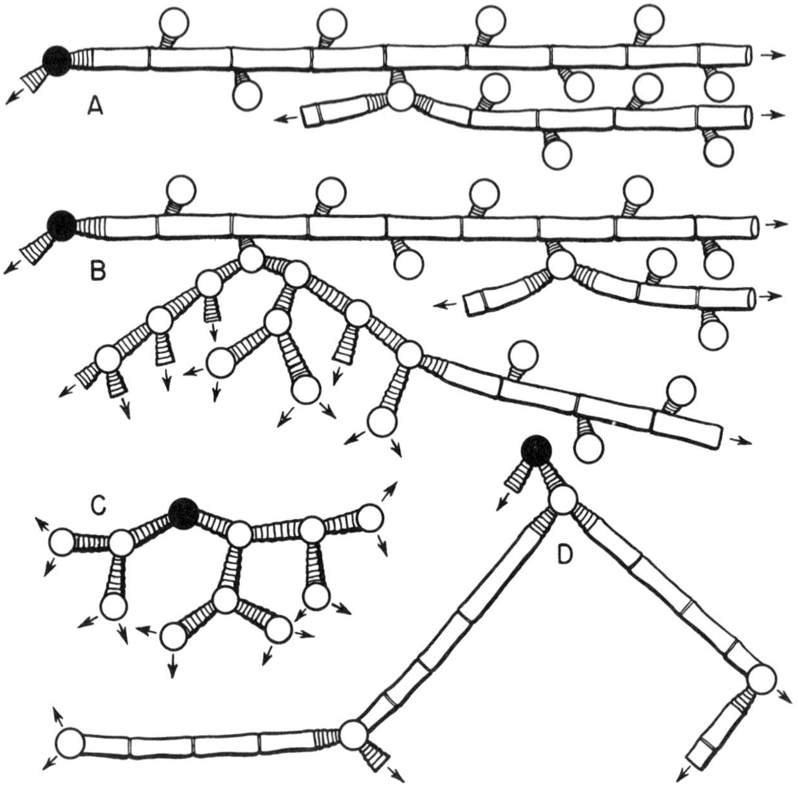

Fig. 9. Types of rhizome system illustrated by Takenouchi (1926:Pl. 3, Figs. i-iv). (*A*) Leptomorph rhizomes with solitary culms, designated by Takenouchi as "Monopodien (*Phyllostachys*)." (*B*) Leptomorph rhizomes with tillering culms, designated by Takenouchi as "Monopodien und Sympodien (*Pleioblastus*)"; this apparently corresponds to the category "amphipodial" of Keng, f. (1948:2). (*C*) Short-necked pachymorph rhizomes, designated by Takenouchi as "Sympodien (*Dendrocalamus, Bambusa*)." (*D*) Long-necked pachymorph rhizomes with tillering culms, designated by Takenouchi as "Sympodien (*Indocalamus*)" [= *Yushania niitakayamensis*]. Takenouchi later (1932:Pl. 4, Fig. 1) referred to the long neck in "Indocalamus" as a "running rhizome." See also Figs. 11, 13, 14, and 3 (insert).

ferent manifestations of the rhizome system. These were then designated as follows: "1. Simple clump-forming; 2. Clump-forming from runners; 3. Lateral clump-forming from rhizomes, and 4. Irregularly branching" (p. 4). Reproduced in my Figs. 11, 13, 14, and 3 (insert), respectively, these are interpreted herein as two typical and more or less distinctly diverse expressions of each of the two basic types. Later in this work, however, Takenouchi shows signs of faltering between the designations "rhizome" and "culm base" in his interpretation of the short, thick, determinate type hereinafter called pachymorph; and in the section captioned "Morphology and characteristics of rhizomes" (Takenouchi 1932: trans., p. 7) he deals concretely only with the long, slender, indeterminate type, hereinafter called leptomorph.

In writing about various Japanese bamboos, Nakai (1933:12), Makino, and others have rather consistently (and with good reason) used the terms "monopodial" and "sympodial" in reference to the manner of origin of the culms rather than to the habit of the associated rhizome. McClure later (1945*b*, 1946) substituted the terms "indeterminate" and "determinate," in the belief that these are more precise and call attention to a more basic difference in the behavior of the growing point of the rhizome prevailing in the two forms respectively.

To the distinctive branching habit in the underground part of the plant which Takenouchi (1931*a*:Fig. 1, *3*; trans., p. 4) calls "lateral clump-forming from rhizomes," Keng (1948:2) gives the name "amphipodial." Its characteristics are described on p. 31.

Shibata (1900:Figs. 1–14) was the first to publish detailed comparative studies of the anatomy of the rhizome in different bamboos. However, he dealt only with leptomorph rhizomes, and made only the briefest reference to their external morphology. He did not differentiate the two basic forms discussed herein. McClure (1963) described and illustrated a hitherto unrecorded anatomical feature, the presence of peripheral air canals in the rhizomes of *Arundinaria tecta.* This feature, paired with the lack of it in the rhizomes of *Arundinaria gigantea,* constitutes a reliable taxonomic character for differentiating the two species.

Reinterpretation of the rhizome system

Further comparative studies of the diverse forms of growth in the underground parts of bamboos should intrigue the physiologist and the anatomist, as well as the morphologist. The taxonomist and the field botanist could use the results to good advantage. Meanwhile, improvements in the differentiation, illustration, description, and terminology of the known modes of expression in the bamboo rhizome system are desirable. The pairs of terms, "caespitose" and "traçant," "sympodial" and "monopodial," "determinate" and "indeterminate" have proved increasingly difficult to use in an unambiguous manner, even with the help of Keng's new term, "amphipodial"—probably because they embrace too many undefined ideas. At any rate, in actual use they all eventually break down in ways that leave either the original concepts or the distinctive features of associated variables somewhat blurred, and they cannot easily be modified. It now appears that a slight shift and a narrowing of the initial focus of attention may improve the interpretation, the terminology, and the descriptions. This changed view will be presented by characterizing the two basic forms of the rhizome proper as pachymorph and leptomorph.

Pachymorph rhizomes

It is proposed to call the rhizome proper pachymorph when it is short and thick, and has the following associated characteristics: a subfusiform (rarely subspherical), usually more or less curved (rarely straight) shape, with a maximum thickness typically somewhat greater than that of the culm into which it is always transformed apically; internodes broader than long, asymmetrical (longer on the side that bears a bud), solid (apparently never hollow); nodes not elevated or inflated; lateral buds solitary, in the dormant state asymmetrically dome-shaped, with a subcircular margin and an intramarginal apex. In pachymorph rhizomes with a horizontal orientation, varying degrees of dorsiventrality are manifested, in (1) a more profuse production of roots on the lower side and (2) dorsiventral flattening of the axis; see Figs. 6 and 7.

Lateral buds of a pachymorph rhizome produce only rhizomes; a culm can arise directly only from the apex of such an axis. A transformation of the internal organization of a pachymorph rhizome axis is prerequisite to culm formation. This feature was the basis for the characterization of this type of rhizome as "sympodial" (McClure 1925) and "determinate" (McClure 1946). In pachymorph rhizomes, the rhizome neck may be short or long.

Leptomorph rhizomes

It is proposed to call the rhizome proper leptomorph when it is long and slender, and has the following associated characteristics: a cylindrical or subcylindrical form, with a diameter usually less than that of the culms originating from it; internodes longer than broad, relatively uniform in length, symmetrical or nearly so, rarely solid, typically hollow, the usually narrow central lumen interrupted at each node by a diaphragm; nodes in some genera usually somewhat elevated or inflated, in others not; lateral buds in the dormant state boat-shaped, with a distally oriented apex; see Figs. 3 and 10, *7–10.*

In leptomorph rhizomes generally (as in the known species of *Phyllostachys*) every node bears a solitary prophyllate bud and a single verticil of roots. In some bamboos, however (as in many arundinarioid species—especially in the genus *Sasa*), buds may be lacking here and there at one to several adjacent nodes, and root development may be very sparse or even lacking entirely at some nodes. No sign of dorsiventrality has been noted, but each rhizome axis shows a strong tendency to orient itself so that the plane of insertion of sheaths and buds is horizontal.

Most of the lateral buds of leptomorph rhizomes are temporarily or permanently dormant. The majority of those that germinate produce culms directly; a few produce other rhizomes. The orientation of the terminal bud in leptomorph rhizomes is consistently diageotropic in most species and under most circumstances. The persistence of this diageotropic character suggested the terms "monopodial" (McClure 1925) and "indeterminate" (McClure 1945*a*:7 and 1945*b*:278). In certain species regularly (as in *Arundinaria tecta,* Fig. 10, *8*), and in others only under certain conditions (*Phyllostachys spp.,* Fig. 4), the leptomorph rhizome turns upward to form a culm. In leptomorph rhizomes, the rhizome neck is always short.

Ueda (1960) has made extensive excavations of the leptomorph rhizome systems of plants of several species of Japanese bamboos, plotting the pattern and rate of their extension in the soil, and calculating the rate of accumulation of total substance.

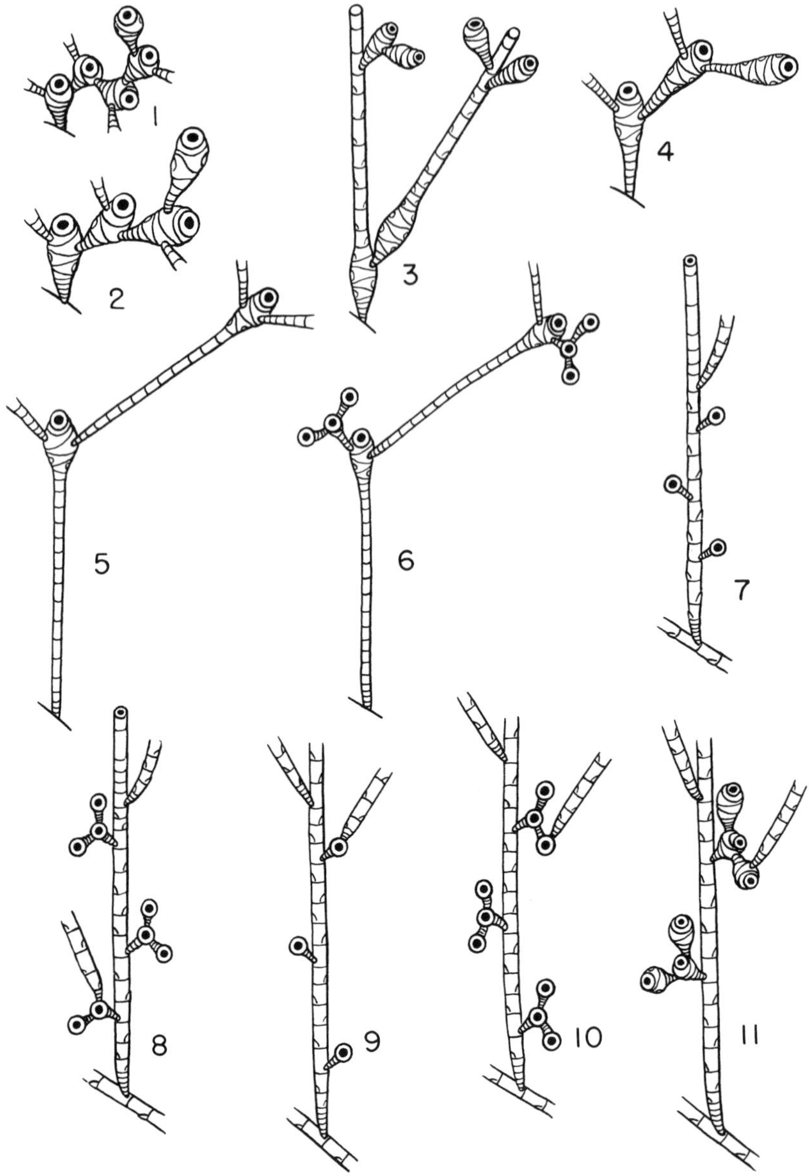

Fig. 10. Known forms assumed by the rhizome system in different bamboos are here illustrated by diagrammatic views from above. Open ends signify the continuation of an axis in a horizontal direction. Double circles represent basal cross-sections of hollow culms. Single circles indicate culms arising by the slow transformation of the terminal bud of a rhizome through a metamorph II axis. Transverse lines indicate sheath scars at nodes; roots are omitted. Rhizomes

proper show buds at their nodes; rhizome necks have no buds. The scale of representation is not uniform. Captions to individual figures follow. See p. 32 for key.

1. Bambusa pachinensis here exemplifies a short-necked parchymorph rhizome in which there is a marked tendency toward wholly erect orientation, that is, negative geotropism. Takenouchi's illustration of this species, showing the growth habit resulting from this orientation of the rhizome, is reproduced in Fig. 11.

2. A typical and widely representative expression of the pachymorph rhizome system is that of *Bambusa tuldoides,* here illustrated. The rhizome neck is shorter than the rhizome proper. The rhizome proper is horizontal in its proximal part, curves upward in its distal part, and terminates in an erect culm. The resulting clump is caespitose, that is, a compact tuft of culms. This is the type of growth first referred to as caespitose (Rivière 1879:312 and Fig. 2), later by McClure (1925) as sympodial, and still later (McClure 1946:107) as determinate. See Figs. 2 and 8, *A.*

3. In *Arundinaria pusilla* the growth of the terminal bud of the short-necked pachymorph rhizome sometimes deviates from the typical expression shown by *Bambusa tuldoides.* Beyond the rhizome proper, the axis describes in this case a broad curve, and its diameter is gradually reduced before it emerges from the ground as a culm. This transitional portion of the axis is typically rootless, but it usually has a bud at each node. Observations made in South Vietnam in 1953 indicate that in this species pachymorph rhizomes with two distinct potentialities emerge from the subterranean buds. In those axes that originate below a certain depth, a long transitional stage (here called a metamorph II axis) intervenes between the rhizome proper and the culm proper. In rhizomes originating above a certain depth, the terminal bud produces a culm directly, without an intervening metamorph II axis. In the Andean highlands of Ecuador, I have observed, in the rhizome system of large, old clumps of *Chusquea scandens,* a growth form similar to the one here described. See Fig. 12 and p. 30.

4. In some bamboos with pachymorph rhizomes the rhizome neck is elongated just enough to give the caespitose clump a noticeably open habit. A typical example of this is afforded by *Sinarundinaria nitida,* here illustrated. *Bambusa vulgaris* shows the same trend. See Fig. 7.

5. Melocanna baccifera, here illustrated, is one of a number of bamboos that are characterized by long-necked pachymorph rhizomes producing solitary culms. Here, and in *Schizostachyum hainanense,* the rhizome neck may be a meter or more in length. In a bamboo native at Tingo Maria, eastern Peru (*Guadua sp.,* McClure 21438-A) the rhizome neck reaches a length of 6 m. Such an elongated rhizome neck is subcylindrical throughout most of its length, except near the distal end, where it expands into a rhizome proper that is terminal to it. It is wholly free of buds, and only exceptionally bears a few roots on the lower side. This type of growth gives rise to a simple, open clump made up of widely spaced solitary culms. See Fig. 82 and p. 36.

6. Long-necked pachymorph rhizomes and culms that tiller are characteristic of *Yushania niitakayamensis,* here illustrated. Buds at the base of culms terminating long-necked rhizomes give rise to axes with two distinct potentialities. They produce culms either (1) by way of long necks, or (2) by tillering, through short-necked rhizomes that often resemble metamorph I axes. The result is a number of spaced-out tufts of culms, joined together by the long-necked rhizomes—a culm habit that may be described as compound-caespitose. See also Fig. 13 and p. 40.

7. Shibataea kumasasa apparently is facultative in respect to the behavior of the growing point of its rhizomes, and in respect to tillering of the culms. The rhizome normally remains diageotropic in its growth, but in an occasional one the growing point becomes negatively geotropic and turns upward to form a culm, through a metamorph II axis. Culms arising from lateral buds of the rhizome normally remain solitary. Occasionally, however—perhaps predominantly in age—they tiller to form tufts of additional culms, by way of metamorph I axes—a condition illustrated in *8.* See also Fig. 14 and p. 36.

8. In *Arundinaria tecta,* shown here, the growing point of the leptomorph rhizome, after carrying the growth of the axis a certain rather uniform distance in a horizontal direction, develops negative geotropism, whereupon the axis turns upward in a broad curve and at the same time gradually modifies its form to that of a culm, by way of a metamorph II axis. This behavior of the rhizome apparently is obligate in *Arundinaria tecta,* but is more commonly facultative elsewhere—as in the example shown in *7.* From their subterranean buds, all culms produce both rhizomes and other culms, the latter by tillering through metamorph I axes. See pp. 34 and 40.

9. A leptomorph rhizome with a normally diageotropic growing point, and culms that normally do not tiller, are here exemplified by *Phyllostachys bambusoides.* Characteristic of all species of *Phyllostachys* that have been studied, this is perhaps the most widely typical expression of the leptomorph rhizome. The diageotropic growth of the apex of the rhizome proper extends the latter steadily in a generally horizontal direction. If and when, for any reason, the apical meristem ceases to funtion—upon the termination of a season's growth, for example—the growth of the plant may be carried on in the same general direction by one or more new rhizomes that emerge from buds near the tip of the old one (Fig. 3). A lateral bud of the rhizome proper may give rise to either a culm or a rhizome. On the periphery of plants of mature stature the culms usually emerge at an open spacing, giving a diffuse clump habit. The culms normally do not tiller, but may do so when damaged, or when growing under unfavorable conditions. In *Phyllostachys,* the apex of the rhizome does not ordinarily develop negative geotropism, but it may do so under exceptional conditions (cf. Fig. 4) as when forced upward by natural obstacles, or when it emerges from the ground upon encountering a steep declivity. See Fig. 16, *2.*

10. In *Indocalamus sinicus,* shown here, the leptomorph rhizome remains diageotropic, and the formation of tufts of secondary culms through tillering by way of metamorph I axes appears to be the normal expression. Takenouchi (1926:Fig. 3, II) illustrates a similar habit in *Arundinaria* (as *Pleioblastus*) *simonii,* and refers to it as typical of the general *Pleioblastus* and *Pseudosasa.* See Figs. 9, *B* and 18, p. 34.

11. The association, in the same plant, of both leptomorph and pachymorph rhizomes, apparently rare among the bamboos, is here exemplified by *Chusquea fendleri.* The swollen character of the pachymorph rhizomes that appear here in association with those of leptomorph habit distinguishes this form of expression from the one illustrated in the foregoing figure, where slender, metamorph I axes appear in their stead. See also Fig. 15 and p. 31.

Metamorph axes

In some bamboos, certain axes or portions of axes of subterranean origin do not fit exactly into the categories "neck," "rhizome,"

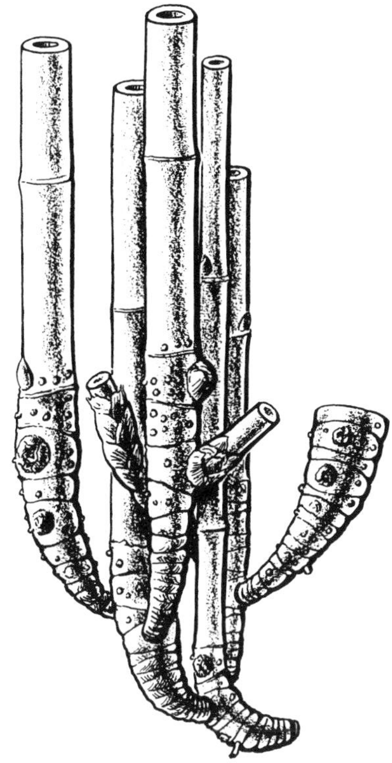

Fig. 11. In some bamboos the pachymorph rhizome shows a marked tendency toward a wholly erect orientation. Takenouchi here uses *Bambusa pachinensis* to illustrate this habit of growth, which he represents as typical for the genus *Bambusa*. Perhaps Takenouchi's inclination (1932:trans., p. 7) to interpret the pachymorph rhizome as a culm base, instead of a rhizome, grew out of his choice of this example as typical of the whole genus. Redrawn from Takenouchi 1932:Fig. 4, *1*. See also Fig. 10, *1*.

and "culm base," as described above. The anomalous form and behavior of these axes, and their position in the plant, suggest the general designation "transitional." For convenience, the term "metamorph" is proposed as a technical designation of such transitional axes. Evidence at hand suggests the characterization of two kinds of metamorph axes: metamorph I and metamorph II. This rough classification is based primarily on the respective positions in which they appear in the plant.

A metamorph I axis occupies a position between the neck proper and its culm. In form, it is intermediate between a typical culm base and a pachymorph rhizome, as contrasted in Fig. 20. Such axes often appear where the culms arising from lateral buds of a leptomorph rhizome tiller to form tufts, as in Figs. 9, *B* and 14.

A metamorph II axis is intermediate in form and position between the apex of a rhizome (pachymorph or leptomorph) and the culm into which the rhizome is transformed apically. It appears where the transformation of the apex of a rhizome into a culm takes place gradually and not abruptly. This gradual change is typical wherever a leptomorph rhizome is transformed apically to form a culm (Figs. 4 and 10, *7* and *8*). The intercalation of a transition zone (metamorph II axis) between a pachymorph rhizome and the culm distal to it, as seen in *Arundinaria pusilla* (Fig. 10, *3* and 12) and some species of *Chusquea*, is rela-

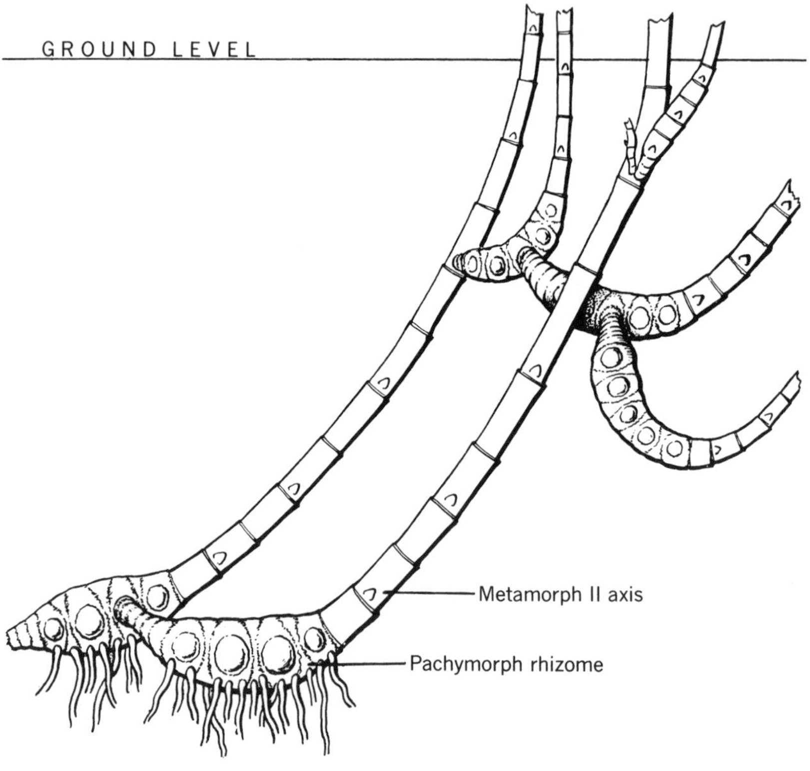

Fig. 12. Metamorph II axes distal to two pachymorph rhizomes, as exemplified by *Arundinaria pusilla*. Field sketch, made en route from Saigon to Dalat, South Vietnam, December 24, 1953. See also Fig. 10, *3*.

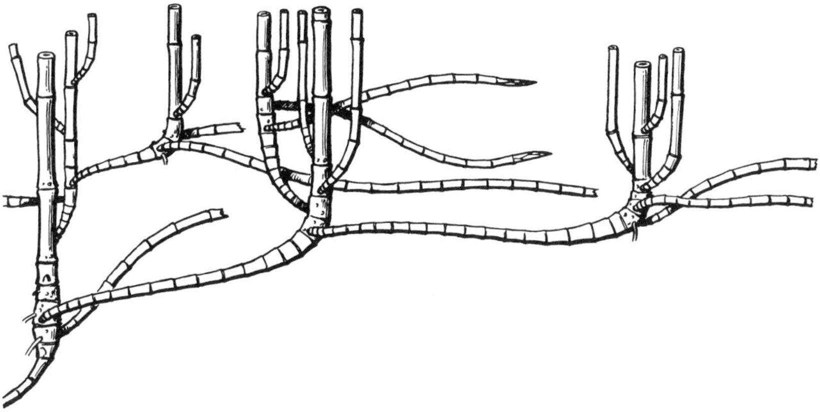

Fig. 13. The combination of long-necked pachymorph rhizomes with culms that tiller was first illustrated by Takenouchi, on the basis of *Indocalamus niitakayamensis,* now known as *Yushania niitakayamensis.* The structure that is interpreted herein as a rhizome neck is in this case called by Takenouchi a "running rhizome." The culms that terminate these long necks tiller by metamorph I axes. Redrawn from Takenouchi 1931*a*:Fig. 1, *2*. See also Fig. 10, *6*.

tively rare, and stands in sharp contrast with the abrupt transformation that is typical of bamboos with pachymorph rhizomes, as illustrated in Figs. 2 and 7.

Other combinations of underground growth forms

As I understand it, the habit of growth designated by Keng (1948:2) as amphipodial is found in the underground parts of certain species of the arundinarioid genera (*Arundinaria, Indocalamus, Pseudosasa, Shibataea, Sasa*). It is the result of the occurrence together, in the same plant, of two distinct kinds of subterranean axes. One of these is of the type described above as leptomorph; the other corresponds to the concept described above as metamorph I. Examples are illustrated in Figs. 9, *B* and 14.

The capacity to produce both typical pachymorph and typical leptomorph rhizomes in the same plant is represented among the buds of the subterranean axes of *Chusquea fendleri* and some other species of this genus. The resulting form of expression is illustrated in Figs. 10, *11* and 15. This potential is not known to exist in any of the bamboos mentioned by Keng.

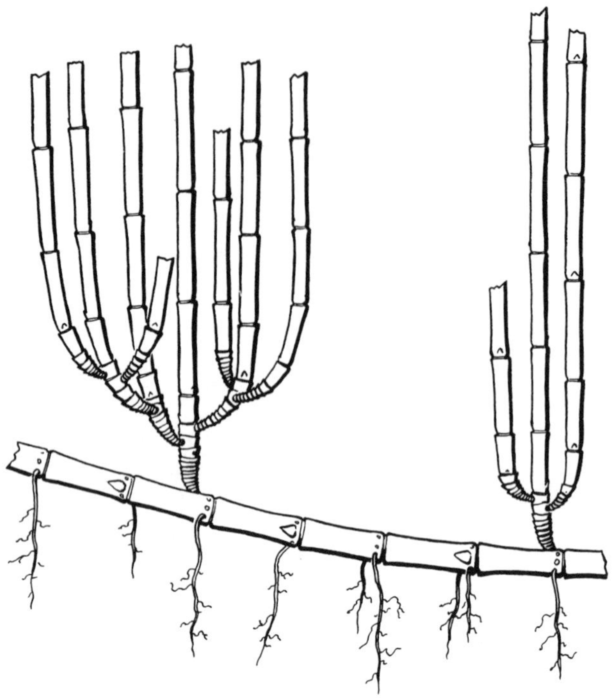

Fig. 14. Tillering of the culms in *Shibataea kumasasa* appears to be a facultative character in this species. Redrawn from Takenouchi 1931*a*:Fig. 1, *3*. See also Fig. 10, *7*.

Key to known forms of the bamboo rhizome system, with examples and illustrations

1. Rhizomes proper all pachymorph; culms each developed from the tip (apical meristem) of a rhizome
    2. Rhizome neck much shorter than the rhizome proper emerging from it
        3. Rhizome proper nearly erect; clump compact-caespitose (Figs. 10, *1* and 11) — *Bambusa pachinensis*
        3. Rhizome proper typically horizontal
            4. Rhizome-culm transition always abrupt; clump compact-caespitose (Figs. 2 and 10, *2*) — *Bambusa tuldoides*
            4. Rhizome-culm transition often prolonged in a metamorph II axis; clump compact- or open-caespitose (Figs. 10, *3* and 12) — *Arundinaria pusilla*

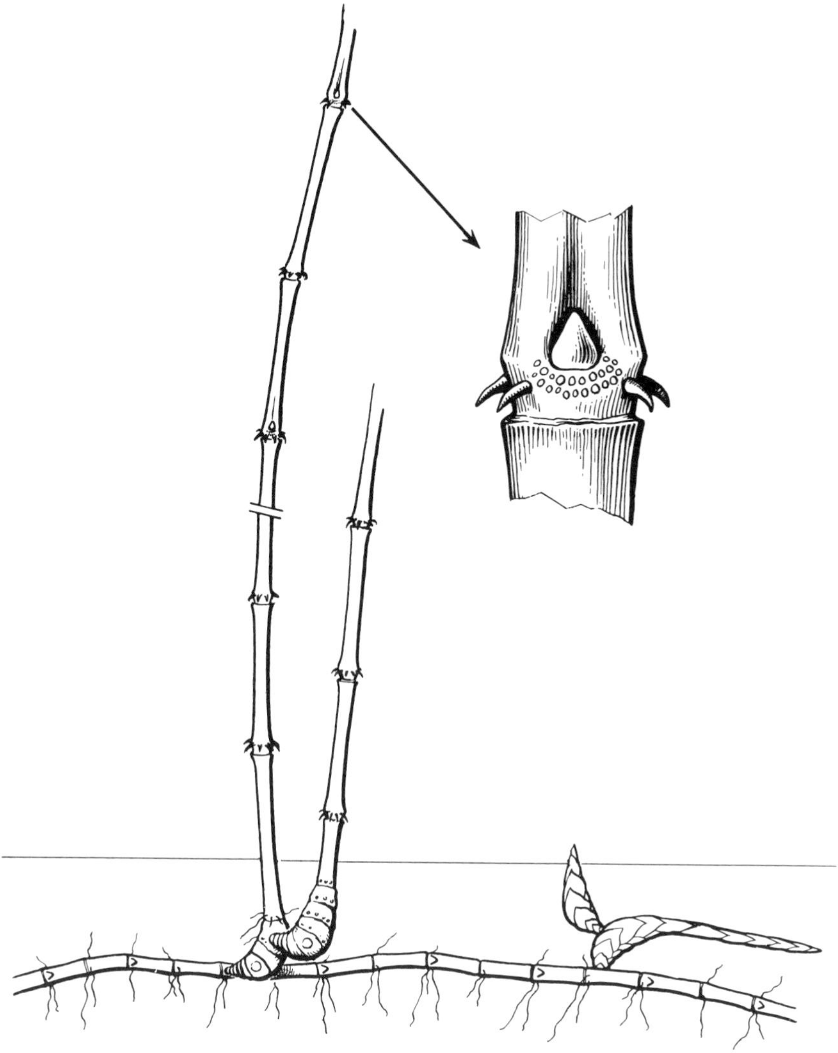

Fig. 15. *Chusquea fendleri.* The capacity to produce either pachymorph or leptomorph rhizomes (or both), is potential in the buds of the subterranean axes of this and some other species of *Chusquea.* The enlarged node above shows the principal branch bud and, below it, a cluster of smaller, subsidiary buds (a character peculiar to the genus *Chusquea*) and also several spinelike aerial roots. Field sketch from plant represented by specimen under McClure No. 21236. See also Fig. 10, *11,* and p. 31.

2. Rhizome neck noticeably elongated; clump open-caespitose (Fig. 10, *4*) — *Sinarundinaria nitida*
2. Rhizome neck greatly elongated
   5. Culms solitary; clump open, diffuse (Fig. 10, *5*) — *Melocanna baccifera*
   5. Culms tillering via metamorph I axes to form tufts; clump compound-caespitose (Figs. 10, *6* and 13) — *Yushania niitakayamensis*

1. Rhizomes proper all leptomorph; rhizome necks all short; culms arising primarily from lateral buds of rhizomes
   6. Growing apex of rhizome becomes negatively geotropic and emerges as a culm, by way of a metamorph II axis
      7. Negative geotropism in rhizome axis obligate, the reaction usually rather prompt; rhizomes show peripheral air canals (Fig. 10, *8*) — *Arundinaria tecta*
      7. Negative geotropism in rhizome axis facultative, the reaction often much delayed; rhizomes lack peripheral air canals (Figs. 10, *7* and 14) — *Shibataea kumasasa*
   6. Growing apex of rhizome typically remains diageotropic; it becomes negatively geotropic and emerges from the ground as a culm only in response to external forces
      8. Culms typically solitary, normally not tillering to form tufts; clump diffuse (Figs. 3, 10, *9*, and 17) — *Phyllostachys bambusoides*
      8. Culms typically forming tufts by tillering through metamorph I axes; clump compound-caespitose (Fig. 10, *10*) — *Indocalamus sinicus*
1. Rhizomes proper of two types, pachymorph and leptomorph, in the same plant; all rhizome necks short; clump compound-caespitose (Figs. 10, *11* and 15) — *Chusquea fendleri*

Clump habit

The nature and meaning of the typical form assumed by the visible part of a given bamboo plant becomes clear only after the form and growth habit of its rhizome system are fully understood. Bamboos with short-necked pachymorph rhizomes grow in discrete, compactly caespitose clumps (*Dendrocalamus membranaceus*, Fig. 16, *1*). Pachymorph rhizomes with a slightly enlongated neck produce a less compact clump (*Sinarundinaria nitida* and *Bambusa*

Fig. 16. *1.* Typical caespitose clump habit of a bamboo with short-necked pachymorph rhizomes, exemplified by *Dendrocalamus membranaceus* at Summit, Canal Zone. U.S.D.A. photo by McClure. *2.* When, in the course of its growth, a leptomorph rhizome encounters a steep declivity, it may emerge into the open. When this happens, the rhizome at first manifests a negative geotropic reaction and, by elongation and curving downward, may re-enter the soil as did the shortest one of the four shown here. If the rhizome is unsuccessful in re-entering the soil, the growing point usually then turns upward in a broad curve, forming first a transitional metamorph II axis, then an erect culm. Two stages of this sequence are here illustrated by a photograph of several rhizomes of *Phyllostachys bambusoides* cv. 'Castillon' that have emerged from the side of a pile of wood chips. See Fig. 4.

vulgaris). Bamboos with leptomorph rhizomes produce culms typically in open array (*Phyllostachys pubescens,* Fig. 17), and the culms of bamboos with only long-necked pachymorph rhizomes (*Melocanna baccifera,* Fig. 82) stand similarly isolated. In writings about the bamboos of India, all bamboos manifesting this open type of clump habit are usually lumped together as "single-stemmed," without distinction and without reference to disparities in subterranean characteristics. Bamboos with long-necked pachymorph rhizomes and culms that tiller via metamorph I axes produce composite clumps made up of compact tufts that appear separate and distinct but actually are connected below the ground by elongated rhizome necks (*Yushania niitakayamensis,* Fig. 13). This same clump habit is found in bamboos that have both leptomorph and pachymorph rhizomes (*Chusquea fendleri,* Figs. 10, *11* and 15) and also in those that have leptomorph rhizomes and culms that tiller via metamorph I axes as in *Indocalamus sinicus* (Fig. 10, *10*), *Semiarundinaria fastuosa* (Fig. 18), and *Shibataea kumasasa* (Fig. 14). In any particular plant of some of these bamboos, deviations induced by any influence that interferes with its free natural growth may obscure the expression of clump habit predicated above.

The culm

Bamboo culms are either lateral branches or apical projections of the rhizome, depending on whether the rhizome proper is leptomorph or pachymorph. In habit, they vary from strictly erect, erect with pendulous tips, or ascending, through broadly arched to clambering (Fig. 19) and from nearly straight to strongly zig zag. Given uniform environmental conditions, each habit character is fairly consistent within a given taxonomic entity, such as a species or a subspecies.

In any given bamboo plant, the culms may originate in one (or sometimes two) of the following ways: (1) as the transformed distal end (apex) of a pachymorph rhizome, either with or without the intervention of a metamorph II axis; (2) from a lateral bud of a leptomorph rhizome; (3) from a bud on the underground base of a culm, by tillering, or by way of a metamorph I axis; (4) as the transformed distal end (apex) of a leptomorph rhizome through a metamorph II axis.

Fig. 17. Open diffuse clump habit in a bamboo with leptomorph rhizomes, exemplified by *Phyllostachys pubescens* in the garden of Mr. Sankichi Ishida, at Komazawa, near Tokyo, Japan. By selective harvesting of the young shoots, the average distance between culms is kept at a level corresponding to optimum yield. U.S.D.A. photo by P. H. Dorsett. See Fig. 3.

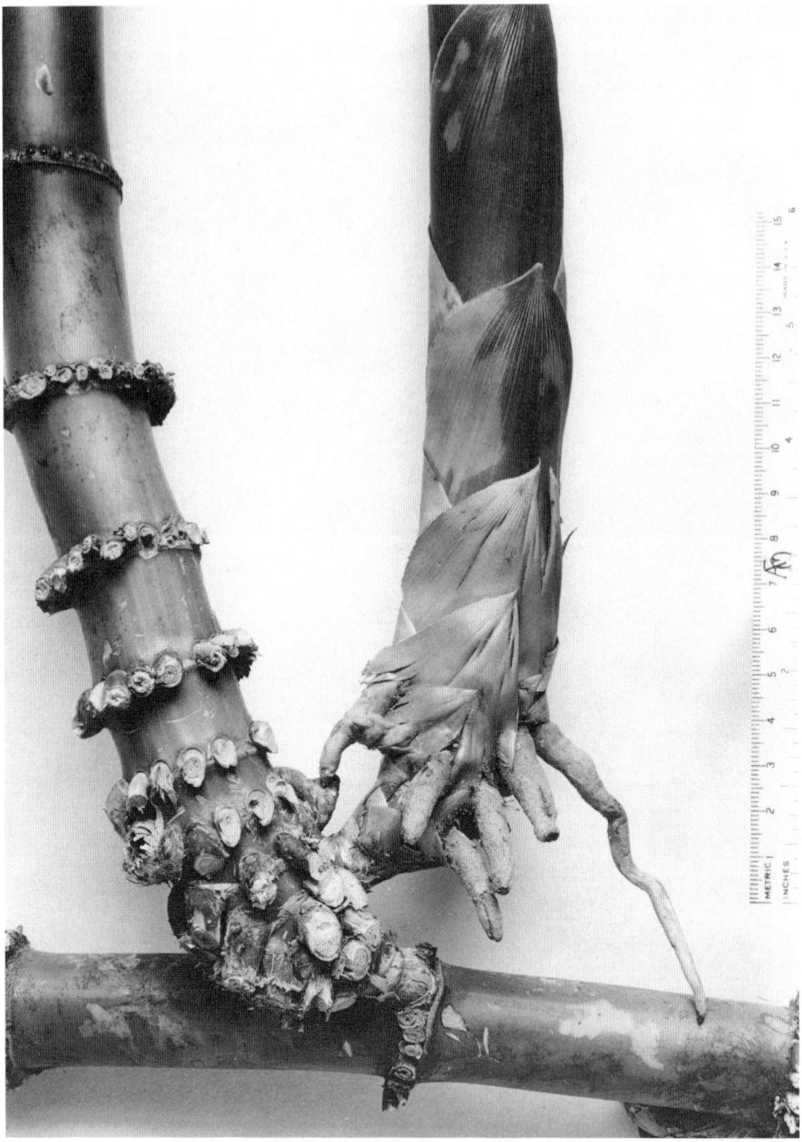

Fig. 18. *Semiarundinaria fastuosa* (P.I. 112080). Underground part of a plant, showing a leptomorph rhizome and the characteristic tillering of a primary culm from a basal bud. U.S.D.A. photo by Robert Taylor. See Figs. 10, *10* and 20.

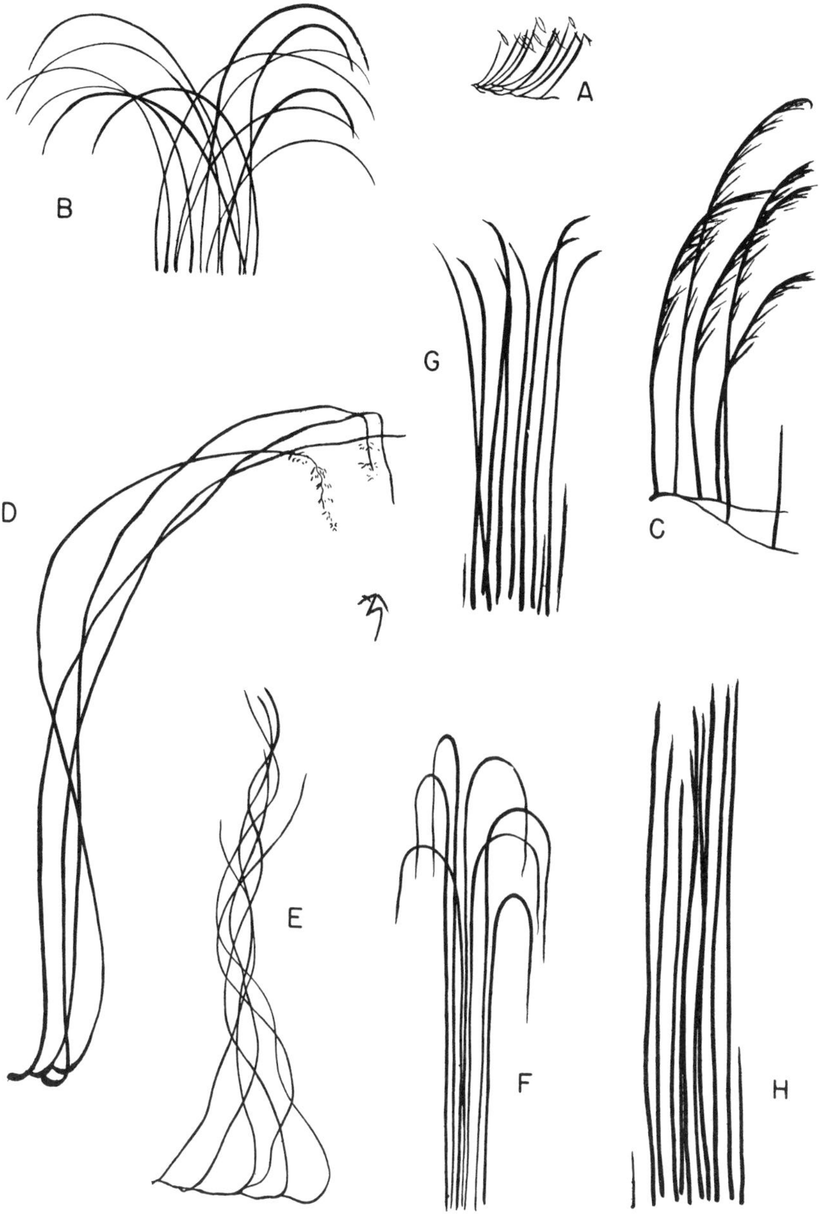

Fig. 19. Selected examples of culm habit, illustrated diagrammatically. (*A*) *Sasa palmata;* (*B*) *Sinocalamus beecheyanus;* (*C*) *Phyllostachys nigra;* (*D*) *Schizostachyum hainanense;* (*E*) *Dinochloa scandens* (auctt. non Kuntze); (*F*) *Sinocalamus affinis;* (*G*) *Bambusa textilis;* (*H*) *Arundinaria amabilis.* Note the result of circumnutation that took place during the growing stage of the culms in (*E*). From McClure and Li 1941:Fig. 1.

A culm that develops from a lateral bud of a leptomorph rhizome (*Semiarundinaria fastuosa,* Figs. 18 and 20, *right*) is an example of the most complete form. It shows two parts: the aerial culm proper and the subterranean culm base. The culm base consists of two structurally distinct parts: the culm base proper and the culm neck. The culm base proper is a narrow, inverted cone with each node marked by the presence of a sheath, a verticil of roots, and, in many cases, a bud. In a plant that tillers, such a bud can give rise to a new culm. The internodes may be solid but generally are hollow and short, increasing in length by small increments from the basal one to the uppermost, which appears at about the level of the surface of the ground. At this point the culm proper begins, as signalized by a more or less abrupt increase in the length of the successive internodes, which here take on a more strictly cylindrical form. At its lower end, the inverted cone of the culm base proper is attached to its point of origin on the rhizome by the slender, curved neck.

A culm arising as the transformed distal end of a leptomorph rhizome (Figs. 4 and 16, *2*) has a metamorph II axis in place of the specially differentiated subterranean culm base and culm neck just described. Even without direct evidence of its actual relationship to the rhizome, the peculiar origin of such a culm is made clear by the character of this basal part. Its lower part is broadly curved in an upward sweep; its lower internodes are solid or nearly so, and shorter than those of culms that emerge from lateral buds of the rhizome. The change from the small-lumened or rarely solid condition of the internodes of the rhizome axis to the large-lumened condition of the internodes of the culm takes place gradually, though in some species (*Phyllostachys bambusoides*) the internodes of such a culm may all be solid.

A culm arising within a pachymorph rhizome system typically originates only as the transformed distal end of a rhizome axis, as in *Bambusa tuldoides* (Fig. 2). In all such culms, the place of the culm base proper is taken by the rhizome itself, to which the culm is here terminal. A neck is found only at the base of the rhizome, and there is no culm base proper and no culm neck (Figs. 2 and 20, left). Exceptionally, new culms may arise in a bamboo with a pachymorph rhizome by tillering, by way of a metamorph I axis, as in *Yushania niitakayamensis* (Fig. 13), or by

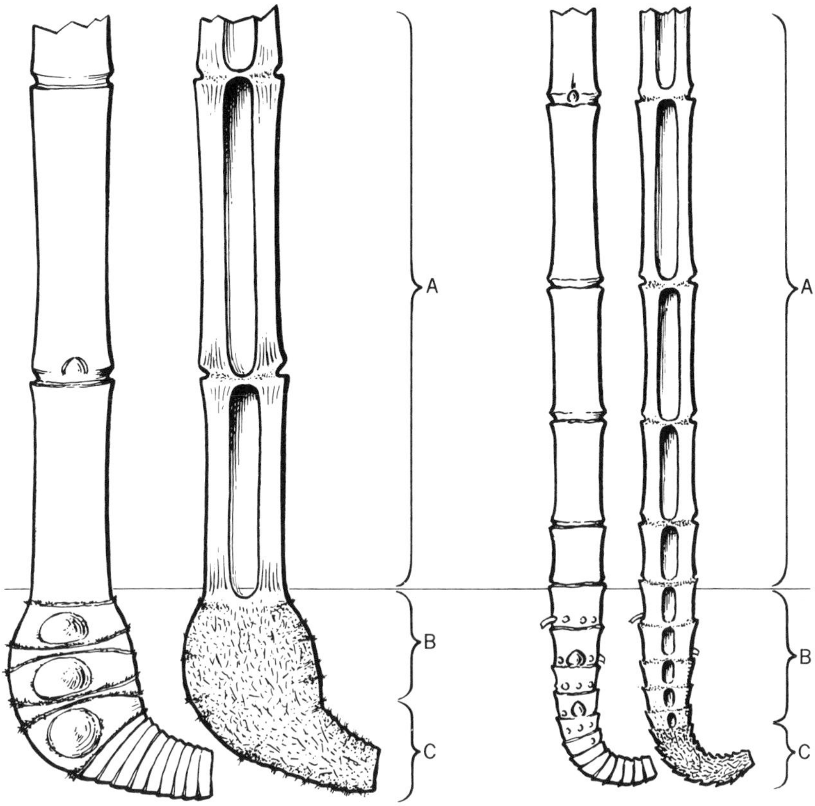

Fig. 20. Lower part of a culm of each of two species—(*left*) *Dendrocalamus latiflorus,* (*right*) *Semiarundinaria fastuosa*—both split in two, to illustrate Takenouchi's interpretation of the structures shown: "Culm with [*left*] and without [*right*] sharp distinction between the true culm and the underground culm base: (*A*) culm proper; (*B*) underground culm base; (*C*) stipe or stalk." As interpreted herein, (*A*) is the culm proper in both; (*B*) in the culm on the left is a pachymorph rhizome, and in the culm of the right it is the culm base; (*C*) in the culm on the left is a *rhizome neck,* and in the culm on the right it is a *culm neck.* The leptomorph rhizome of the culm on the right, not shown here, may be seen in Fig. 18. The discrepancies between the two interpretations given above illustrate the observation made elsewhere herein (p. 6) that "some of the outward forms and behavior of the bamboo plant are not easily amenable to the rigid application of a hierarchy of technical terms or reasoned concepts." Redrawn from Takenouchi 1931*a*:Fig. 4.

way of a metamorph II axis that intervenes between the rhizome and the culm proper as in *Arundinaria pusilla* (Fig. 12).

Small culms generally taper gradually from base to tip. In larger ones, the lower half or so is roughly cylindrical or imperceptibly tapered, and the rest of the culm more sharply so. In the largest ones from vigorous plants of many species (in the genus *Phyllostachys,* for example) there may be an appreciable increase in the diameter for some distance, beginning at or near the base, then a gradual decrease in diameter, and finally, in the upper $\frac{1}{3}$ or $\frac{1}{4}$ of its length, the culm is more sharply tapered.

In the cycle of increase and decrease in length, and in the progressive change of shape, of the successive internodes and their associated sheaths (a combination of structures illustrated by Porterfield, 1930*b*, Fig. 4, as phytons) bamboo culms and their branches manifest a periodicity that is well described and documented by Porterfield (1933). Shigematsu (1958) has developed mathematical formulas, based on detailed measurements, to elucidate the patterns of change in length and diameter of culm internodes and thickness of culm wall characteristic of 15 species representing 8 genera of Japanese bamboos.

The march of increase and decrease in internode length, as found in self-supporting bamboo culms, may be graphically represented by plotting the length of each internode against its serial number, beginning with the lowest. The graphs so constructed (Figs. 21–23) generally take the form of the profile of an asymmetrical mountain, with a single acute or more or less rounded crest, or with two or more, usually less well-marked, peaks in the median region. It will usually be steeper on the "west" side (as a result of larger length increments in successive internodes in the lower part of the culm) and more gently sloped on the "east" side. In scandent bamboos, of which apparently there are no culm-internode measurements on record, the curve of internode length would embrace a broad, nearly flat middle section, representing a series of mid-culm internodes of more or less equal length.

Schromburgk (1841) brought to light the occurrence of a greatly elongated first above-ground internode in culms of a bamboo now known as *Arthrostylidium schromburgkii.* In a culm 15 m tall the first internode may be 5 m long. Another feature of this plant (McClure 1964:2) is the occurrence, immediately above the long

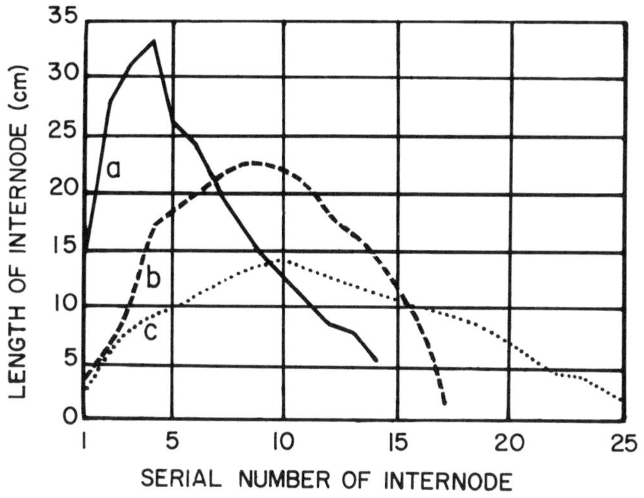

Fig. 21. Graphic representation of the march of increase and decrease in internode length in the culms of three bamboos: (*a*) *Bambusa multiplex* (as *B. nana* var. *typica*); (*b*) *Sinobambusa tootsik;* and (*c*) *Arundinaria* (as *Pleioblastus*) *simonii.* Redrawn from Takenouchi 1932:Fig. 16.

first internode, of two or more nodes without an intervening internode, intercalary growth having been suppressed locally. Here the consecutive sheath scars are usually less than 1 inch distant from each other. The insertion of branch complements and the subtending branch sheaths at such congested nodes is distichous, and the development of the nodes is normal in other respects as well. The congestion of nodes (suppression of internodes) also occurs, in conjunction with a greatly elongated first internode, in a number of other species of this and other bamboo genera of the Western Hemisphere, including *Glaziophyton* and *Myriocladus.* Since these elongated first internodes, followed by congested nodes, appear regularly in all culms of the species of which they are characteristic, it seems clear that they are not teratic in nature, but have a stable genetic basis.

The shape of the component segments (nodes and internodes) of bamboo culms varies from species to species, and in some cases from genus to genus. Differences in the internodes include the degree and extent of the depression (groove) immediately above the insertion of a bud or a branch complement. In all bamboos of the genus *Phyllostachys,* for example (Fig. 27) each culm inter-

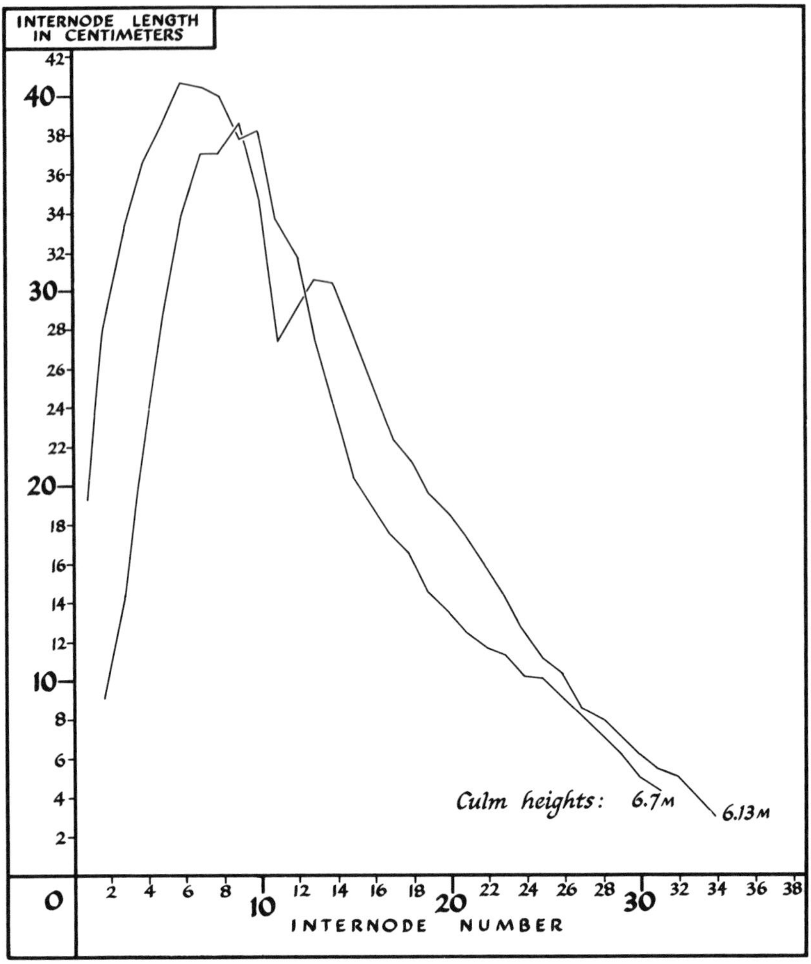

Fig. 22. Pattern of increase and decrease in internode length in two culms from the same clump of *Phyllostachys nidularia* (P.I. 67399), based on measurements made at the U.S.D.A. Barbour Lathrop Plant Introduction Garden, Savannah, Georgia, by McClure and Ditmeyer, August 12, 1942.

node is grooved (sulcate) throughout its length above the insertion of a bud or branch complement; in *Shibataea* the internodes are only flattened. The internodes of the culm branches and of the rhizome in these genera show this same characteristic.

The buds at culm nodes generally protrude more or less perceptibly above the level of the surrounding culm surface. In some bamboos whose culms have prominent culm nodes ("supranodal ridges"), buds at lower nodes are so deeply sunken that they do not project above this ridge (*Phyllostachys arcana; Sasa longiligulata*). It has been suggested (Takenouchi 1932, trans., p. 56) that pressure exerted by the bud upon the contiguous developing internode, while both are closely confined by the enveloping sheath, is responsible for the channeling of the internode at and above the insertion of a bud. But this channeling is so diverse in its expression, and so completely lacking in some bamboos (known species of *Melocanna* and *Schizostachyum*, for example) where such pressure can scarcely be wholly absent, that it must have some other cause as well. Unpublished studies by Frank Venning on the young shoots of bamboos of several genera indicate that there may be fundamental differences in the manner in which the bud primordia of the culm and its branches are initiated, and that these differences may be correlated with the absence, or the presence and extent, of channeling. Takenouchi (1931*b*:154) relates the deep

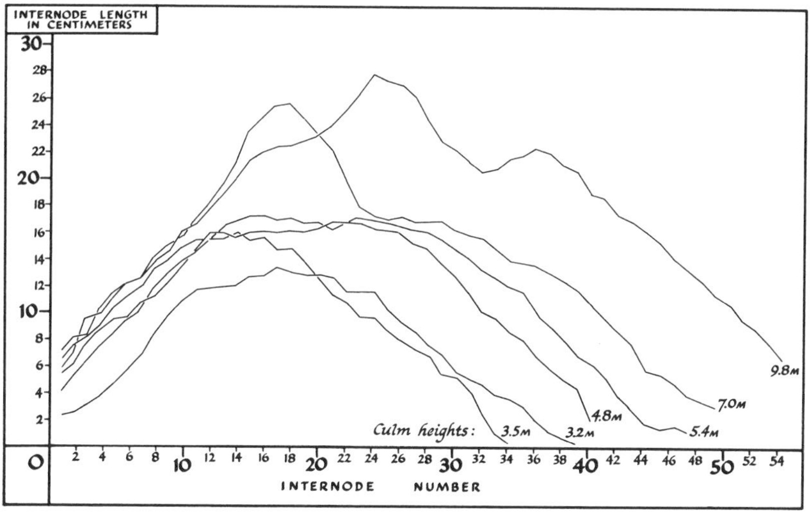

Fig. 23. Pattern of increase and decrease in internode length in six culms from the same clump of *Phyllostachys pubescens* (P.I. 80034), based on measurements made at the U.S.D.A. Barbour Lathrop Plant Introduction Garden, Savannah, Georgia, by McClure and Ditmeyer, August 12, 1942.

channeling throughout each culm internode in *Phyllostachys* to the simultaneous development of all of the internodes (and buds) of a young shoot in this genus, while in those bamboos where channeling does not continue throughout the internode, the internodes and buds are said to develop successively and acropetally.

A characteristic of many bamboos is the appearance of a white exudate on the surface of the culm internodes. This varies from a barely perceptible bloom (similar in superficial appearance to that familiarly seen on some kinds of plums and grapes) to a conspicuously abundant, fluffy, flourlike deposit that more or less completely conceals the green surface of the internodes, as in *Lingnania chungii* (Fig. 24). The time of the appearance of the exudate in relation to the development of the culm, and its texture, abundance, distribution, and persistence, are variables that may, in conjunction with other internode characters, be useful in the identification of some species of bamboos in the field. Kurz (1876:252) mentions the "white-pruinous" culms of *Dendrocalamus giganteus* and of an unidentified bamboo, "booloo idyooh," and the occurrence of "a white fugaceous meal" on the culms of "booloo kenneng," a yellow-culmed form of *Bambusa vulgaris.* The exudate collected from young culms of *Lingnania* (as *Bambusa*) *chungii* was analyzed at Harvard University by Chang (1938). One component, comprising about 25 percent of the powder, was identified as a triterpenoid ketone, identical with or closely resembling friedelin, a constituent of the wax of corks. Two similar compounds, bambuselin I and bambuselin II, were later described from the same source (Chang 1941).

Among other conspicuous internode characters are the surface texture, the combination and pattern of colors, and the type of pubescence. The pigskinlike surface of the lower mid-culm internodes in *Phyllostachys makinoi* and *Ph. viridis;* the papillate surface of the lower and mid-culm internodes of *Chimonobambusa quadrangularis;* the color pattern yellow-with-a-green-sulcus of the internodes of the whole culm in *Phyllostachys bambusoides* cv. 'Castillon'; the velvetlike pubescence of the internodes in young culms of *Ph. pubescens;* and the hard, rough, siliceous surface of the lower culm internodes, often clothed with pale appressed hairs, in species of *Schizostachyum,* are examples. Kurz (1876:251–252) lists others.

Fig. 24. *Lingnania chungii.* Plants of this species are made conspicuous by a layer of flourlike exudate that more or less completely conceals the green surface of the internodes in culms of the current year's growth. Culms in plants of mature stature may lack buds and branches in the lower $^{2}/_{3}$ or $^{3}/_{4}$ of their height. "Gray bloom on green bamboo culms adds an indescribable beauty to this forest cathedral." Au Tsai, Kwangsi Province, China. Reproduced from McClure 1935:203.

Arber (1934:77, Fig. 27) describes and figures as "spinous hairs" the characteristic retrorse hooks—outgrowths of the epidermis—that appear on the slender upper internodes of young culms in a climbing bamboo, *Arthrostylidium multispicatum.*

The culm nodes also show more or less strongly marked variation in appearance as between different species. An important feature of culm nodes is the sheath scar—which marks the sheath-node of Hackel (1887, trans., p. 3). The sheath scar is a transverse circumaxial offset in the surface of the culm, marking the circumaxial locus of insertion of the culm sheath. The sheath scar may be thin and inconspicuous, without any outgrowths, as in *Melocanna,* or it may be thick, with a strongly flared callus (see p. 67) and conspicuously fringed with brown hairs, as in *Lingnania chungii* (Fig. 25, *A*), *Phyllostachys nidularia,* and *Sinobambusa tootsik.* The sheath scar may be symmetrical and level, or it may take a strong dip on the side that passes under the point of insertion of the branch complement, as it does in *Bambusa arundinacea*

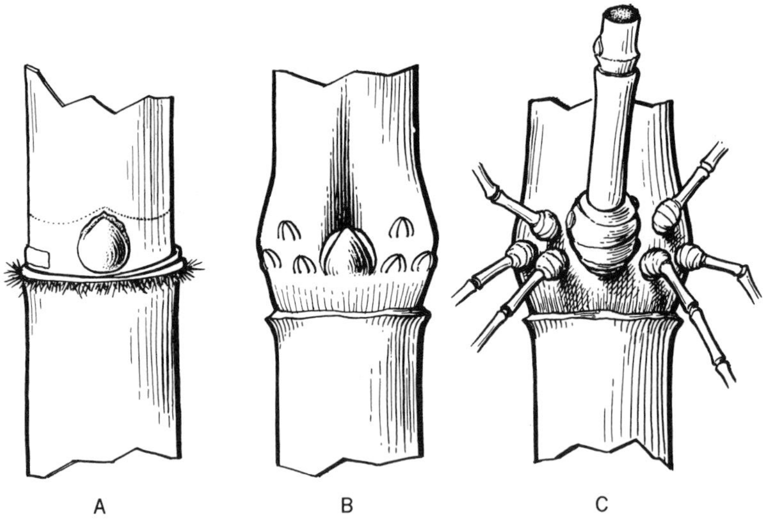

Fig. 25. (*A*) Mid-culm node of *Lingnania chungii,* showing sheath scar with conspicuous callus and fringe of hairs and typical solitary branch bud; for an illustration of the branch complement of this species, see Fig. 26, *4.* (*B*, *C*) Mid-culm node of *Chusquea sp.,* showing (*B*) multiple-bud (constellate) complement peculiar to this genus, and (*C*) branch complement; both somewhat diagrammatic; not drawn to scale.

and in many species of *Chusquea*, especially those in the section *Rettbergia*.

Velasquez and Santos (1931) published an excellent portrayal of the culm anatomy of five tropical bamboos.

Branch buds at culm nodes

Branch buds emerge on alternate sides of the culm just above the sheath scar at successive nodes. Each bud stands in a position median to the base of the sheath that subtends it. With the exception of those of the genus *Chusquea*, all bamboos studied have the primary buds solitary at culm nodes, as in *Lingnania chungii* (Fig. 25, *A*). In *Chusquea* the principal bud at each culm node (above ground) is flanked by two to many smaller ones. These represent independent branch primordia, each enclosed in a separate prophyllum (Fig. 25, *B*). In exceptional cases (sometimes, for example, in *Ch. pittieri*) the small buds emerge in a continuous narrow band that extends completely around the culm. The unusual nature of the branch buds, and the course of development of the branch complement, shown in *Merostachys* and in some species of *Arthrostylidium* will be described below (pp. 53ff).

Takenouchi (1931*b*:143) was the first to call attention to the order of breaking of branch buds on the young culm. In some species these buds awaken in acropetal order, that is, each bud breaks as soon as the internode to which it is basal has completed its growth; examples are typical *Arundinaria gigantea* and all species of *Phyllostachys*. In certain other species the order is basipetal. Here, some or all of the branch buds of the full-grown young culm remain dormant for a while (sometimes for several weeks, as in *Bambusa textilis*) before breaking. In small plants of *Semiarundinaria viridis* the mid-culm buds break first, and the wave of branch growth then spreads simultaneously upward and downward in the culm. In some bamboos, and under certain conditions, both the awakening of the primary branch buds of the culm and their subsequent proliferation may be irregular both in time and in order of incidence. In mature bamboo plants one or more of the lowest above-ground buds on each culm may remain dormant indefinitely. In most bamboos, immature plants may bear buds at all of the culm nodes. In many bamboos, culms of progres-

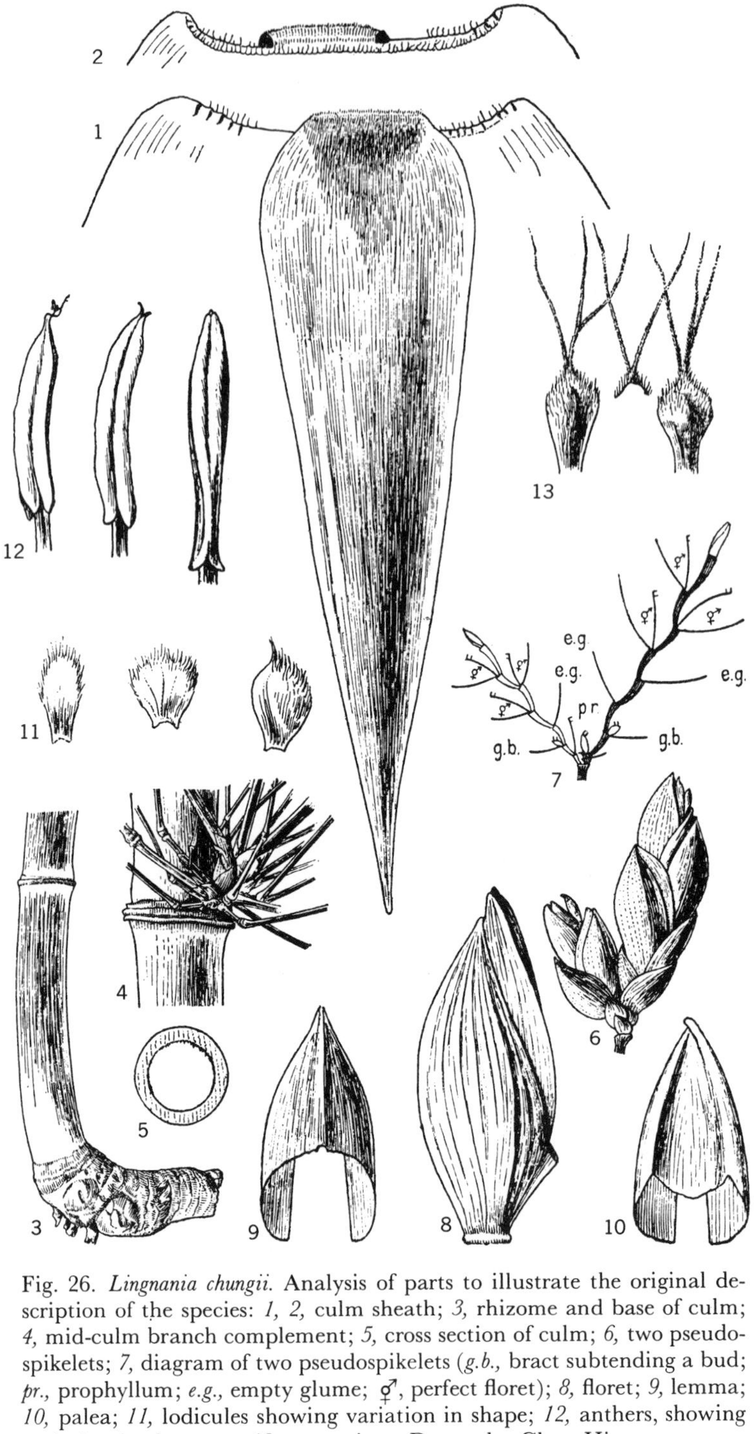

Fig. 26. *Lingnania chungii.* Analysis of parts to illustrate the original description of the species: *1, 2,* culm sheath; *3,* rhizome and base of culm; *4,* mid-culm branch complement; *5,* cross section of culm; *6,* two pseudospikelets; *7,* diagram of two pseudospikelets (*g.b.,* bract subtending a bud; *pr.,* prophyllum; *e.g.,* empty glume; ⚥, perfect floret); *8,* floret; *9,* lemma; *10,* palea; *11,* lodicules showing variation in shape; *12,* anthers, showing variation in the apex; *13,* gynoecium. Drawn by Chan Hin-yau.

sively larger sizes (those approaching mature stature) have a progressively longer series of budless lower (above-ground) nodes. In *Bambusa textilis* and *Arundinaria amabilis,* for example, culms from plants of mature size may lack buds and branches in the lower $\frac{1}{2}$ to $\frac{2}{3}$ or even $\frac{3}{4}$ of their length. This is a desirable aesthetic feature in bamboos planted for ornament—especially such species as *Lingnania chungii* (Fig. 24). In culms used in handicrafts or in industry, the absence of the knots that mark the insertion of branch complements makes for economy and ease in working the material. In the sterile condition, culms commonly remain unbranched above the ground in bamboos of the genera *Glaziophyton, Guaduella, Neurolepis, and Puelia.* The incidence of buds at subterranean nodes of culms arising from lateral buds of leptomorph rhizomes may differ from species to species in the same genus (*Phyllostachys*), or within the same species as the plant advances toward maturity (*Arundinaria amabilis*).

The branch complement

In bamboos whose culms are branched (the condition prevailing in most genera) it is at the mid-culm nodes that the typical form of the branch complement appears. The lowermost one or more branch complements in the series on a given culm are generally less well developed than those at mid-culm nodes, and so cannot be taken as typical examples. In most bamboos the branching at the uppermost culm nodes is usually not strongly distinctive of a genus or species. However, bamboos with a fixed number of branches at each culm node, as in *Sasa* and *Phyllostachys,* are exceptional in this respect.

In a few bamboo genera, the number of axes in a mid-culm branch complement is fixed at a characteristic total. In *Sasa,* the single primary branch remains solitary because it possesses no basal buds. In *Phyllostachys,* the typical primary mid-culm branch complement is binary, the two axes being more or less strongly unequal (subequal in *Ph. arcana*), and subsidiary branching is limited to a single, very much smaller axis that sometimes emerges between them from a bud at the base of the smaller of the two (Fig. 27).

In many bamboos the branch primordium at each mid-culm node ramifies precociously before it ruptures the prophyllum. The

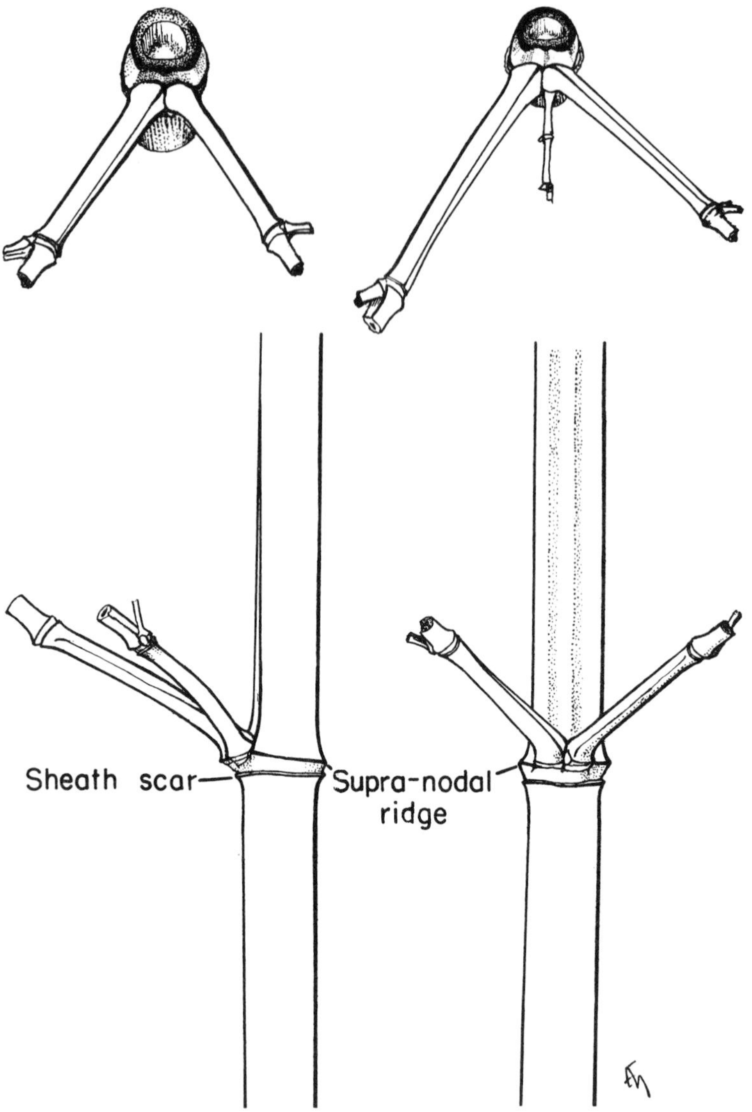

Fig. 27. Branch complement from a mid-culm node of *Phyllostachys elegans* (P.I. 128778). The features shown characterize the genus in the following respects: the branch-bearing nodes of the culm are marked by a prominent ridge (sometimes referred to by Hayata as a pulvinus, and by others as a supranodal ridge) above the sheath scar; the internodes of the culm are flattened throughout by grooves above the point of insertion of each bud or branch complement (all branches are similarly flattened); and the primary branches are borne typically in twos at each node, but sometimes a third one develops between the other two (*upper right*). From McClure 1957:Fig. 2. See p. 13, and *Sheath node* and *Culm node* in the Glossary.

generic name *Pleioblastus* (Nakai 1925:145–146) was based on this character. However, Nakai's description may leave the incorrect impression that the bamboos included by him in this genus actually have more than one primary bud at each culm node. The extent of this precocious ramification of the original branch primordium varies from species to species and from genus to genus. Usui (1957*b*) illustrates the phenomenon in several of its manifestations (Fig. 28). In *Melocanna baccifera,* and the known species of *Schizostachyum,* the degree of precocious branching is very pronounced, and the branch complement emerges from the prophyllum a sunburst of juvenile axes. In these genera, the primary branch is indistinguishable in size from the subsequent orders of axes, and the result is a dense tuft of slender subequal branches. In these and other bamboos, the growing out of buds at proximal nodes of branches of successively higher orders augments still further the number of component axes in a given branch complement. The tempo of this augmentation, and the length of time during which it continues, determine the composition and nature of the mid-culm branch complement found in any culm of a given age. In many bamboo genera (*Bambusa, Dendrocalamus, Gigantochloa*) the primary branch emerges, and remains, strongly dominant. Subsequent orders of axes that develop from its basal buds have progressively diminished diameters and lengths (Fig. 30).

In bamboos of the genus *Chusquea,* the branch that arises from the large central bud at each mid-culm node is strongly dominant—sometimes nearly equal to the mother culm in diameter and length—and is often tardily or not at all rebranched at its basal nodes. However, this large central bud often remains dormant, and may become completely obscured by the dense cluster of slender, subequal branches that develop from the small adventitious buds at the same culm node (Fig. 25, *B*).

The branch complement in bamboos of the genus *Merostachys* and in some species of *Arthrostylidium* bears a superficial resemblance to that in *Schizostachyum* and *Melocanna,* in that it consists of a large number of slender, subequal leafy axes that usually branch from their basal buds only. Closer examination, however, reveals a basic difference. In *Schizostachyum* and *Melocanna,* the branches at mid-culm nodes radiate from a common point, the locus of insertion of the primary branch primordium. In *Mero-*

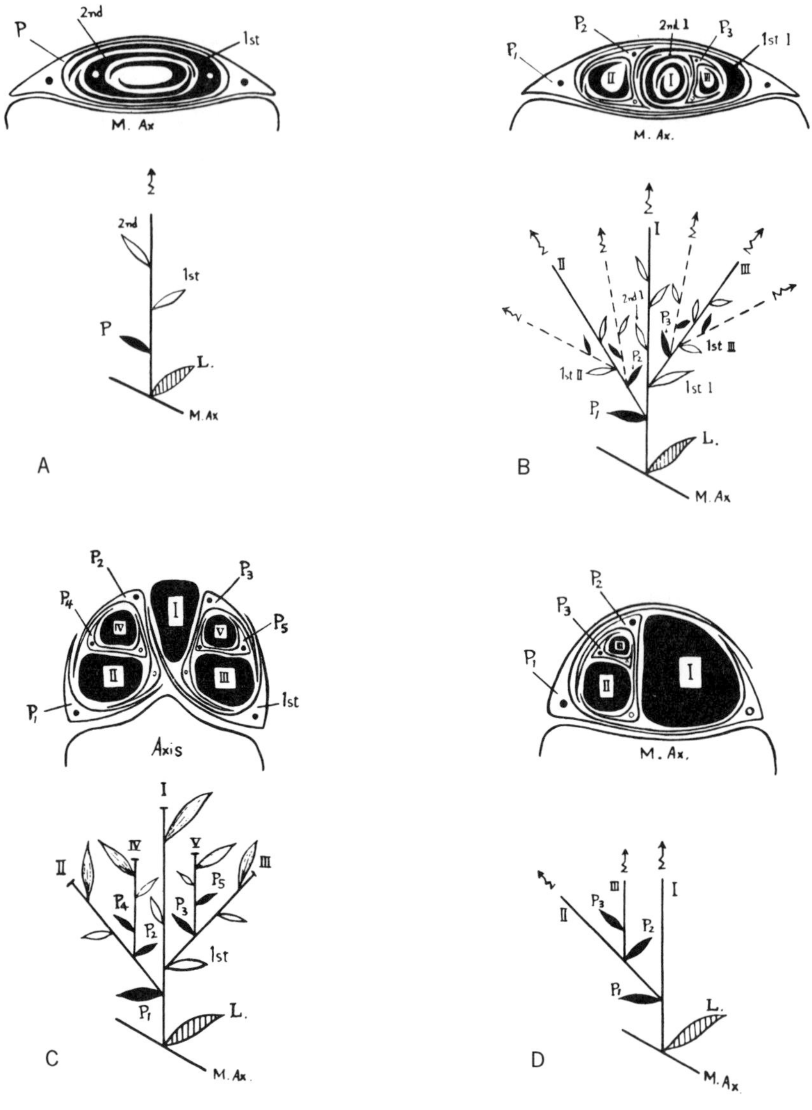

Fig. 28. Diagrammatic cross sections of four types of mid-culm branch buds, and diagrams of corresponding branch complements, presented by Usui (1957:Figs. 1–4) as typical of bamboos of six genera. (*A*) The genus *Sasa;* here the primordial branch axis remains solitary, both before and after the rupture of the prophyllum that enclosed it in the bud stage. (*B*) Represents three genera: *Pleioblastus* (merged with *Arundinaria* in the present work), *Semiarundinaria,* and *Sinobambusa;* the broken lines ostensibly represent branches produced after the rupture of the prophyllum. (*C*) The genus *Shibataea.* (*D*) The genus *Phyllostachys;* here the branches emerge as a pair of subequal axes, the basal bud on the smaller one of which may remain dormant, or develop as shown here, to form a third, very much smaller one between the other two. The captions given above are somewhat modified from the originals. The author's legend: "*P,* prophyll; *1st,* the first foliage leaf; *2nd,* the second foliage leaf [on a given axis]; *M. Ax.,* main axis; *L,* a leaf on the node of main axis; the black leaf shows the prophyll and the white leaf shows the foliage leaf. Branch internodes not drawn to scale." See Figs. 27 and 29.

stachys (Fig. 31) and in some species of *Arthrostylidium* (*A. harmonicum,* Fig. 32), the focus of radiation of the mid-culm branches falls within a vacant area. This hitherto undescribed character has its origin in the peculiar nature of the branch bud, the stage of development of the internode when the bud is formed, and the behavior of the primary branch primordium (primordia?) as subsequent development takes place. The material at hand for

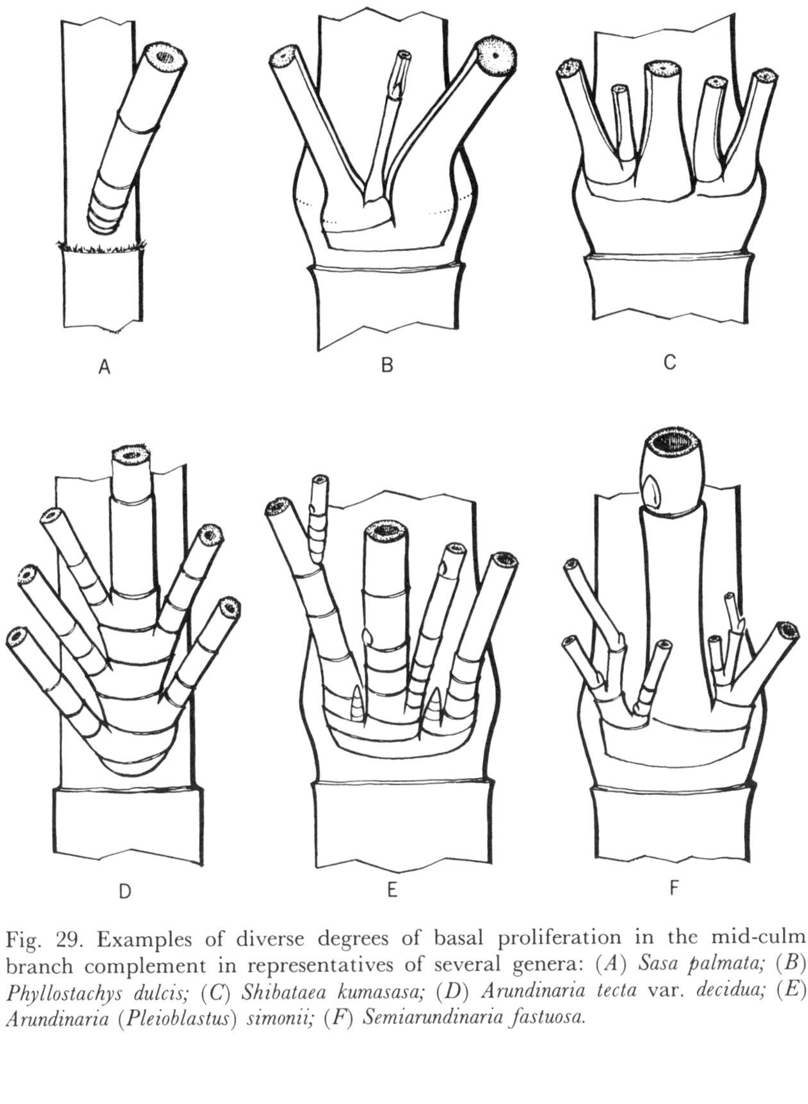

Fig. 29. Examples of diverse degrees of basal proliferation in the mid-culm branch complement in representatives of several genera: (*A*) *Sasa palmata;* (*B*) *Phyllostachys dulcis;* (*C*) *Shibataea kumasasa;* (*D*) *Arundinaria tecta* var. *decidua;* (*E*) *Arundinaria* (*Pleioblastus*) *simonii;* (*F*) *Semiarundinaria fastuosa.*

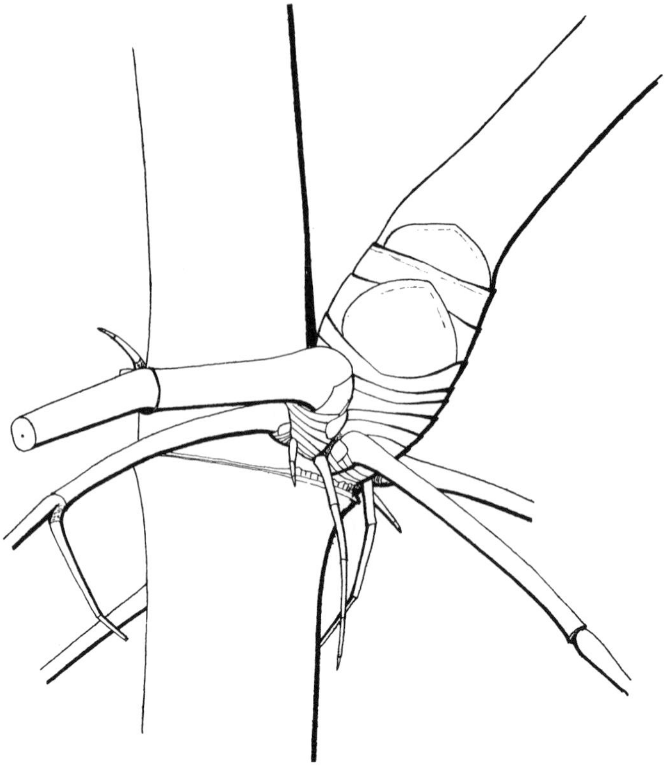

Fig. 30. Branch complement of *Bambusa rutila,* illustrated from a mid-culm node of a 3-year-old culm in the Lingnan University Bamboo Garden (LUBG 1476) showing a strongly dominant primary branch and thornlike branches of higher orders. Drawn by M. K. Hoh, 1939.

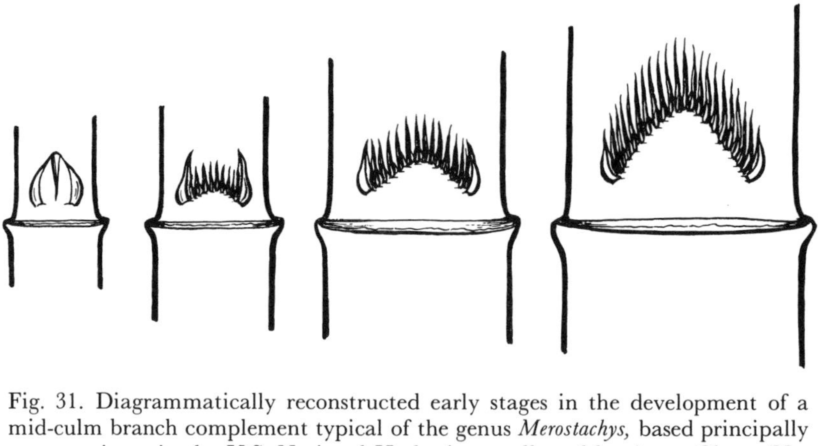

Fig. 31. Diagrammatically reconstructed early stages in the development of a mid-culm branch complement typical of the genus *Merostachys,* based principally on a specimen in the U.S. National Herbarium, collected by Agnes Chase (No. 9463). The development of the branch complement in *Arthrostylidium harmonicum* (Figs. 32 and 33) is similar.

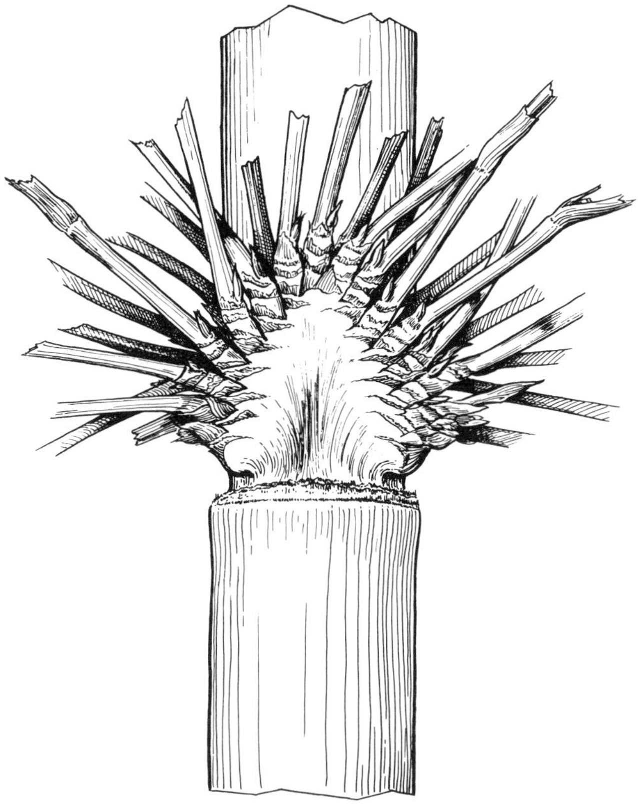

Fig. 32. *Arthrostylidium harmonicum.* Basal portion of a fully developed mid-culm branch complement, showing the relation of the component branches to each other at the locus of their emergence. Based on a specimen in the U.S. National Herbarium, collected in Ecuador by W. H. Camp (No. E-1613). Drawn by Elmer W. Smith. See Figs. 31 and 33.

study is inadequate to a complete elucidation of the development of this type of branch complement. The earliest stage available to me is found in a sterile specimen of *Merostachys sp.* accessioned in the National Herbarium under US 562138 (Agnes Chase 9458). A slightly later stage is shown in US 1257740 (Agnes Chase 9466). A solitary two-keeled prophyllum encloses the primary branch primordium (primordia?). Several primordia appear later in alternate, distichous insertion (emergence), enclosed individually in small, bladeless sheaths. The spatial pattern of their insertion suggests separate origins, though this can only be determined by anatomical studies. In the early stages, the bud lies within a few millimeters of the level of insertion of the culm sheath by which it is subtended, and the surrounding culm tissue is still in a meristematic condition. When the branch complement has reached its full development, however, the spot originally occupied by the bud is vacant, and the primary branches are seen to be inserted in an arch above its periphery. The fragmentary evidence at hand suggests that, by the elongation of the culm internode, the several branch primordia are carried forward acropetally and fanned out along the surface of the culm. The branches arising from these primordia finally emerge at separate loci arranged in a tight arch. To characterize branch complements with this type of insertion the term apsidate is proposed. The manner of origin of branch complements of this type remains to be elucidated by anatomical studies initiated at early stages of their formation.

Branch complements originating in this manner generally comprise numerous slender, subequal axes. In *Arthrostylidium harmonicum* (Figs. 32 and 33) the pattern of development appears to be basically the same as that typical in most of the known species of *Merostachys.* Exceptionally, however, one or two of the axes at the top of the arch of insertions may elongate and branch from buds other than their basal ones. This has been observed frequently in *Arthrostylidium capillifolium* and *A. cubense,* and once in a sterile leafy specimen of an unidentified species of *Merostachys* (US 1910797; Froes 20060, sheet 2).

In branch complements where the primary axis is dominant, this branch resembles the culm itself in varying degrees, and usually on a very reduced scale. This resemblance is especially noticeable in many bamboos that have pachymorph rhizomes. Here the base of the primary branch at each mid-culm node incorporates a structure somewhat resembling the pachymorph rhizome (Figs. 25, *C* and 30). The internodes of this basal part of the branch are solid, and often bear roots or root primordia, as in *Bambusa tulda* and *B. textilis* (Fig. 97), *B. vulgaris,* and *Gigantochloa apus,* for example. The transition to hollow, elongated internodes is as abrupt here as is the transition from rhizome to culm that is typical in bamboos with pachymorph rhizomes (Fig. 20, *left*). In some bamboos with scandent culms (*Chusquea* and *Dinochloa*) the principal branch at lower or mid-culm nodes may approach

Fig. 33. *Arthrostylidium harmonicum.* Lower and mid-culm nodes, showing early stages in the development of the absidate, pleioclade branch complement. Documented by specimens in the U.S. National Herbarium collected in Ecuador by McClure (No. 21416). U.S.D.A. photo by James Mitchell. See Figs. 31 and 32.

the culm itself in diameter and length as well as form. In some bamboos with pachymorph rhizomes, the resemblance of the primary branches to the culm is much less spectacular. This is particularly true where the primary branch is not appreciably larger than the secondary ones at mid-culm nodes, as in *Lingnania chungii* (Fig. 26, *4*). In *Melocanna baccifera* and *Schizostachyum spp.*, the branch complement is made up of numerous, widely radiating, small, slender, subequal axes. Such branches generally do not develop root primordia. However, an exceptional example of the spontanious production of roots from the basal nodes of such branches, *in situ,* is shown in a sterile specimen of *Schizostachyum sp.* collected by N. W. Simmonds (*s.n.*) in Western Samoa, October 13, 1954, and preserved at the Kew Herbarium.

In self-supporting culms the pattern of the length of the primary branches at successive nodes, from the bottom to the top of the series, follows the pattern of culm-internode length. Of this correspondence, Porterfield (1933:362) says: "The relation between the length of [culm] internodes and branch length is significant, and strongly suggests a connection with functions of growth." However, certain scandent bamboos show quite a different pattern. In *Chusquea simpliciflora* and *Melocalamus compactiflorus,* for example, the principal branch at mid-culm nodes may approach the culm itself in diameter and length, but the internodes of the mother culm where these branches originate are not correspondingly elongated.

In some bamboos, certain branches of the several orders become dwarfed, indurated, sharp-pointed, and curved, in degrees that generally correspond to the species. Extreme degrees of these modifications result in a local or more or less general manifestation of thorniness in the plant. Branches in the lower part of the culm are usually the most thorny ones. All of the known species of *Guadua,* and a number in the genus *Bambusa* have thorny basal branches. *Bambusa arundinacea* is a classic example. The degree of thorniness may vary within a species. For example, in its most northerly distribution, the typically thorny *Guadua amplexifolia* is represented by a thornless form. Intermediate degrees of thorniness characterize other forms of this species. In some cases, the extent to which branches are invested with thornlikeness is so slight that the character may be overlooked (*Bambusa malingensis*).

Again, *Bambusa dissimulator* shows an inconspicuous but appreciably more pronounced thorniness, and in *B. rutila* (Fig. 30) the spines are still more strongly developed. Referred to by Arber (1934:77) as "spinescent buds, which are either dormant or permanently arrested" (cf. Kurz 1876:236), these thorns actually show all of the characteristic marks of segmented axes. The nodes bear simple, scalelike sheaths, and these sheaths often subtend buds. The buds on primary thorns sometimes produce other, smaller thorns, or leafy twigs, or even inflorescences. Seibert (1947:272) refers to bamboo branch thorns as "short shoots."

The sheathing organs

Every node of every segmented vegetative axis of a bamboo plant bears a sheathing organ, which embraces the developing internode(s) distal to its insertion. This sheath is often referred to loosely as a leaf. The sheath proper should, however, be distinguished from its blade, which is the only leaflike part. In the bamboos, a foliage leaf (a leaf proper) is an appendage to a leaf sheath proper (cf. *Leaf* and *Sheath* in the Glossary). Sheath primordia are the first appendages to emerge behind the apex of the growing point of a segmented axis (Porterfield 1930*a*:Fig. 1). Next to emerge is the primordium of the bud subtended by each sheath. Greatly in need of further study is the question of the possible influence of substances elaborated in the developing sheath upon the initiation of bud primordia and upon ensuing events. These later events include the development of segmentation in the elongating axis, the formation and transverse anastomosis of fibrovascular bundles, intercalary growth, and even the initiation of root primordia, each of which takes place only at or just above a node (see pp. 12 and 123).

With reference to the relation between a bud and its axillant leaf, Arber says (1950:125):

> It is generally agreed that the vast majority of buds arise in connection with an 'axillant' leaf; but to what, exactly, this amounts remains vague, and Warming's criticism, though made in the eighteen-seventies, is still pertinent. He wrote [in French]: "It appears that no one has clarified the relation between a bud and the leaf called mother-leaf [axillant leaf]."

Porterfield (1930*a*) and Sharman (1945) trace the anatomical development of axillary buds and of the corresponding axillant (subtending) "leaves" in a bamboo and in several nonbambusoid members of the Gramineae, respectively. Garrison (1949) and others have made developmental studies of the anatomical relation between bud and axillant leaf in several species of woody dicotyledonous plants. However, it appears that the respective origins of the two associated classes of structures, and their relation to each other, still await effective illumination from the point of view of physiology and biochemistry.

The term "sheath" is commonly used in a comprehensive sense, to include both the sheath proper and its appendages. In the bamboos the principal appendages of the sheath proper are the blade, the ligule, and the auricles (Fig. 35). These appendages may assume very different sizes and forms in the sheaths of different parts of the plant. The blade, the ligule, or both may be very small or even rudimentary, and the auricles may be lacking entirely, as in the sheaths of the rhizome, for example (Fig. 34).

The sheath proper is the basal, and basic, part of the structure. By their close, firm application to the axis that bears them, and by their form and structure, the sheaths proper are specially fitted to perform the function of protecting, supporting, and stiffening the internodes of the young aerial axes during the period while these are actively elongating. Without this support during their critical period, the tender meristematic tissues of the intercalary zones of a growing culm or branch would be incapable of resisting the transverse mechanical forces that impinge upon them. In the rhizome, however, the most obvious function of the sheaths proper is the formation of a sharp-pointed resistant shield that protects the tender apical meristem as it is pushed through the soil by the elongation of the rhizome axis. All bamboo sheaths have one character in common: the smooth, lustrous inner (adaxial) surface of the sheath proper. This feature facilitates the adjustments made necessary by the rapid elongation of the young internodes while still tightly embraced by the sheaths. The white exudate mentioned earlier (p. 46) may function here as a lubricant.

The strong disparity that prevails between the sheaths found on different axes of a bamboo plant, in respect to (1) their size,

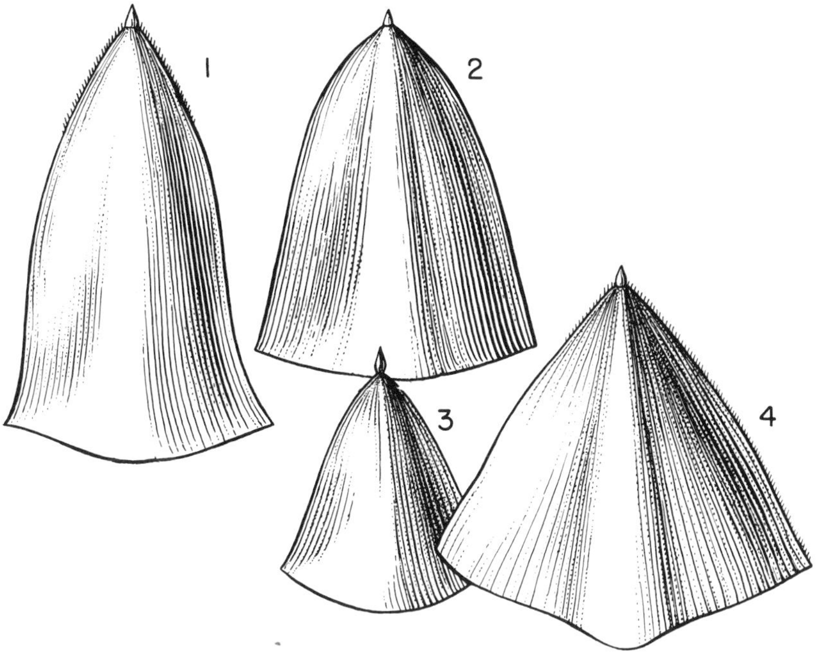

Fig. 34. Rhizome sheaths, showing typical extreme reduction of the sheath blade, illustrated by Takenouchi (1931*a*:Fig. 2): *1, Semiarundinaria viridis; 2, Sasa* (as *Pseudosasa*) *kurilensis* var. *nebulosa; 3, Arundinaria* (as *Pleioblastus*) *variegata* var. *viridis* f. *major;* and *4, Phyllostachys bambusoides* (as *Ph. reticulata*) var. *marliacea.* See pp. 75f.

shape, and texture, and (2) the relative development of the blade and other appendages, constitutes an important basis for differentiating the bamboos from the rest of the Gramineae. The dimorphism between the culm sheath and the leaf sheath is especially conspicuous in the bamboos (Fig. 35).

Taxonomists and anatomists frequently fail to identify for the reader the axis, or the point of origin, of the sheath whose features are being described. Too often this structure is referred to indiscriminately either as a "leaf" or a "sheath." For clarity, the following types of sheath found in the bamboo plant should be differentiated: culm sheaths, branch sheaths, leaf sheaths, rhizome sheaths, neck sheaths, and prophylla. The last-named are often referred to ambiguously as bracts, or bracteoles, or bud scales.

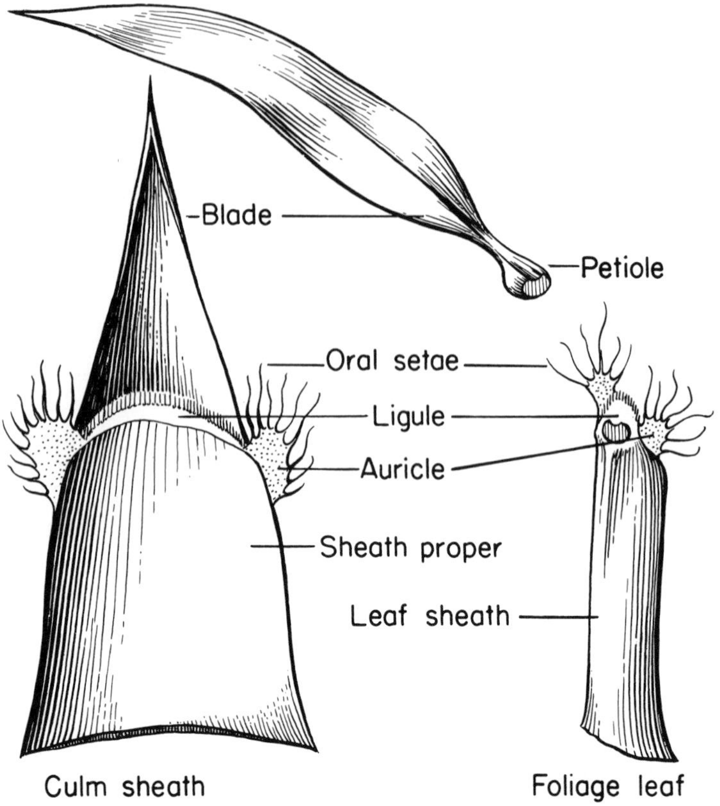

Fig. 35. Diagrammatic representation of culm sheath and foliage leaf sheath in the genus *Bambusa,* showing the several component parts; not drawn to a uniform scale. The dimorphism that prevails as between these two types of sheath is generally much more marked in the Bambuseae than in the other tribes of the Gramineae. Oral setae are sometimes referred to as shoulder bristles, when auricles are absent.

Culm sheaths

The series of sheaths that clothe the successive internodes of a developing bamboo culm show a progressive change, acropetally, in the size, shape, substance, and vesture of both the sheath proper and its appendages. A given culm sheath is therefore in these respects characteristic of the position of its origin. In a culm of mature stature the culm sheaths at nodes in the mid-culm range show the development of their various parts and appendages that is most highly characteristic of a given species (Fig. 36). In sheaths

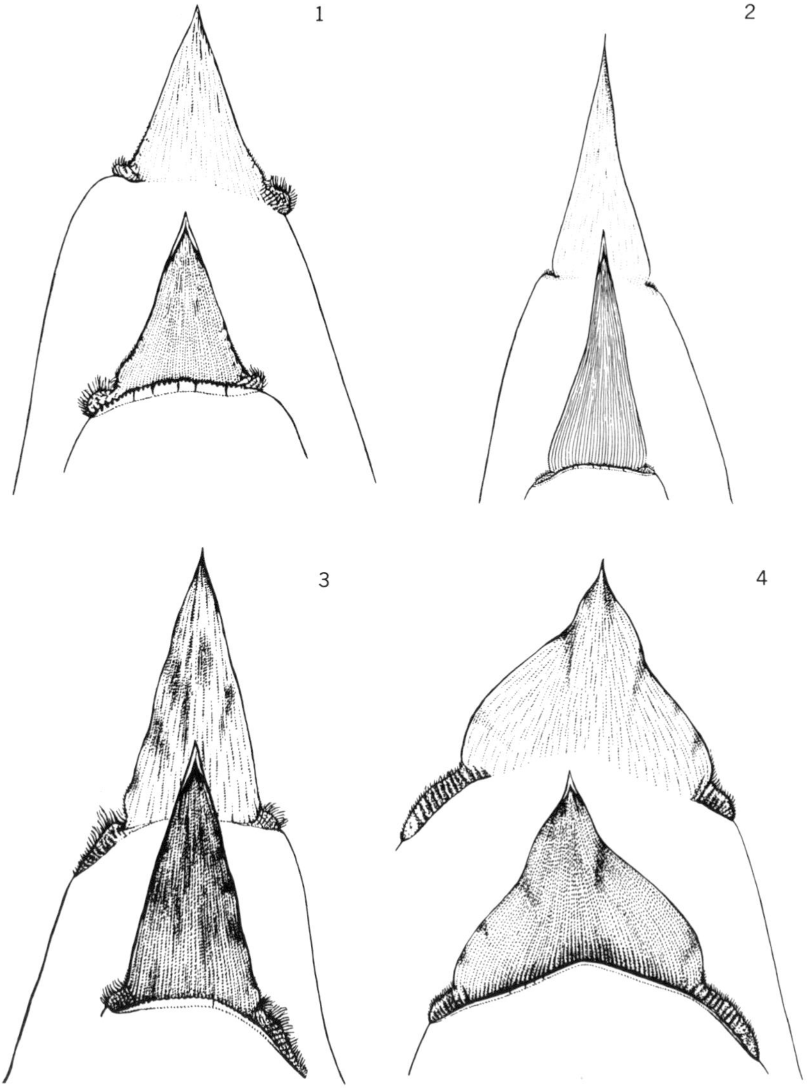

Fig. 36. Culm sheaths of four congeneric species: *1, Bambusa tuldoides,* LUBG 2664; *2, B. textilis,* LUBG 1842; *3, B. pervariabilis,* LUBG 1226; and *4, B. tulda,* LUBG 1350; all from the mid-culm level. As a rule, differences between species are shown by the culm sheaths more dramatically in bamboos of large stature, such as these, than in very small species. As demonstrated here, the auricles often contribute useful characters. The main drawing in each case shows the outer (abaxial) aspect of the whole sheath; the insert shows the inner (adaxial) aspect of the apical portion, including the ligule at the junction of the blade with the sheath proper. The original drawings (here reproduced at about $\frac{1}{5}$ natural size) where made by Miss Y. M. Lau, from specimens grown at the Lingnan University Bamboo Garden, under the accession numbers indicated.

from near the base and tip of a culm, however, the features typical of the species are as a rule somewhat obscured by a kind of generalization of their expression. At the base of the culm the sheaths approach in form those of the rhizome or the neck. At the uppermost several nodes the sheath and its blade, by a rather sudden change of proportions, take the form of leaf sheath and petiolate leaf blade. The progressive change in the form and dimensions of sheath proper and sheath blade characteristic of a series of sheaths from successive nodes of a single culm of each of two species is shown in Fig. 37.

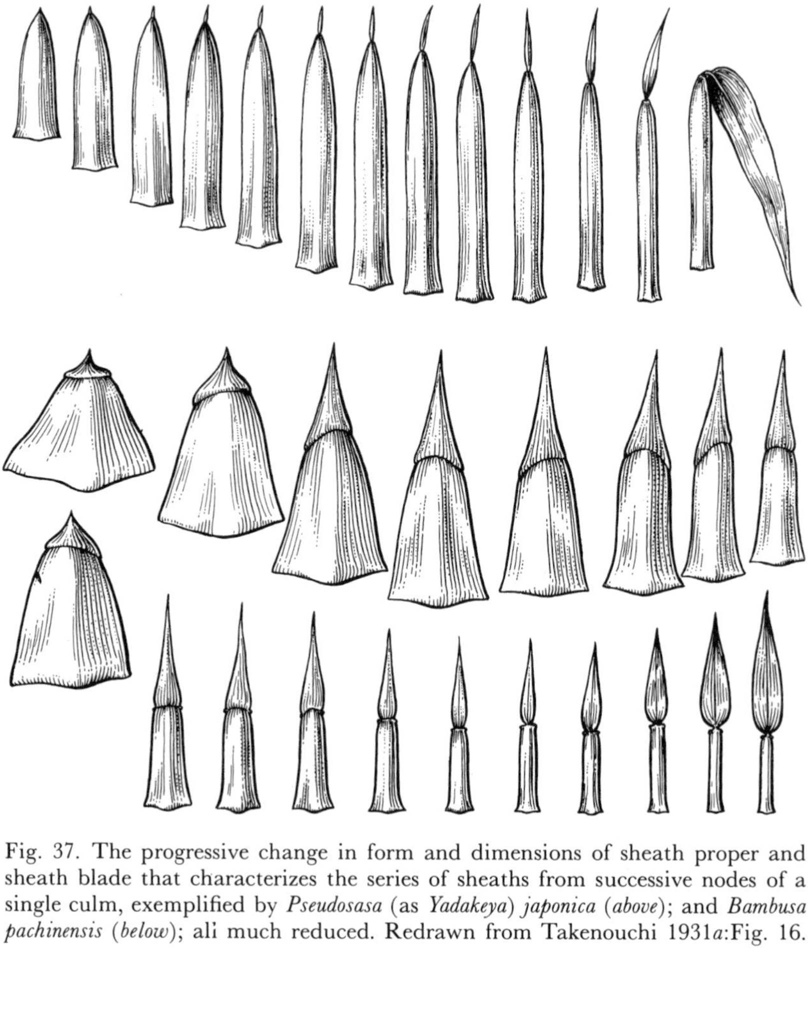

Fig. 37. The progressive change in form and dimensions of sheath proper and sheath blade that characterizes the series of sheaths from successive nodes of a single culm, exemplified by *Pseudosasa* (as *Yadakeya*) *japonica* (*above*); and *Bambusa pachinensis* (*below*); all much reduced. Redrawn from Takenouchi 1931*a*:Fig. 16.

In a given species, the culm sheaths may be persistent, tardily deciduous, or promptly deciduous. In some species (*Guadua amplexifolia*) the sheaths at the lower culm nodes may be persistent while those above are promptly deciduous. In *Semiarundinaria fastuosa* abscission is incomplete, and a culm sheath usually hangs for a time attached by a narrow sector of tissue at the middle of its base. In bamboos of the genus *Phyllostachys* the culm sheaths are promptly and completely deciduous; in *Sasa* they are persistent. The incidence of deciduousness may vary from species to species within a given genus, as in *Arundinaria.* As a rule, the development of a branch complement dislodges the subtending sheath partially or completely. Where the culm sheaths are stubbornly persistent, and particularly where they are closed, as in some species of *Chusquea,* by having the outer edge fused to the outer surface of the sheath, the emergence of the branch complement ruptures the sheath locally (Fig. 38).

In many bamboos, such as *Phyllostachys nidularia, Sinobambusa tootsik* and *Lingnania chungii* (Fig. 25, *A*), for example, we find at the base of each culm sheath in the lower and midculm range a ring of corky tissue properly called a callus or sheath callus (see p. 48). Upon the abscission of a sheath most of this tissue remains attached to the sheath node. Arber (1934:309) attributes to Godron (1880) the description of a similar structure in the leaf sheath of *Chimonobambusa* (as *Arundinaria*) *falcata,* and proposed for it the term collaret. Actually, it is rather at the base of the culm sheaths, and of the branch sheaths that are proximal to the leaf sheaths, that this structure is conspicuously developed in *Chimonobambusa falcata.*

In *Melocalamus compactiflorus* and in some scandent species of *Chusquea* (Fig. 38) the base of each mid-culm sheath consists of, or is inserted upon, a conspicuous horizontal band of tissue distinct in structure from that of the rest of the sheath proper, and probably related to the callus anatomically. It is expansible in all directions as long as the corresponding intercalary zone of the embraced internode is in active growth, and it takes on an asymmetrical form at geniculate nodes. Upon the abscission of the sheath, this band remains attached to the sheath node. Gamble (1896:111 and Pl. 98, Fig. 10) characterizes this structure as "a broad leathery ring left by the sheath." The term sheath girdle seems eminently suitable for it.

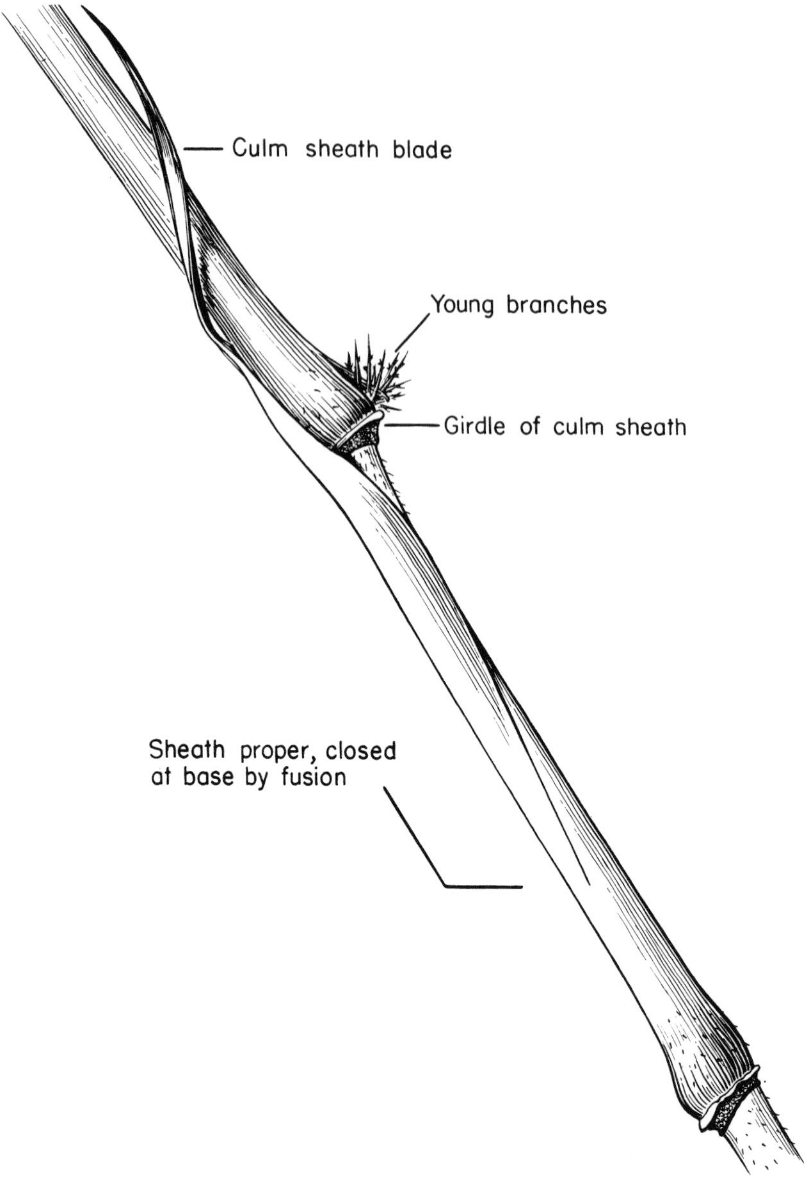

Fig. 38. Lower mid-culm nodes of young culm of *Chusquea sp.*, showing the well-developed "girdle" of the culm sheaths. The lower sheath shows the fusion of the outer margin to the sheath itself. This closure results in the persistence of the sheath, a condition related to the extravaginal eruption of the branch complement shown above. Field sketch, documented by herbarium specimens (McClure No. 21440) collected in Ecuador.

Culm Sheath Blade. A culm sheath blade is sessile, and may be either persistent, incompletely abscissile, or completely abscissile.[2] In the lowermost sheaths the blade is usually "reduced" to a sturdy mucro and the rest of the appendages are lacking or very rudimentary.

Branch sheaths

The series of sheaths clothing each branch of the culm recapitulates, on a reduced scale, that clothing the culm. The dimensions are progressively reduced, and the form of the sheaths is progressively generalized, in the successive orders of branches. In some bamboos with scandent culms, where the principal branch at mid-culm nodes may approach the culm itself in diameter and length, its sheaths may be indistinguishable for those of the culm proper (*Chusquea simpliciflora; Melocalamus compactiflorus*). Branch sheaths may be persistent or deciduous. The blade of a mid-branch sheath may be persistent but more often it soon abscisses. In sheaths at the basal nodes of a branch the blade is usually represented by a short mucro.

Leaf sheaths

The sheaths that clothe the distal nodes of all culms, and of branches of all orders, are properly called leaf sheaths. Their blades, called leaf blades, or simply leaves, are the plant's principal source of elaborated food. The sheath of a foliage leaf is distinguished from other kinds of sheaths by having a petiolate (stalked) blade, and by being relatively small and inconspicuous in relation to its blade (Fig. 35). Each one being much longer than the internode it embraces, leaf sheaths are deeply imbricate, that is, overlapping, each one nearly covering the one above it, except in rare cases where, under weak illumination, the axis bearing them is atypically elongated. Takenouchi illustrates the variation in auricles and ligules of leaf sheaths as between several Japanese bamboos in different genera (Fig. 39).

[2]A new feature of vascular anatomy, in the form of transverse bars that take the place of the end walls of the component cells of the annular vessels in a "sheath leaf" (culm sheath blade) from a young culm shoot of *Phyllostachys nidularia* is described and illustrated by Porterfield (1923).

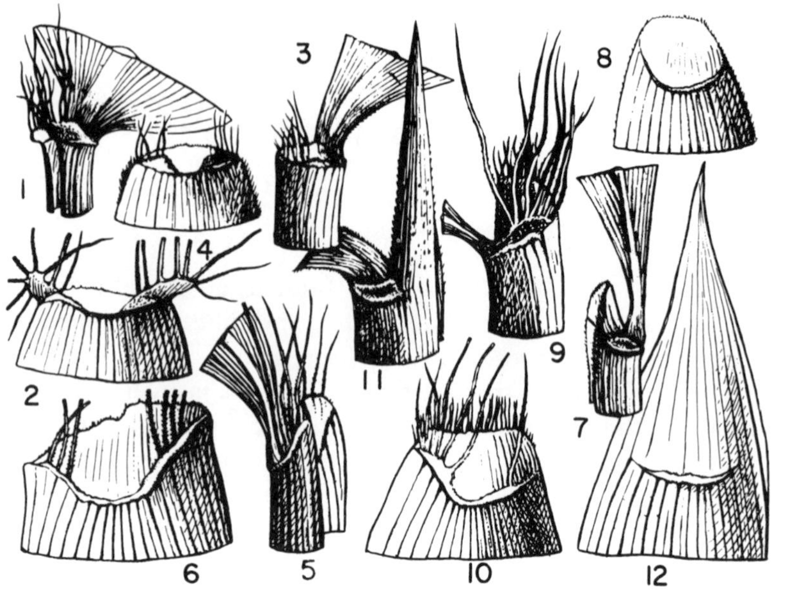

Fig. 39. Diverse forms assumed by the ligule and auricles of the bamboo leaf sheath, as exemplified by: *1, 2, Bambusa shimadai; 3, 4, Sasa tsuboiana; 5, 6, Arundinaria* (as *Pleioblastus*) *variegata; 7, 8, Phyllostachys pubescens* (as *Ph. edulis*); *9, 10, Phyllostachys formosana; 11,* 12, *Shibataea kumasasa;* all drawn at same scale. From Takenouchi 1931*a*:Fig. 15.

The Ligule. The ligule, in the conventional sense of the term, is an upward extension of a sheath proper, behind and beyond the locus of insertion of the sheath blade. The ligule is to be recognized by its position rather than by a common shape or texture. In size, texture, and substance, the bamboo ligule ranges from a small, thin, and scarious structure similar to that commonly found in the herbaceous grasses (see Jackson 1949:215) to a structure of substantial size and thickness. The "leaf scar" left by the abscission of the leaf stalk (petiole) is bounded on its adaxial side by the conventional ligule, and is surrounded on the outer side by an upward extension of the leaf sheath. The two structures combined make a little cup, referred to earlier (McClure 1941:33–34) as the "cupule" (Lat., *pocillum*). Hackel (1899:716 and elsewhere) refers to the outer rim as an "external ligule" (Lat., *ligula externa* and *ligula exterior*), a designation I have occasionally used. Holttum (1958:15, Fig. 3) refers to this structure as a "callus." In describing *Guadua perligulata,* Pilger (1937:57) refers to it as a "collar"

(Lat., *collum*) 5 mm long, and characterizes the ligule proper (the *ligula interior* of Hackel) in the same species as "lanceolate" and 5–7 cm long.

The Leaf Blade. Leaf blades (see *Leaf* in the Glossary) differ from culm sheath blades and branch sheath blades in being stalked or petiolate. Leaf blades are generally thinner than culm sheath blades, and often show more marked dorsiventrality. In many bamboos the petiole has a pair of rather strongly developed pulvini at its base (*Dendrocalamus latiflorus; Indosasa gibbosa*). By means of these, the petiole orients the blade in response to the dominant incidence of light, but the mechanism of the functioning of pulvini has not been studied in the bamboos.

The order of development of leaf blades is acropetal. They emerge, singly, through the "mouth" of the preceding leaf sheath. As it emerges, each one is rolled up tightly along its long axis (Fig. 40, *left*). Brandis (1907:72) makes reference to the convolute

Fig. 40. (*Left*) Cross section of a developing leaf blade of *Phyllostachys pubescens*, showing the convolute vernation characteristic of the foliage leaves of most bamboos; greatly enlarged. From Takenouchi 1931*a*:Fig. 12 (as *Ph. edulis*). (*Right*) Diagrammatic magnification of features of the lower (abaxial) surface of a leaf of *Bambusa dolichoclada*, showing: (*K*) short spinelike hairs; (*St*) long spinelike hairs; (*S*) stomata; (*G*) vascular bundles, which appear as longitudinal veins. The transverse veinlets are here not visible, being submerged in the mesophyll tissue. From Takenouchi 1931*a*:Fig. 17.

vernation of bamboo leaf blades. *Chusquea pinifolia,* which Brandis says may possibly be an exception, conforms in this respect.

The foliage leaves of all bamboos have more or less clearly manifest parallel veins of three orders: primary (the midrib), secondary, and tertiary (in descending order of size). In those bamboos whose leaf anatomy has been adequately studied, mutually adjacent veins of all orders are connected with each other by transverse veinlets (Fig. 41). In some bamboos (principally the hardy species), these transverse veinlets may be clearly seen to form, with the parallel veins, a pattern conventionally called tessellation, or tessellate venation. In the leaves of most of the physiologically tender, so-called "tropical," species, on the other hand, tessellation is not at all or, generally, only barely detectable externally. Mitford (1896:55) calls attention to the remarkably wide incidence, among the bamboos, of a correlation between cold-hardiness in the plant and strongly marked tessellate venation in the leaves. Wisely refraining from drawing a conclusion with respect to all bamboos, Mitford terminates his observations with these words.

> It would be, of course, idle to assert that every bamboo which has tessellated leaf veins is hardy; indeed we know that there are many bamboos with tessellated venation which, from their habitat, cannot be grown in this country [England]. Only one thing is certain, viz., that no bamboo introduced up to the present has proved hardy that has not such tessellation.

Comparative anatomical studies should reveal structural differences that explain the strong diversity of external expression that characterizes the transverse venation in bamboo leaves of different genera. Brandis (1907:Pl. 12, Fig. 18) presents clear evidence that in the tropical bamboo *Dendrocalamus giganteus* the transverse veinlets may be obscured from superficial view by thick layers of tissue both above and below them, in addition to the epidermal layer. On the other hand, in the hardy bamboo *Pseudosasa japonica* (*ibid.:* Pl. 11, Fig. 3, as *Arundinaria*) the transverse veinlets impinge upon the epidermis on both surfaces of the leaf blade. Such structural differences are not often brought into focus where the leaves being compared are sectioned only in a plane at right angles to the long axis (see Prat 1936:Fig. 21, *A* and Metcalfe 1956:Figs. I and II).

Takenouchi (1931*a*:Fig. 17) illustrates variations in the size, shape, and distribution of the short, spinelike, antrorse processes

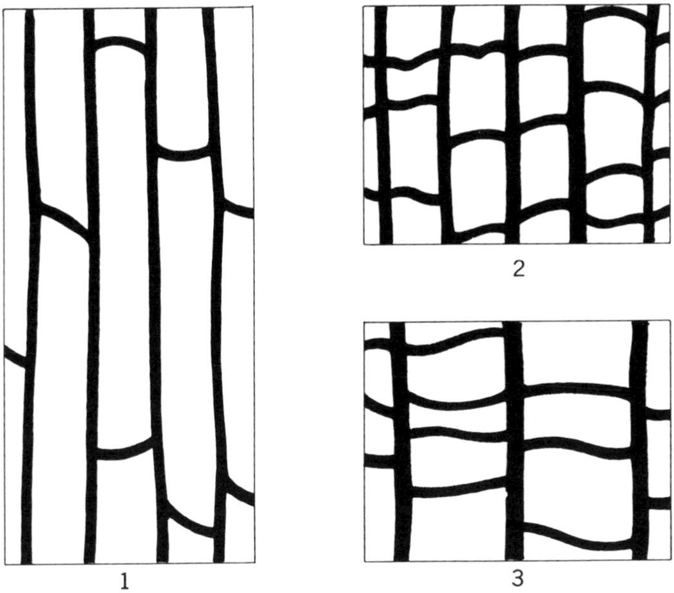

Fig. 41. Patterns made by transverse veinlets with the longitudinal ones, as exemplified in leaf blades of: *1, Bambusa stenostachya; 2, Phyllostachys makinoi; 3, Sasa veitchii* (as *S. albo-marginata*); all greatly enlarged. From Takenouchi 1931*a*:Fig. 20. The transverse veinlets are usually not manifest to superficial examination in unskeletonized leaves of the genus *Bambusa* and most other tropical genera with pachymorph rhizomes.

that are found, usually in unequal abundance, on the two edges of each leaf blade in most bamboos. These and other microscopic features of the leaf surface, which he illustrated diagrammatically, are reproduced in Fig. 40, *right.* Ohki (1932) made a systematic study of these features of the epidermis in the common bamboos of Japan by means of spodograms of the leaf blade, a method first demonstrated by Molisch (1920:262ff, Pl. 2, Fig. 9). Ohki (p. 4) recognizes eight "types" of spodograms: "1. the *Shibataea* type; 2. the *Dendrocalamus* type; 3. the *Bambusa* type; 4. the *Sasa* type; 5. the *Semiarundinaria* type; 6. the *Sinobambusa* type; 7. the *Pleioblastus* type; and 8. the *Phyllostachys* type." His analytical key (p. 3) to the Japanese genera of bamboos, based on characters drawn from his leaf spodograms, follows:

1 { Silica cells are not found in the epidermis above the assimilation tissue on the under surface of a leaf . . . . . . 1. *Shibataea* Makino
Silica cells are found in the epidermis above the assimilation tissue on the under surface of a leaf. 2

2 Silica corpuscles (Kieselkörper) are found on either or both sides of the longer rectangular cells of a leaf . . . . 2. *Dendrocalamus* Nees ab Esenbeck
Silica corpuscles are hardly ever found on either or both sides of the longer rectangular cells of a leaf 3

3 Silica cells in the upper epidermis above the veins are comparatively short, and are nearly dumb-bell shaped or rectangular in form . . . . 3. *Bambusa* Schreber
Silica cells in the upper epidermis above the veins are nearly rectangular, ovate or elliptical in shape 4

4 When an articulation band consists of 3 cell-rows, the length of the same cells in the middle row is nearly equal to the width, or may be shorter or longer . . . . 4. *Sasa* Makino et Shibata
5. *Pseudosasa* Makino
6. *Sasaella* Makino
When an articulation band consists of 3 cell-rows, the articulation cells in the middle row are mostly measured more in their length than in their width. 5

5 When an articulation band consists of 3 cell-rows, the articulation cells in the middle row are generally 11.6–24.9μ wide, even in an exceptional case measure at least 10μ in width. . . . . 7. *Semiarundinaria* Makino
When an articulation band consists of 3 cell-rows, the articulation cells in the middle row are generally 3.3–16.6μ wide; but some of them are sometimes less than 10μ in width 6

6 Silica corpuscles in the articulation cells are especially numerous . . . . 8. *Sinobambusa* Makino
9. *Chimonobambusa* Makino
Silica corpuscles in the articulation cells are few in number or are not found 7

7 Silica cells in the upper epidermis above the veins are mostly 2.5–10μ long; and in some species, in addition to such a short silica cell stomata measuring more than 28μ in length are found in the under epidermis. . . . . 10. *Pleioblastus* Nakai
Silica cells in the upper epidermis above the veins are about 4.9–10μ or more long, and stomata in the under epidermis are shorter than about 27μ in length . . . . . . 11. *Phyllostachys* Siebold et Zuccarini

At the conclusion of his paper (p. 128) Ohki has the following to say about the diagnostic value of leaf spodograms:

Generally speaking, there can be no doubt but that leaves have some characters which may be regarded as criteria for the determination of different species. Yet it may well be said that they never possess all the characteristics for the classification of species . . . Moreover, some genera such as *Sasa, Sasaella* and *Pseudosasa* bear so much resemblance as to their spodograms that I have found it extremely difficult to distinguish one from the other. This, I think, is most likely due to the fact that they came from the same phylogenetic branch.

The results of Ohki's survey suggest that the epidermis of the leaf blade in some bamboos offers characters which, *when used in conjunction with other morphological features,* may be of diagnostic value for the recognition of genus or species (see Prat 1936; Jacques-Felix 1955; Metcalfe 1956, 1960).

Brandis (1907) produced the first comprehensive comparative study of the anatomy of gramineous leaves, having examined material of "122 species in 21 genera." Of these, 40 species representing 15 currently recognized genera of bamboos are discussed or illustrated by him. Brandis concludes (p. 87) that "the uniformity of the structure of bamboo leaves is very remarkable, considering the great morphological differences in the order. In spite of the great variety in regard to important morphological characters, the structure of the leaves is the same in all genera and species, as far as they have been examined."

Metcalfe (1956) made a comparative study of transverse sections of the foliage leaves of 45 species, representing 28 currently recognized genera, and illustrated the central portion of sections taken at mid-point of the leaf blades of 14 species. This author concludes that "although [in their leaf anatomy] the bamboos exhibit combinations of characters that appear to be diagnostic for the species in which they are exemplified, there are no clear anatomical lines of demarcation between the genera." Six references that represent the principal previously published contributions to the knowledge of this subject are listed here by Metcalfe (see also Metcalfe 1960: lvii, 583–584, and 586–589; M. V. Brown 1958; Jacques-Felix 1958, 1962). Porterfield (1937) described and illustrated the stages in the development of stomata in the "sheath leaf" of *Phyllostachys nigra* and *Ph. pubescens.*

Rhizome sheaths

Of all the kinds of sheaths borne by a bamboo plant, those of the rhizome proper vary least from node to node within a given axis. A series of sheaths from the successive nodes of a pachymorph rhizome shows a perceptible progressive change in size that corresponds to changes in the diameter of the axis. In a leptomorph rhizome proper, however, the sheaths are remarkably uniform in all respects. Examples of sheaths from leptomorph rhizomes of several species of bamboos are shown in Fig. 34. All of the appendages of rhizome sheaths proper are depauperate; auricles are completely suppressed; the blade is commonly reduced to a hard mucro. Rhizome sheaths are usually strongly imbricate, and persistent or tardily abscissile.

Neck sheaths

Neck sheaths clothe the basal part of every segmented vegetative axis in all bamboos (p. 12f). Neck sheaths have the simplest of the various forms assumed by the sheath on any of the axes of the bamboo plant; moreover, they never subtend a bud. The neck sheath proper is small, broadly triangular in shape, typically hard and tough in texture, and generally smooth and lustrous on the inner surface. The blade is obsolete or reduced to a minute, usually hard mucro; auricles are lacking and the ligule is obsolete. Neck sheaths are strongly imbricate and generally absciss very tardily if at all.

Prophylla

The distinctive sheath at the basal node of each vegetative branch is properly called a prophyllum, or prophyll. Some authors refer to it by a general term, such as bract, bracteole, or bud scale. By its position, orientation, shape, and function, this sheath is placed clearly in a class by itself. It is addorsed (that is, its back is appressed) to the axis from which the branch that bears it emerges. In its rudimentary state, the branch that bears the prophyllum is completely enclosed by the latter to form a bud. In contrast with other types of sheaths described, the prophyllum lacks the conventional appendages, blade, auricles, and ligule. The characteristic shapes of prophylla representative of the buds at culm nodes in different species are illustrated in Fig. 42. Typical prophylla have in common two salient vascular bundles (veins), which mark the "keels" along which the margins are folded forward, and which set off the "margins" from the "back." The prominence of each keel is generally accentuated by a wing of thin tissue. The

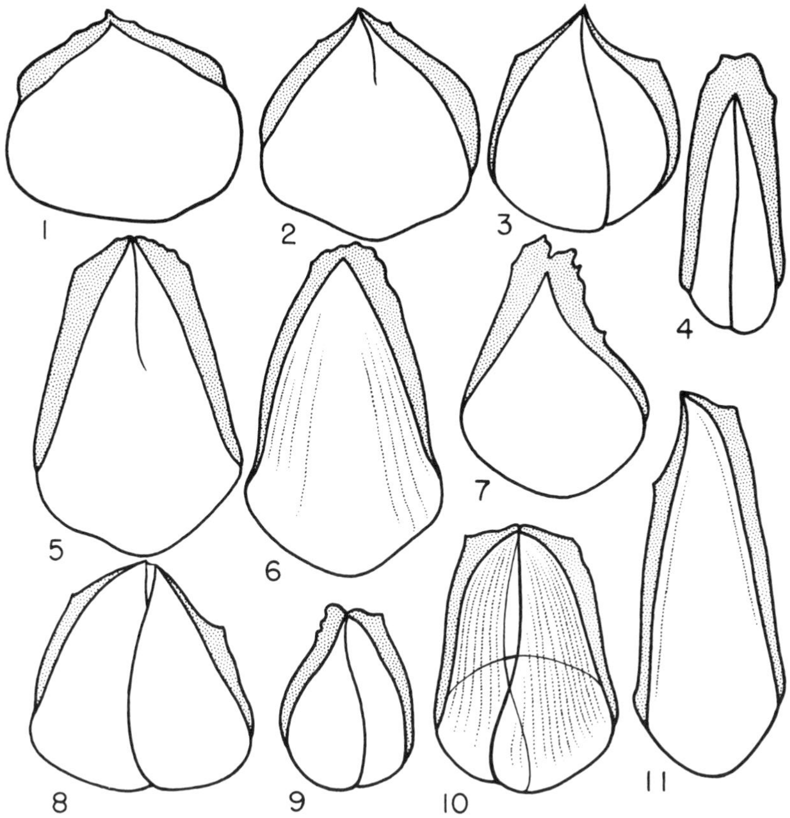

Fig. 42. Prophylla of buds at culm nodes, showing variation in form, fusion of the margins, and shape of the wings, from species to species, as exemplified by: *1, Sinocalamus latiflorus; 2, Bambusa multiplex* (as *B. nana* var. *typica*); *3, Phyllostachys makinoi; 4, Arundinaria* (as *Pleioblastus*) *simonii; 5, Sasa paniculata* var. *paniculata; 6, Pseudosasa japonica; 7, Shibataea kumasasa; 8, Semiarundinaria fastuosa; 9, Sinobambusa tootsik; 10, Chimonobambusa quadrangularis;* all much enlarged. From Takenouchi 1932:Fig. 58.

wings may be entirely separate from each other or they may be joined at the apex of the prophyllum. When the bud is dormant these wings are appressed to the axis on which the bud is borne. The margins of the prophyllum may be either free or fused together in varying degrees. Some prophylla are in this manner closed all the way from base to tip while others are only partly closed (Fig. 43).

As in other types of sheaths, the size, shape, vesture, and other

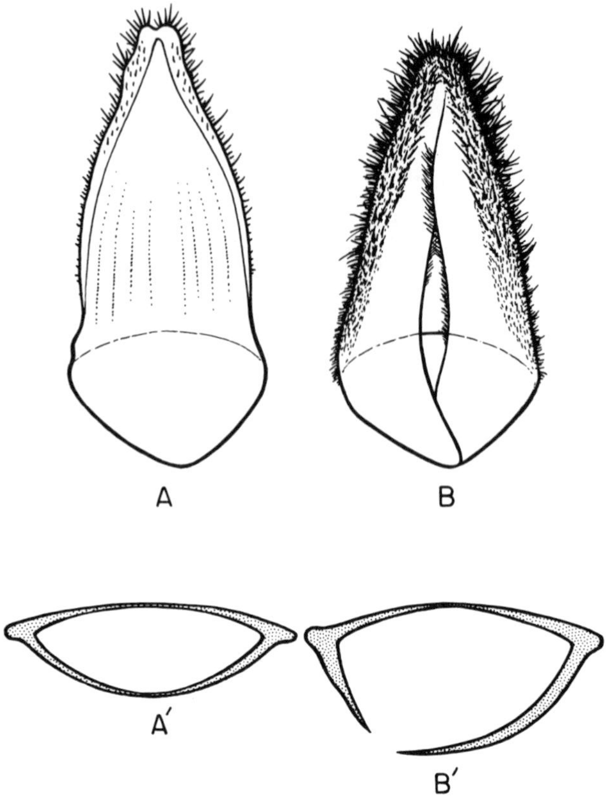

Fig. 43. Prophylla of buds at culm nodes, closed and open forms, each shown in abaxial view and in cross section: (*A, A′*) *Sasa veitchii* (as *S. albo-marginata*) with the prophyllum closed on the abaxial face by the fusion of the margins; (*B, B′*) *Yushania* (as *Arundinaria*) *niitakayamensis,* with the prophyllum open on the abaxial face, the margins not fused, greatly enlarged. From Takenouchi 1932:Fig. 59.

features of the prophyllum vary more or less from species to species and with its position on the plant. In buds at successive nodes of a leptomorph rhizome, the prophylla are always very uniform, but in pachymorph rhizomes they may show appreciable changes in size and shape from node to node. In culms and their several orders of branches, the size, shape, and vesture of the prophylla are modified progressively from the base to the tip of a given axis (Takenouchi 1931*a*:Pl. III). They are smallest in size on leafy twigs, and largest on pachymorph rhizomes. For a discussion of the prophyllum in inflorescence branching, see p. 108.

Roots

Roots are the only unsegmented vegetative axes in a bamboo plant. They conform to the popular characterization "fibrous" (in a sense of the term not found in Jackson) which probably is best described as slender, roughly cylindrical, and apparently not increasing in diameter with age. In some large bamboos, as in *Guadua angustifolia,* subterranean root axes of the first order may be relatively massive, with a diameter of nearly a centimeter.

The ephemeral primary root of the seedling bamboo is soon overshadowed by the progressive development of adventitious roots, which take over the functions of anchorage and absorption. Adventitious roots in the bamboos are initiated only at a level immediately distal to a sheath node. They develop chiefly at the nodes of the rhizomes proper and at the underground nodes of the culms (Fig. 44, *A*). In certain species with pachymorph rhizomes root primordia form on the swollen base of primary branches at lower midculm nodes (see text, p. 244). In many species, roots also appear normally in a primordial or partially developed state on nodes at varying levels of the aerial part of the culm (Fig. 44, *B*). In *Bambusa vulgaris, Gigantochloa verticillata, Chusquea pittieri,* and several species of at least one genus with leptomorph rhizomes (*Chimonobambusa*) root primordia may form at nodes in the lower half or more of the culm length. These primordia may remain dormant, or they may develop to a length of one to several centimeters. The degree of their development at the successive nodes always diminishes acropetally. In *Chusquea pittieri* and *Chimonobambusa quadrangularis,* and in other species of both genera, these aerial roots typically become dry, somewhat indurated, and thin at the tip, assuming the appearance of thorns, though actually they are of rather brittle consistency (see Arber 1934:Fig. 28, *B*).

Published information about the subterranean root system in bamboos is very scarce. Apparently unique is an exploratory study by White and Childers (1945) of the extent and subterranean distribution of the root system in *Bambusa tulda,* with particular reference to its potential stabilizing effect on the soil where erosion is a problem. The root system of a well-established clump growing in a heavy clay soil on a hillside of unrecorded gradient was studied by digging a broad trench 4 ft deep extending radially from the base of the clump to a point more than 17 ft away. The

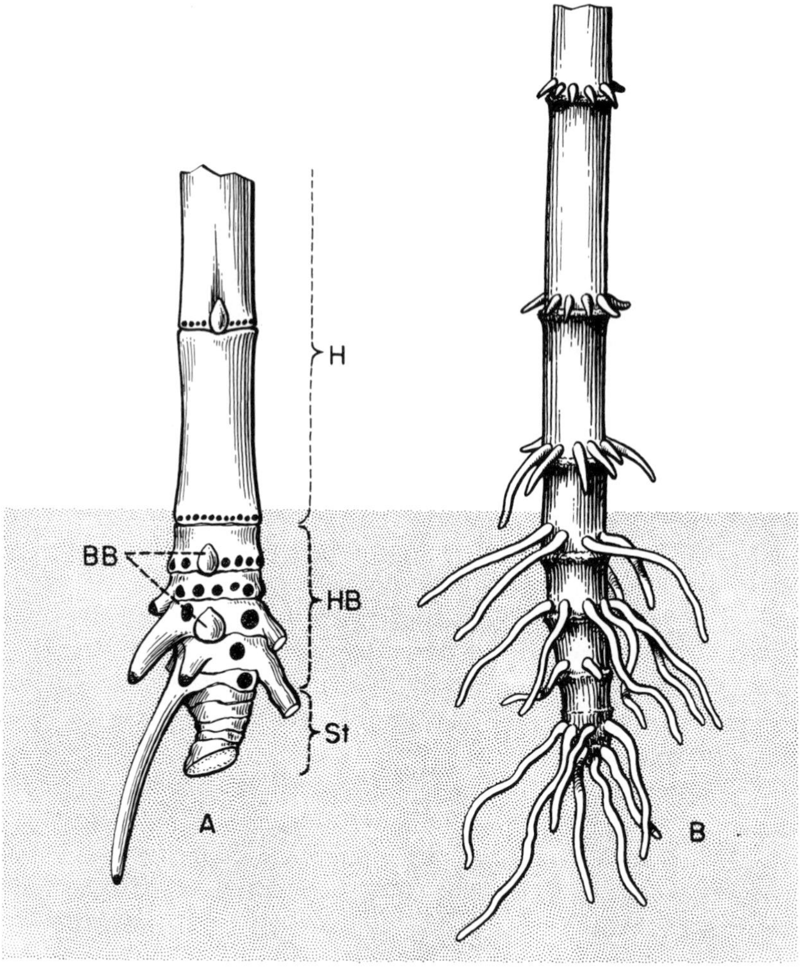

Fig. 44. (*A*) "The base of a shoot of *Phyllostachys nigra* from which the sheaths have been removed, showing the young roots in various stages of development. *BB,* basal buds; *H,* culm; *HB,* culm base; *St,* 'Steil' or connecting bridge to rhizome" (see "The neck," pp. 13f). From Porterfield (1935:Fig. 1). (*B*) Culm base of *Chimonobambusa quadrangularis,* showing adventitious roots: normal subterranean roots and spinelike aerial roots. Redrawn from Takenouchi 1931*a*:Fig. 4.

number of roots exposed in each square foot of the surface of the trench wall was counted. Eighty-three percent of the roots were present in the upper foot of the soil; 12 percent between 1 and 2 ft; 4 percent between 2 and 3 ft; and 1 percent between 3 and 4 ft from the surface. The longest roots extended 17 ft from the base of the clump. The growing tips of the uppermost root branches extended above the soil into the accumulated mulch of fallen bamboo leaves.

As a general rule, in bamboos with pachymorph rhizomes the underground roots are much more concentrated, both in their place of origin and in the space they occupy, and the radial extension of individual roots is much greater, than in bamboos with leptomorph rhizomes. However, the more widely ranging growth of the leptomorph rhizome effects a certain compensation by dispersing its sparser array of shorter roots more rapidly into the previously unoccupied soil.

Shibata (1900:Figs. 23–47) illustrated anatomical features of the roots of ten different species, representing six genera. In this work, and later in collaboration with Makino (Makino and Shibata 1901:Figs. 15–19), Shibata brought into focus a fundamental anatomical difference between bamboos of the genus *Bambusa* and those of *Sasa* and other arundinarioid genera. In the endodermis of the roots of *Bambusa,* exemplified by *Bambusa multiplex* (as *B. nana*) and *B. stenostachya,* the cell walls are shown to be thickened symmetrically on all sides, while in the arundinarioid genera, exemplified by *Sasa borealis* and members of *Arundinaria, Chimonobambusa, Phyllostachys, Sasa,* and *Shibataea,* the cell walls of the endodermis are conspicuously more heavily thickened on the side toward the center of the root.

This and other, apparently correlated, root differences noted by Shibata may now be recognized as associated with gross morphological differences, such as type of rhizome, branching habit in vegetative and reproductive axes, the relative salience of transverse veinlets in the leaf blades, and even differences in ecological adaptation, that separate two large groups of genera.

Further comprehensive and detailed comparative studies are greatly needed to improve our perspective on external and internal features of the bamboo root system and the range of variation within and between both species and genera. See also p. 122, where the primary root is discussed.

2 Reproductive Phase

The duration of the vegetative state and the incidence of flowering and fruiting vary from one species to another. In some bamboos the vegetative state shows signs of persisting indefinitely, the plants producing new growth with undiminished vigor year after year, for example, *Arundinaria variegata, Bambusa vulgaris,* and *Sasa tessellata,* as known in cultivation. Whether the plants under discussion represent physiological strains distinct in this one character from the rest of the population of the species to which they belong is not known. Perhaps such bamboos may, in some cases, possess a physiological stability which, under most conditions, keeps them perpetually in a vegetative state, and which permits individual plants under the impact of certain unknown circumstances—even pure accident of a physiological nature—to produce gametes. At any rate, the behavior in *Bambusa vulgaris* suggests such a condition. Ever since it was described and given a scientific name (now more than 150 years ago) by Wendland (1810), *Bambusa vulgaris* appears to have been the most generally recognized of all known bamboos, and it has become the most widely distributed species in cultivation. Apparently the general population of this bamboo has remained steadily in a vegetative state as long as the plant has been under scientific observation. Each occurrence of flowering (as reported, and as observed personally) has involved at most only a few plants (see Hildebrand 1954:50). It appears that these always die upon flowering. I have found no published record of bamboos of this species having produced fruits (see Moebius 1898:86). If they really do not do so, then no seedling progenies are being produced to perpetuate or accentuate, in this species, the tendency to enter the flowering state. However, in spite of its apparently not having been "rejuvenated" in recent times by sexual reproduction, *Bambusa vulgaris* remains to this day one of the most vigorous of the known bamboos. No evidence of systemic senescence in sterile plants of this species has come to my attention.

In contrast with this apparent potential permanence of the purely vegetative state in some bamboos, we find in others a persistent tendency to flower. *Bambusa lineata,* for example, is often described as "constant flowering" (Gamble 1896:47). Plants of this bamboo have been under observation upwards of a hundred years in cultivation at Bogor, Calcutta, and Peradeniya. No record of the production of fruit, or of the death of the flowering plants, has been published. Holttum (1946:343, and personal conversation) states that at Singapore *Schizostachyum brachycladum* and some plants of *S. longispiculatum* flower continuously, while *S. gracile* and *S. grande* constantly flower and fruit a little, and all of the plants retain their vegetative vigor apparently undiminished.

A few bamboos are designated as flowering annually. *Indocalamus wightianus* (as *Arundinaria wightiana*), *Bambusa lineata,* and *Ochlandra stridula* are so listed by Gamble (1896:viii); others are said to flower constantly. In most records there is no clear evidence to serve as a basis for distinguishing the behavior described as "flowering annually" from that called "constant flowering." Since, at low elevations in the deep tropics, temperature and light values vary only slightly during the year, it seems likely that in both cases the active production of new flowers follows the yearly pattern of rainfall. Dr. Holttum observes that in Singapore, where there is no dry season, flowering is continuous, not "annual," while in places with a definite dry season flowering may occur once a year.

The majority of the known bamboos for which records are available fall between two physiological extremes (constant sterility and constant flowering) by manifesting a cyclic recurrence of the flowering state at intervals of several to many years. Individual kinds are not always consistent in their flowering habits, but two general modes of behavior are distinguishable. Bamboos of some species generally die within one or two years after flowering (*Bambusa arundinacea*) while others do not die, but their vegetative growth slows down during the flowering period. This latter group includes all species of the genus *Phyllostachys* and several species of *Arundinaria* whose behavior is a matter of record. In order to think clearly about the ontogeny of a bamboo plant, therefore, it is necessary to distinguish between the life cycle and the flowering cycle, since the two may not be coextensive.

In most bamboos the flowering cycle is much greater than one year. Because of its great length in many cases, and because of frequent conflict in details between different accounts, and since the flowering and fruiting of bamboos has in many cases (especially in India) relieved a local famine brought on by drought (see Munro 1868:4), the flowering cycle in the bamboos has been subject to much popular interest and speculation. Published accounts designed to fix the length of the flowering cycle for particular species are rather numerous. Suessenguth (1925) and Blatter (1929–1930) have published the most complete available summaries of such records. These relate principally to Asiatic species, with extended discussions of the circumstances, patterns, and possible causes of flowering in bamboos. The evidence presented by such records usually consists primarily of a series of dates at which a given species flowered gregariously in a given region. From this evidence, many species of bamboo appear to have a characteristic and more or less sharply defined flowering cycle, of the order of roughly 1, 3, 7, 11, 15, 30, 48, 60, or possibly (Kawamura 1927:341) even 120 years, at the end of which time all plants of a given seedling generation flower gregariously. However, the records on which the flowering dates assembled by these authors were based are in no case documented by first-hand personal observation of the complete cycle in a single hereditary line from seed to seed. Parodi's documentation (1955), based on his personal observation of one full flowering and fruiting cycle in *Guadua trinii,* appears to be unique. Fuller details are given on p. 275. (see also *Bambusa copelandii,* p. 269). A scientific approach to the study of the question of flowering cycles in bamboos calls for the frank recognition of one hard fact. Existing records do not provide a positive basis for the generally held assumption that some bamboos have, as an innate character, a flowering cycle of precise and invariable length.

On the basis of their flowering behavior, Brandis (1906:662) defined three groups of bamboos. These are briefly interpreted by Blatter (1929–1930:136) as follows: (1) those that flower annually or nearly so; (2) those that flower gregariously and periodically; (3) those that flower irregularly. However, a given species may manifest behavior that places it in more than one of these groups. *Dendrocalamus strictus,* for example, falls within both the second and the third groups. Ghinkul (1936:26) conceives three groups as follows: (1) mostly small species, whose life span is undetermined, flowering and bearing fruit repeatedly at short intervals without dying; (2) plants dying regularly after the first and only fruition; (3) plants dying in part, and renewing the dead parts after fruition.

It is clear that gregarious flowering is not the universal rule; nor is it invariable in those species where it is the general rule. Flowerings, called sporadic, of only a portion of the plants within a given population known or presumed to be of the same age are frequently recorded. The flowering of some of the culms, and not others, in a single clump has been recorded. And I have in my own records a note (dated Canton, China, June 19, 1938) registering the flowering of a single twig on a plant of *Bambusa eutuldoides* (Lingnan University Bamboo Garden accession number LUBG 1206). No further flowering was seen in the following 18 months during which I was able to observe the plant. Holttum (1955:408) says, "I know of no recorded case of mature Malayan bamboos flowering gregariously and then dying, as is the habit of many bamboos in the monsoon region (a plant may be considerably weakened by the full flowering of several culms, but may yet gradually regenerate from small basal buds)."

Fruiting behavior in different bamboos

It is rarely possible for even the most interested and dedicated observer to assemble a complete account of the full course and the outcome of flowering in a given bamboo. Two important items of information most often lacking in available reports are: whether, and in what quantity, mature or viable fruits were produced; and whether any plants survived the flowering period.

The Indian Forester (1875–) contains rather frequent notices, communicated by members of the Indian Forest Service and others, of the availability, in quantity, of the seeds of certain bamboos. As a result of these notices and published accounts of the relief of famine conditions in India by heavy yields of edible fruits following the flowering of bamboos such as *Bambusa arundinacea* (Munro 1868:4) and *Dendrocalamus strictus* (Arber 1934:95), there is abroad a rather widespread impression that heavy fruiting automatically follows gregarious flowering in the bamboos as a general rule. This impression may have been strengthened by accounts of the plagues of rats that develop periodically in southern Brazil, when local populations of certain bamboos (*Merostachys spp.*) flower gregariously and fruit heavily (Pereira 1941; Giovannoni *et al.* 1946, see also Udagawa 1958). However, a different impression has been derived from certain published records and from personal observations made during a period of more than a third of a century. These observations cover flowering plants from twenty of the recognized genera, and thousands of preserved flowering specimens of bamboos representing, in aggregate, all of the known genera. They add up to the impression that the incidence of maturation of fruits is relatively low in the majority of known bamboos. It appears that abundant yields occur in only a relatively few species out of the hundreds that have been observed in flower; and these few are principally bamboos not under cultivation. However, this impression should be disciplined by a recognition of the diverse nature of the evidence. In bamboos that have a long flowering period, the number of fruits discoverable at a given moment is usually very low. This suggests that either the set of fruit, or its maturation, is relatively infrequent, in both time and space. Again, in most bamboos the florets bearing mature fruits are more or less promptly released by the disarticulation of the rachilla, and fall to the ground. Wild creatures, particularly certain birds and rodents, harvest them either from the plant or from the ground. Hughes (1951) states that in his experience at Plymouth, North Carolina, the incidence of destruction of the immature fruits by insects is so high in flowering natural stands of *Arundinaria* [*tecta*] that in order to get any mature seeds he found it necessary to dust the inflorescences regularly with DDT. Of *Guadua amplexifolia,* which flowered in 1954 at the Federal Experiment Station in Puerto Rico, Kennard (1955:194) says: "In spite of the fact that many thousands of flowers were produced and that the pollen apparently was viable, fruit set was very low. Examination of thousands of spikelets yielded only 1003 fruits." Only 1 percent of the seeds that were planted germinated.

For circumstantial records partially documenting the flowering and fruiting behavior of selected bamboos of different genera and species, see Chapter 6.

Flowering habits of bamboos as related to taxonomy and the problem of field identification

To the taxonomist and to the plant collector, the bamboos present special difficulties. These difficulties have their origin in several characteristics common to most species. Rarity of flowering is one of them. Another is the more or less severe suppression of vegetative activity which generally prevails during the flowering period, and which in some species usually results in the death of the plant. In some tropical bamboos, a failure to initiate new vegetative growth at the usual season may give advance warning of the impending onset of flowering. This means that intact examples of the perishable culm sheaths will usually not be available when flowering takes place.

Since the taxonomic differentiation and description of all angiosperms have been, from the time of Linnaeus, based primarily upon the reproductive structures, it is only natural that bamboos not in flower or fruit would be passed over by most collectors. It is also quite natural that little or no effect would be made to incorporate a representation of the vegetative parts of the plant, since systematic procedures traditionally concede only minor taxonomic value to characters from these structures. Moreover, the most neglected vegetative parts of a bamboo plant are usually bulky and require special treatment.

However, certain vegetative structures may be used effectively for the recognition of bamboo entities, without reference to the flowers, after the range of their variation in the plant has been studied. Prominent among these vegetative structures is the culm sheath. Since the production of new culms, and therefore of culm sheaths, generally ceases entirely during the flowering, and since the old sheaths usually deteriorate rapidly, these structures are rarely found associated in herbarium specimens with flowers from the same plant. Branching habit and other features of the several categories of segmented axes are also useful for taxonomic purposes. It is important therefore, in the study of the bamboos, to build up independently for each recognizable entity a complete representation (specimens, notes, sketches, and photographs) of the vegetative structures of the sterile plant. Then, when flowering finally occurs, the association of characters from the two phases of a given entity will be facilitated. In this connection, it is important to be aware that the morphological expression of component parts varies within certain limits, in correlation with the relative age of the different parts of the plant, and that the mode of expression found in the youngest (currently produced) culms reflects the relative maturity and the current stature of the plant as a whole (see McClure 1935:193–197, and Holttum 1958*b*:1–2). The importance of the foregoing considerations relates primarily to the fact that, as encountered in the field, most bamboo plants are not in a flowering condition; their vegetative structures provide the only means of identifying them.

Franchet (1889:281) in his remarks following the characterization of a new genus of bamboos asserts that

> it is especially in its vegetative characters that *Glaziophyton* is clearly distinguished from all of the known Bambuseae, and perhaps even more so from most of the other genera of the group to which it belongs. It was, of course, a long time ago that Kunth noted that, in the Gramineae, most of the genera could be clearly separated only by their vegetative organs, as the floral organs furnish only vague elements for distinguishing the plants, and even these [elements] are of doubtful value.

Franchet's assertion with regard to *Glaziophyton* is correct. But the statement attributed to Kunth (I have not been able to find the original source of it) goes beyond what can be supported by available evidence. While it is possible to recognize and identify individual species of bamboos, once they are thoroughly known, by an adequate array of vegetative characters, it is not always possible, given the present state of our knowledge, to determine the generic affinities of an unknown bamboo with certainty, by means of vegetative structures alone.

During the first half of the present century, and particularly in Japan, the description of new species of bamboos on the basis of vegetative structures alone was practiced by taxonomists on an increasing scale. Where conscientiously executed, this procedure can be sanctioned on the same premises as the naming and describing of Fungi Imperfecti from sterile material. This expedient is in general more satisfactory for bamboos of large stature than for small ones, when characters are drawn from gross morphology alone. In any case, the original description should be full enough to set forth reliable specific characters which, in combination, are truly distinctive.

The inflorescence

The earliest description of the bamboo inflorescence as a distinct structural unit apparently is that of Nees (1841:460). This was quoted verbatim by Munro (1868:11) as a part of his characterization of the "Bambusaceae." Freely translated it reads as follows: "The inflorescence [in Tribe xvii, the Bambuseae of Nees] is a panicle, very rarely with strongly compound branching, more commonly smaller and depauperate; in some bamboos it is reduced to the form of a raceme or a spike (in some verticillate, in others simple); it may even be reduced to a peduncle bearing a single spike." Gamble (1896:1), in his characterization of the tribe, simplified it to the following phrase: "Inflorescence various, usually a large, compound panicle with spicate branches." These definitions approximately cover the usage still generally accepted. The situation is not much helped by the adoption of such borrowed terms as "thyrsoid" and "capitate." The "inflorescence" of *Thyrsostachys* is not really like a thyrse. And the term "capitate" is applied, without discrimination or qualification, to both the globose indeterminate inflorescence of *Guadua capitata* (a cluster of pseudospikelets) and the subglobose determinate inflorescence of *Athroostachys capitata,* which latter is really a short-branched paniculate raceme.

In the introductory chapter of his monograph of the bamboos, Ruprecht (1839:15) called attention to the difference between *Arundinaria* and *Bambusa* in respect to the "disposition of the spikelets." However, in setting up the two major divisions of the bamboos known to him, Ruprecht differentiated them not on this character (that is, the branching habit of the inflorescence) but on the number of stamens in the androecium, thus: "*Bambusae verae*" with 6 stamens, and "*Arundinariae*" with 3 stamens.

In the introduction to his description of the genus *Schizostachyum,* Nees (1829:534, Obs. III) called attention to the "peculiar structure" (*singularem structuram*) of the "spikelets" in the type species of this genus. From his description of this structure, it is evident that the inflorescences in the specimen he had before him consisted of pseudospikelets in an advanced state of development (see pp. 65 and 66). Ruprecht (1839:44; 1840:134), in his own description of the genus *Schizostachyum,* used the following words (translated from his Latin): "The glumes of Nees's description are [labeled] in our illustration as lemmas [but they are] really bracts subtending spikelets or buds." Here the matter rested until Moebius (1898:Pl. IV, Fig. III) figured an inflorescence of *Bambusa vulgaris* and, during the course of a discussion of morphological details, made several references to the continuing nature of its development.

Pilger (1927:Figs. 1 and 2) illustrated an inflorescence of *Guadua tessmannii* and presented a diagram purporting to show the details of its organization. Pilger's description of the dynamic aspects of the development of this inflorescence is difficult to follow. Moreover, the diagram of its organization (*ibid.,* Fig. 2) is defective in respect to a fundamental detail. It fails to show that all of the branches of the indeterminate inflorescence characteristic of this species recapitulate the organization and the dynamic aspect of the primary axis. To be explicit, the ultimate branches are not shown to have the basal buds (mentioned in the text) that are themselves potential pseudospikelets, and capable of developing into reproductive branches exactly like the ones on which they are borne. For this reason, Pilger's diagram of this indeterminately branching inflorescence portrays it as exactly like the fundamentally different inflorescence of *Glaziophyton mirabile,* in which, while each branch is prophyllate, and subtended by a bract, the branching is determinate. Also noteworthy is the fact that, in confining his comparisons to grasses of nonbambusoid genera, Pilger failed to place this indeterminate type of inflorescence in perspective among the bamboos. Pilger's important contribution received no published notice from students of the bamboos until McClure (1934) presented a study of the inflorescence in the Chinese species of *Schizostachyum,* having arrived at the concept "pseudospikelet" to distinguish a whole (spikeletlike) branch of an indeterminate inflorescence from the spikelet proper that terminates it. The following statement is an attempt to elucidate more fully some hitherto generally neglected basic features, particularly the branching, of the bamboo inflorescence. The extreme forms of expression are described and given distinctive names.

The bamboo inflorescence is an axis, or a system of axes (associated branches) emanating from a common axis, the primary rachis. The primary rachis ends in a spikelet; and so does its every branch of every order. Two basically different forms of inflorescence may be distinguished: indeterminate and determinate (Figs. 45–53). These two forms manifest distinctive characteristics related to fundamental differences in their manner of development.

The indeterminate inflorescence

An indeterminate bamboo inflorescence is one the course of whose development is prolonged indefinitely by the progressive elaboration of its branching. A separate "grand period of growth"[1] is initiated and completed, independently, in each flowering axis of each successive order of branches. Each flowering axis is spikeletlike in appearance, and it terminates in a spikelet. However, its basal part is a very short rachis, clothed with lemmalike bracts, each of which subtends a prophyllate branch bud instead of a flower. For this reason, these spikeletlike branches have been given the distinctive name, pseudospikelets (McClure 1934:Fig. 1). Being repositories of meristem, the buds at the base of a pseudospikelet make possible the continued expansion of the indeterminate inflorescence as long as the physiological state of the meristem and adjacent tissues favors the breaking of their dormancy and the continued development of the new axes and their appendages.

The determinate inflorescence

A determinate bamboo inflorescence is one the course of whose development is strictly limited to a single "grand period of growth" that encompasses the elaboration of the whole complement of branches of a solitary rachis. Terminal growth ceases in all branches of the inflorescence within a limited time. Each branch terminates in a conventional spikelet; no meristem remains afterward in the form of dormant lateral buds.

Other contrasts

The expansion of an indeterminate inflorescence follows very closely the pattern of development characteristic of the vegetative phase of growth. The expansion of the determinate inflorescence,

[1]According to Porterfield (1928:327), "The grand period of growth covers not only the enlargement of some part already in existence, but the production of a fully formed organ or series of organs with definite parts and specific characters."

Fig. 45. *Bambusa multiplex* var. *riviereorum:* (*A*) base of plant (showing pachymorph rhizome and caespitose clump habit); (*B*) leafy branches terminating in primary pseudospikelets (young inflorescences of indeterminate branching). Drawn by Elmer W. Smith.

on the other hand, manifests a high degree of simplification, as in the panicle of *Sasa veitchii* (Fig. 46), for example. This change marks an evolution in the direction of patterns common in many of the herbaceous grasses of nonbambusoid subfamilies (*Poa pratensis*, for example).[2] The diversity of form manifested by the determinate inflorescence is discussed on pp. 101f.

The course of development of the indeterminate inflorescence

The following description of the development of the indeterminate inflorescence departs, in some minor details, from that of Holttum (1956*a*:82; see "spikelet tufts") and an earlier one (McClure 1934) and it is more complete than either (see also Nees 1829; Moebius 1898; and Pilger 1927). The basic structural unit of the indeterminate inflorescence is the pseudospikelet. For clarity it is necessary to think in terms of two categories of pseudospikelets: primary and secondary. Primary pseudospikelets have their origin on segmented axes that are themselves vegetative, in a morphological sense at least. Secondary pseudospikelets originate from other pseudospikelets.

Primary pseudospikelets differ in minor structural details which are correlated with their respective positions on the vegetative axis that bears them. Lateral ones are sessile; a terminal one is made pedicellate by the distal internode of the vegetative axis that bears it. Closer examination reveals further that, whereas the lateral ones have a two-keeled prophyllum as the first (proximal) foliar appendage, the terminal one has an unkeeled bract in place of this two-keeled structure. (Fig. 47, *A*). The prophyllum that embraces this axis is located far below, at the base of the vegetative branch itself. This is the only structural difference between a terminal pseudospikelet and a lateral one. Secondary pseudospikelets of all orders are of the sessile category. A lateral primary pseudospikelet is therefore the prototype of all secondary pseudospikelets. And since secondary ones are by far the most numerous form in any flowering specimen, a description of the development and structure of its prototype will be most generally useful.

Initiating the sequence of events, a solitary prophyllate inflo-

[2]No report of the occurrence of indeterminately branching inflorescences in any grass outside the subfamily Bambusoideae has yet come to light.

Fig. 46. *Sasa veitchii.* Portion of a plant, showing leptomorph rhizome, diffuse clump habit, and leafy culm bearing solitary branches (monoclade branch complements) with determinate inflorescences. Adapted from Makino and Shibata (1901:Pl. 1).

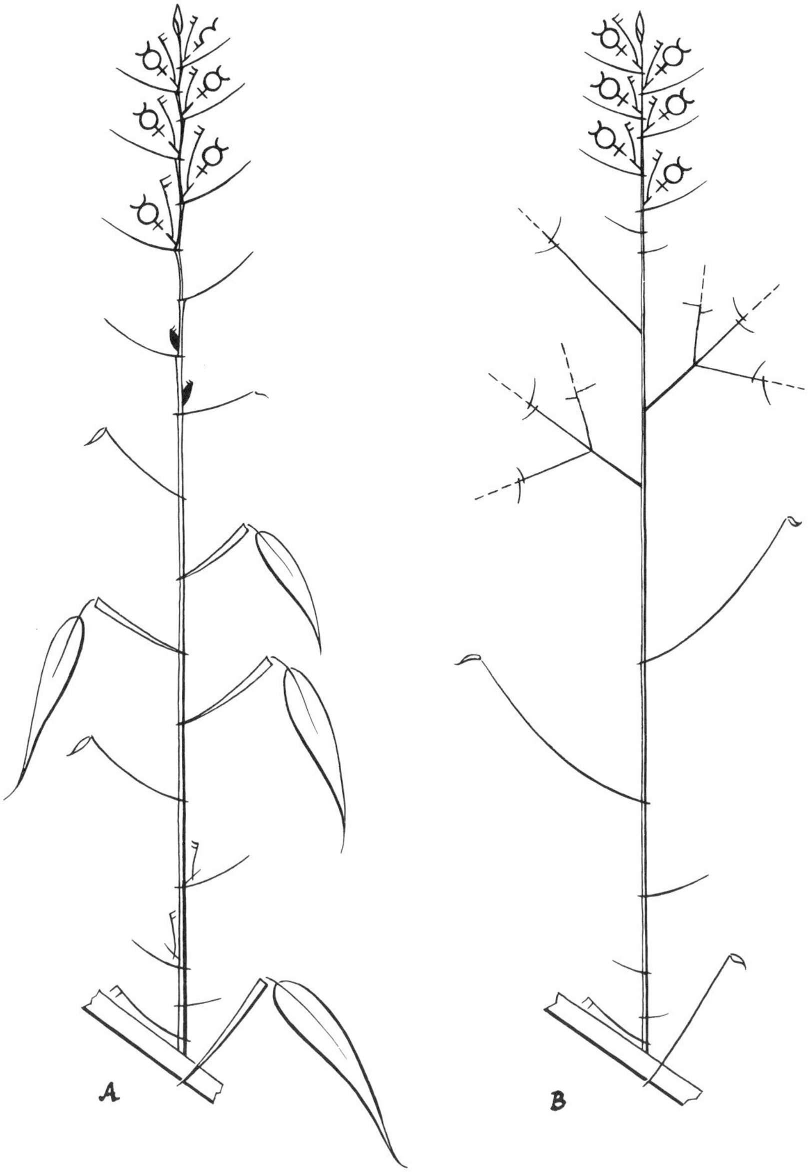

Fig. 47. Diagrammatic comparison of a branch of (*A*) *Bambusa multiplex* terminating in an *indeterminate* inflorescence with a branch of (*B*) *Sasa veitchii* terminating in a *determinate* inflorescence. The two dark objects at the base of the inflorescence proper in (*A*) are prophyllate buds, each capable of duplicating the pseudospikelet of which they are a part. This is a basic distinguishing feature of the *indeterminate* inflorescence. The broken lines in the inflorescence proper in (*B*) indicate fully developed spikelets like the apical one shown. Here no buds or other repositories of meristem remain to initiate additional branches. This is a basic distinguishing feature of the *determinate* bamboo inflorescence. Cf. Fig. 48 (*A, B*).

rescence bud lateral to a segmented vegetative axis develops into a short primary reproductive branch. This branch is spikeletlike in superficial appearance, but may be recognized as a pseudo-spikelet by the presence of a two-keeled prophyllum at its base, followed by one or more bud-subtending bracts (Figs. 48, *B* and 49, *B*). The presence of branch buds (revealed only by careful dissection) characterizes this part of the axis of the pseudospikelet as a rachis; the distal part is the rachilla of the spikelet proper. The transition between the lower, bud-bearing part of the pseudo-spikelet and the spikelet proper that follows is marked (1) by the initiation of the development of abscission layers at the nodes of the axis and (2) often by the intercalation of one or two empty glumes (see p. 109). With or without intervening empty glumes, the transformation of the rachis to a rachilla is made plain by the appearance, at the next level, of a floret—the beginning of the spikelet proper.

Sometimes, as in *Schizostachyum terminale* (teste Holttum 1958:51, 52), *Bonia tonkinensis*, and *Arundinaria prainii* (firsthand observation), the buds subtended by bracts on the rachis of the primary pseudo-spikelet may remain dormant. In such cases, the individual inflo-rescence consists of but a single pseudospikelet. In most bamboos, however, these buds break more or less promptly, and each re-sulting branch becomes a secondary pseudospikelet, whose basal buds produce tertiary pseudospikelets, and so on. The promptness with which the successive orders of pseudospikelet buds form new pseudospikelets is characteristic and relatively uniform in each bamboo. In *Oreobambos buchwaldii*, an embryo pseudospikelet cluster forms precociously, emerging from the primary prophyllum with several orders of branches already well advanced. In *Dendrocalamus strictus*, the development from the primary pseudospikelet to a full-blown, spherical head, consisting of many pseudospikelets, proceeds at a spectacular rate, but apparently no record of the duration of the active growth of these quickly maturing pseudo-spikelet clusters has been published. In *Bambusa vulgaris*, inflo-rescence clusters have been observed to continue to produce new pseudospikelets related to the same primary axis through two full years (Moebius 1898), while in *Bambusa multiplex* cv 'Alphonse Karr' they apparently may remain active for a still longer period.

The internodes of the rachis of the pseudospikelet are very

short—often not more than 1 or 2 mm in length. This being the case, the inflorescence cluster tends to become congested as the number of orders (successive "generations") of pseudospikelets increases. When the spikelets are all persistent, a globular head is formed, as in *Dendrocalamus strictus.* Commonly, however, the spike-

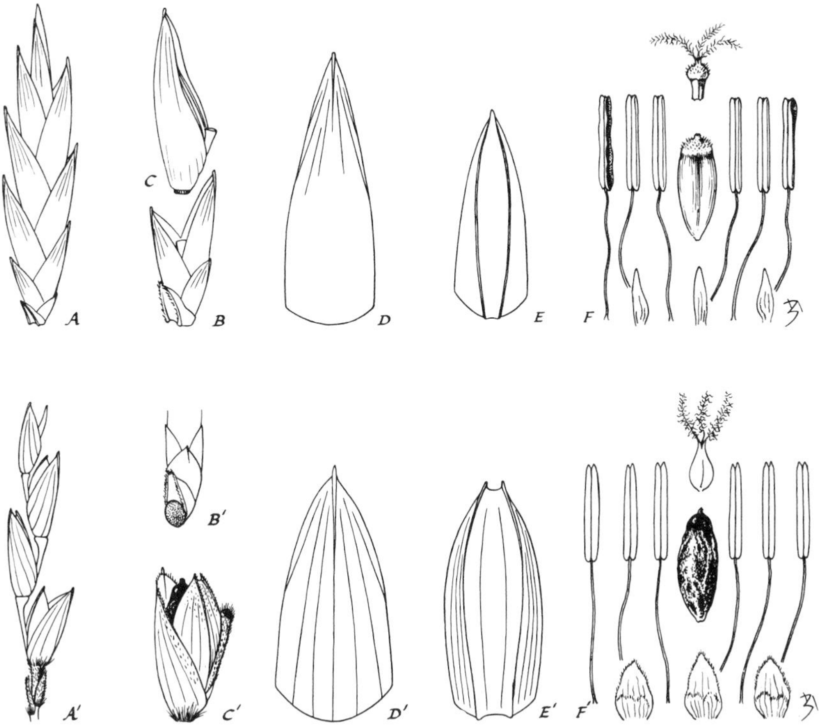

Fig. 48. Components of the inflorescence in *Bambusa multiplex* (*A-F*) and *Sasa veitchii* (*A′-F′*). (*A*) Branch of the indeterminate inflorescence—a pseudospikelet with a 2-keeled prophyllum at its base. (*A′*) Branch of the determinate inflorescence—a spikelet with 2 empty glumes at its base. (*B*) Base of (*A*), showing the prophyllum, the two bud-subtending bracts, and the empty glume that precede the first lemma of the spikelet (not shown). (*B′*) Base of the flowering branch that bears (*A′*), showing the 2-keeled prophyllum—here (in contrast) far removed from the spikelet. (*C, C′*) Floret. (*D, D′*) Lemma. (*E, E′*) Palea. (*F, F′*) the component parts of the flower: lodicules, stamens (androecium), pistil (gynoecium), and fruit. Based on specimens deposited in the U.S. National Herbarium. (*A-F*) US 516519 collected by G. T. Lane (sn) February 25, 1898, in The Royal Botanic Garden, Calcutta. (*A′-F′*) US 2241716, collected by T. Makino (sn) June 1909, near Ihagi-mura, Musashi, Japan.

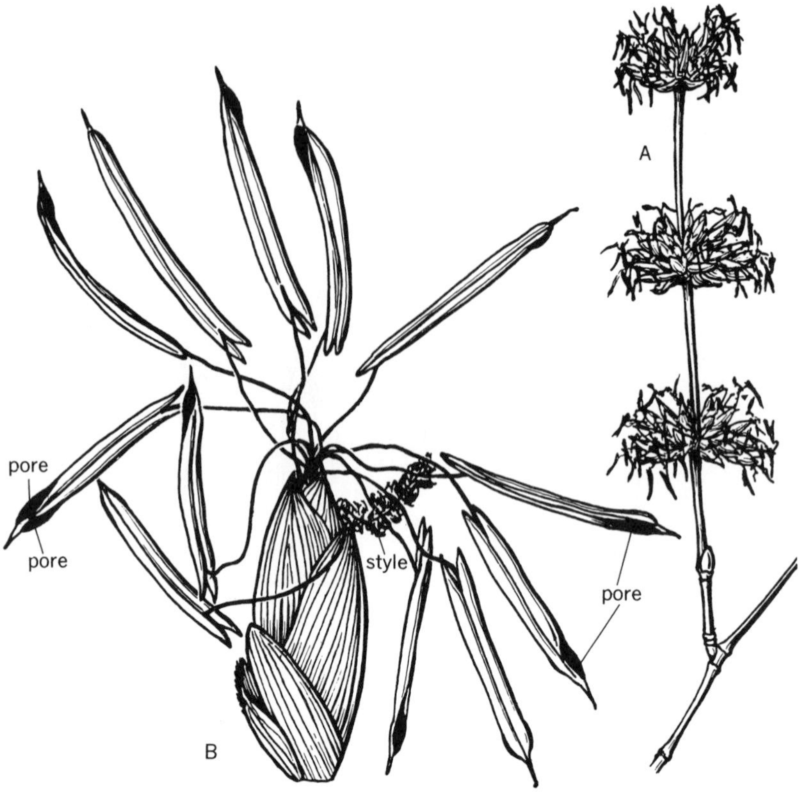

Fig. 49. The indeterminate inflorescence in *Dendrocalamus sikkimensis:* (*A*) flowering branch "about one month after anthesis," with a cluster of pseudospikelets at each node; (*B*) pseudospikelet (recognizable as such by the prophyllum at its base on the left) terminated by a spikelet with three florets. A solitary stigma (tip of the style) is seen projecting from the apex of the lowermost floret. Six stamens have emerged from each of the upper two florets. Pollen is discharged from the anthers through apical pores. From Arber 1934:Fig. 32*A*.

lets terminating the earlier pseudospikelets in a given cluster fall away more or less promptly as they mature, leaving the cluster of pseudospikelets sparser, and irregular in shape. This is characteristic of some bamboos of the genus *Bambusa,* all known species of *Elytrostachys,* and many species of *Schizostachyum.* To the experienced eye, the naked rachis tips that remain afford a clue to the loss of spikelets that has occurred, and an intimation of the relative age of the inflorescence.

Porterfield (1926) illustrated the inflorescence of *Phyllostachys nidularia* in two stages of development: "before anthesis" and "after anthesis." The capitate form resulting from the progressive branching from the basal nodes of short axes in the successive orders of branching is strikingly shown (Fig. 50) but the author makes no reference to the indeterminate (continuing) nature of the branching habit of this inflorescence. Porterfield (p. 259) says only that "the characteristic capitate cluster [inflorescence] of

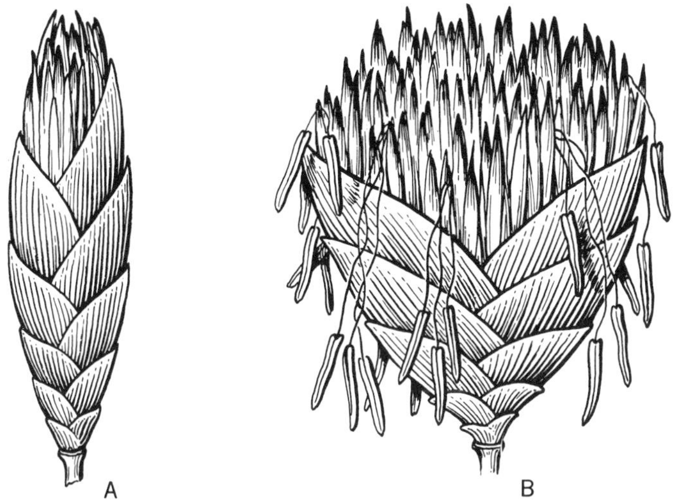

Fig. 50. The inflorescence of *Phyllostachys nidularia,* as illustrated by Porterfield (1926:Figs. 1, 2). (*A*) An early stage, "before anthesis," showing the form of the primary branch (a pseudospikelet) with the spikelet tips of pseudospikelets of higher order protruding beyond the uppermost bracts. The primary pseudospikelet in this illustration appears to be terminal to the internode shown below it, but the sheath scar that precedes the first bract suggests that it may be a lateral branch at the node that terminated the internode shown. If it actually was subtended by a leaf sheath (now fallen away) of which the scar is shown, then a prophyllum preceded the small bracts shown at its base. This prophyllum could not be shown in this view because it would lie behind the pseudospikelet as shown here. It would be addorsed to the next internode of the axis on which the pseudospikelet is borne, and which is hidden by the latter. (*B*) A later stage, "after anthesis," showing the capitate form attributed by Porterfield (p. 259) to the "telescoping" of the inflorescence. Actually, this is a tuft of pseudospikelets, an indeterminate form of inflorescence. Its congested form results from the rapid consecutive branching of the very short pseudospikelet axes of successive orders arising from buds in the axils of basal bracts of each pseudospikelet. It appears that the indeterminate nature of the inflorescence in bamboos of the genus *Phyllostachys* has not hitherto been reported in the literature.

Ph. nidularia is in effect a cluster of spikelets in which the branching inflorescence has been telescoped."

Variations and "anomalies" in the indeterminate inflorescence

The pseudospikelets of successively higher orders may become progressively depauperate. In the spikelet terminating a primary pseudospikelet the flowers are generally perfect and, potentially at least, functional. In pseudospikelets that are branches of successively higher orders, the florets and flowers may become smaller, and the development of the sexual organs may become progressively restricted. The gynoecium is the first part affected, and purely staminate flowers may result. Other changes may be noted, as well. In *Schizostachyum lima,* for example, fully developed perfect-flowered spikelets of the primary pseudospikelets are rather promptly deciduous, while those bearing rudimentary perfect flowers absciss much more tardily, and those bearing purely staminate flowers are persistent. In plants that have been flowering for some time, this behavior results in a growing preponderance, in pseudospikelet clusters, of spikelets with rudimentary perfect flowers and rudimentary staminate ones (McClure 1934:544ff). When preserved in herbaria, specimens with such a history may give the impression that the species they represent has unisexual flowers and the sexual pattern called polygamy (see p. 117), both as typical expressions.

Again, the inflorescence branches (either those of higher order or those higher on the rachis) may fail to develop buds at basal nodes. This puts a stop to the further production of pseudospikelets. Such a movement away from the typical indeterminate form of expression (clustered pseudospikelets) might be construed as a step in the direction of (or a sign of the potential for) evolution of the bamboo inflorescence toward a determinate form (raceme or panicle). The essentially determinate inflorescences in *Glaziophyton mirabile* and *Greslania montana* appear to be the result of a strong trend in this direction. Here, the inflorescence branches bear well-developed prophylla and are subtended by well-developed bracts. The occurrence of an occasional dormant prophyllate bud on the ultimate branches of an inflorescence, especially in the latter genus, suggests a relict tendency toward indeterminateness.

The combination of prophyllate lower branches subtended by bracts with upper branches neither prophyllate nor subtended by bracts, found in the inflorescences of *Thamnocalamus spathiflorus,* suggests a stage of evolution perhaps still further removed from the indeterminately branching condition.

Diversity of form manifested by the determinate inflorescence

The brief general statement (p. 91) characterizing the determinate bamboo inflorescence as having but a single grand period of growth is elaborated here by a few additional observations. Its form may be analyzed for diversity in terms of several features, the more important of which are the following: the presence or absence of a strong central axis (rachis); the spatial disposition of the loci of insertion of the rachis branches; the dimensions, physical proportions, and habit of the rachis and its branches; the number of orders of branches; the absence, or presence and degree of development, of pulvini at the base of branches; the absence, or presence and degree of development, of bracts subtending branches; and the presence or absence of prophylla.

The rachis may be strong and excurrent (*Sasa veitchii*), or it may be deliquescent (*Aulonemia queko*). A strong central axis with a single order of widely spaced, very short, solitary branches produces an elongate spicate or subspicate raceme in *Arthrostylidium cubense.* In such inflorescences the pattern of emergence of the primary branches is usually somewhat obscured by torsion of the rachis. This twisting produces a secund, subspicate raceme in most known species of *Merostachys.* In *Merostachys retrorsa* and *M. pauciflora* most of the spikelets are retrorsely oriented. Each reflexed pedicel bears a pair of minute pulvini inserted adaxially at its base. In other species of this genus the spikelets are antrorsely oriented and pulvini appear to be lacking. In *Athroostachys capitata* the inflorescence assumes a capitate aspect, but on dissection it is seen to be a very compact raceme which may be designated as "paniculate" because the rachis bears its branches mostly in fascicles of two or three, of which each primary one appears to have given rise, subcutaneously, to one or two secondary ones. Each fascicle is subtended by a single, narrowly acuminate bract bearing a long awn. In *Arundinaria tecta* the usually racemose inflorescence may occasionally combine the branching habits of both a raceme and a panicle or it may assume a typically paniculate form.

A strong, elongate, excurrent central axis with more than one order of branches produces a "typical" panicle. The primary branches of a panicle may be solitary, as in *Sasa veitchii* (Fig. 46), or two to several branches may emerge from a single node of the rachis, as in *Arundinaria amplissima.* Branches of higher orders are nearly always solitary. All orders of branches may remain more or less appressed or only slightly divergent (*Arthrostylidium subpectinatum*). In *Indocalamus sinicus* the individual branches of the panicle are very long and stiff; as they mature, they are given a wide angle with the rachis by the operation of basal pulvini. In all but one of the species mentioned, as in most bamboos with determinate inflorescences, the bracts subtending branches of the inflorescences are rudimentary or absent, often being represented only by a mere line or ridge below the locus of insertion of primary branches. As a rule, however, the bract subtending the lowest primary branch of determinate inflorescences is more noticeably developed than any of the others.

The bamboo inflorescence from the point of view of physiology

The results of determinateness and indeterminateness that appear in the bamboo inflorescence are commonly interpreted by the taxonomist only in terms that will most simply describe the superficial aspect of the resulting growth forms. These may be more fully understood when viewed also through the eyes of the physiologist, particularly if attention is given to the extreme range of expression often to be found within the inflorescences of a single plant.

It sometimes happens (now and then in *Arundinaria simonii,* for example) that a determinate inflorescence is reduced to a single spikelet borne at the apex of the rachis. Reference was made earlier (p. 96) to a similar reduction in inflorescences whose branching is indeterminate (that is, normally of a continuing nature). The use of the term "reduction" in this sense is traditional in taxonomy that is based solely on gross morphology. However, the physiologist probably would see behind the visible phenomenon (extremely early cessation of growth) the premature development of a physiological state that corresponds either to dormancy or to the loss of the meristemmatic potential—or possibly to the exhaustion of stored nutrients. In the case of an indeterminate inflorescence comprising but a single pseudospikelet, the potential for subsequent branching might remain, in the form of dormant buds at the base of the primary pseudospikelet (as in *Bonia tonkinensis* and *Arundinaria prainii*). On the other hand, when a determinate inflorescence stops growing after producing one spikelet, that usually is the end of the story. The development of additional spikelets is no longer possible, since no meristem remains. However, it sometimes happens in an inflorescence with determinate branching that the spikelets themselves do not stop growing at the usual point. The continued activity of the apical meristem may extend the length of some spikelets to almost a foot, as I have observed in flowering plants of *Arundinaria dolichantha.* This suggests the steady persistence, in the meristem at the tip of a spikelet, of a physiological state corresponding to active reproductive morphogenesis.

As long as the signal that determines the length of the spikelet is thus inhibited, the spikelet continues to grow apically. H. Fung 695, a specimen of *Arundinaria dolichantha* collected at Hoh Tung in Tonkin (now North Vietnam) shows another striking morphological anomaly. After producing flowers for a length of about 3 in., the rachilla in some of the spikelets reverted to a purely vegetative state and terminated in a leafy twig.

From the point of view of morphology, the component parts of the bamboo inflorescence are conventionally seen as the result of a more or less radical "modification" of the structures (nodes, internodes, and sheathing appendages) characteristic of the segmented axes of the plant in the vegetative state. The specialized (or "reduced") sheathing organs of the inflorescence—and even the androecium and gynoecium—are commonly referred to as "leaves" or "sheaths" in modified form (see Watson 1943 and Arber 1950:68). In some bamboos with indeterminately branching inflorescences (*Bambusa multiplex* and *Phyllostachys bambusoides,* for example) the transition from the vegetative phase to the reproductive phase may be marked by such gradual changes in the form of the sheathing structures that the correspondence is, indeed, striking (Figs. 45 and 51). The indeterminately branching inflo-

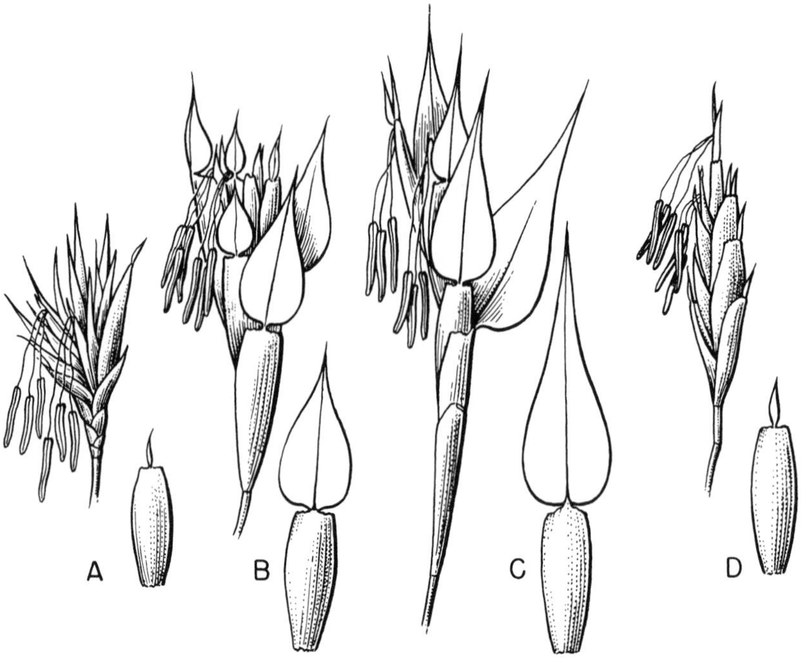

Fig. 51. In some bamboos with indeterminate inflorescences, the bracts that precede primary pseudospikelets or subtend branches thereof bear more or less strongly developed, leaflike blades. This condition prevails in different degrees in four species of *Phyllostachys,* as illustrated by these figures: (*A*) *Ph. nigra* cv. *henon* (as var. *henonis*); (*B*) *Ph. bambusoides;* (*C*) *Ph. aurea;* (*D*) *Ph. pubescens* (as *Ph. mitis*). This leaflike character of the bracts is lacking in some species of this genus, as in *Ph. nidularia* (Fig. 50) but is sometimes seen in *Bambusa multiplex* (Fig. 45). From Kawamura 1927:Fig. 6.

rescence is, in fact, like the vegetative system in being made up essentially of ramifying axes that are all clearly segmented and bear subtending sheathing organs and prophyllate primordia (buds) at all points of branching. The determinate inflorescence, on the other hand, in the extreme form of its expression, as in *Sasa veitchii* (Figs. 46 and 47, *B*), has the rachis of the inflorescence and its branches of all orders below the spikelet level segmented only weakly if at all. These axes are often more strongly parenchymatous in substance than the corresponding branches of an indeterminate inflorescence. Subtending foliar organs (bracts) are usually rudimentary or absent, and amplectant prophylla are typically lacking entirely in determinate inflorescences. Exceptionally, however, the subtending bracts, and prophylla, may be well developed (as in *Glaziophyton mirabile* and *Greslania spp.*; see Pilger 1945:22, 24, on *Greslania*).

From the physiological point of view, the similarities between component elements of vegetative and reproductive shoots, especially in respect to the form and pattern of insertion of certain of the foliar organs, do not indicate the lingering influence, in the inflorescence, of the *form* in which the corresponding (homologous?) organs were cast during the vegetative phase of the plant's life. They appear to be rather a sign, or the result, of features common to the corresponding *inner* (physiological) *states* of the respective bodies of meristem. The distinctive forms and functions that characterize the inflorescence, particularly at the level of the androecium and gynoecium, speak eloquently of strong monitoring factors of a physiological nature peculiar to the flowering state (cf. Heslop-Harrison 1959:269 *et passim* and Wardlaw 1961). Wilson (1942:759) says, "It is becoming widely recognized that the generally held concept of the essential organs of the flower as modified foliar appendages rests upon a most insecure foundation." Nozeran (1955:13) credits Thompson (1944) with having proposed the following ideas, based on studies of ontogeny: (1) the parts of the flower have no apparent relation to leaves; (2) they are outgrowths that cannot be equated to leaves; (3) all of the primordia of the floral parts are mutually homologous; and (4) their differentiation is monitored solely by the physiological conditions that prevail in the floral axis at the time of their development. In this perspective, the plausible traditional designation of all of the floral parts as "modified leaves" loses some of its appeal. This whole subject is admirably treated from the historical and philosophical points of view by Arber (1950:chap. V; see also Watson 1943; Wardlaw 1956; and Takhtajan 1959:chap. 1).

The physiological point of view also places in fresh perspective several fundamental aspects and features of the reproductive phase: (1) the sudden change in phyllotaxy, from the distichous expression that prevails elsewhere in the plant (except in the branching of some determinate inflorescences), to the trimerous arrangement of parts that prevails within the flower; (2) the radical and abrupt changes that take place in the very short axis of the flower, giving rise to forms so highly specialized, and so diverse, as palea, lodicules, stamens, and pistil within the space of 1 or 2 mm; (3) the "anomalous" manifestations in the bamboo flower that appear in the form of teratic structures intermediate between contiguous categories of the floral structures just mentioned (cf. Munro 1868:152 *et passim;* Arber 1934:Fig. 202–210); (4) the general lack, in the determinate inflorescence (as contrasted with the indeterminate form), of well-developed subtending bracts and of prophylla at the points of branching, and the corresponding loss (below the spikelet level) of both clearly marked segmentation and distichous branching; (5) the diverse behavior of different species and strains of bamboos in respect to the initiation, patterns of development, and fruition of the reproductive phase; (6) the corresponding patterns of behavior in the vegetative life of the plant, both during and after flowering; and (7) inflorescences intermediate between the two extremes described above as determinate and indeterminate. Studies of the bamboos, along the lines illustrated by Zimmerman (1961), should improve our understanding of the genesis of some of these manifestations.

The development of a comprehensive perspective with regard to the characteristics of the intermediate (transitional) morphological expressions just mentioned requires a close examination of the whole range of the known genera of bamboos. Such a broad perspective is not yet available in published dissertations, partly because of the hitherto incomplete coverage of the known forms, but chiefly because the close examination of the materials has generally been confined to the level of the flower and the spikelet. As a rule, published observations on individual species have not penetrated deeply enough to reveal the basic nature and origin of the branching habit of either the vegetative or the flowering axes.

Comparative studies of the branching habit of the bamboo inflorescence carried out personally indicate that the evolution, by stages, from the less specialized form of the typical indeterminate inflorescence, as seen for example in *Bambusa multiplex* (Figs. 45 and 48, *A–F*), to the more specialized form of the typical determinate inflorescence, as seen in *Sasa veitchii* (Figs. 46 and 48, *A′–F′*), corresponds to the general pattern hypothesized by Holttum (1958:18–19) in the following words:

Now how can we rationalize a comparison between the group of short flowering branches at one node found in a bamboo with a panicle of spikelets as found in a grass? If all the lower internodes of a tuft of short flowering branches [pseudospikelets] in the bamboo were much elongated, as far as the empty glumes in each case, we should have something very like a panicle of spikelets, each spikelet on a stalk and either terminal on a branch or in the axil of a bract with a prophyll immediately above the bract. If we eliminate these branch-bearing bracts and the prophylls adjacent to them and reduce the empty glumes at the base of each spikelet to two, we have exactly the condition of a grass panicle.

As Pilger did earlier (see p. 90), Holttum here sets off the indeterminate inflorescence ("short flowering branches") of a bamboo, against the determinate inflorescence ("panicle of spikelets") of a nonbambusoid grass. Holttum states elsewhere (1956*a*: 85) that

the primitive condition would seem to be that now normal in several species of Malayan *Schizostachyum,* in which every branchlet is leafy (with normal blades), the spikelet tufts occurring at the few most distal nodes on many branches, flowering being often continuous. This condition could easily be modified to that of a terminal panicle on each leafy branchlet.

It appears that conventional preoccupation with observations made at the spikelet and flower levels has diverted attention from important characters that can be brought into focus only by examining the structure of the inflorescence as a whole, and studying its development in the light of facts relating to physiology and morphogenesis (cf. Thompson 1944, especially pp. 66 and 68, and the discussion that follows). Systematic studies carried out in the present century have resulted in the description of hundreds of new species and a number of new genera, but the fundamental nature of the branching habit of the bamboo inflorescence as pictured herein has been generally neglected, in spite of its phylogenetic significance, and its taxonomic value. Arber, whose numerous anatomical studies have so greatly helped to elucidate the structural diversity manifested in the spikelet and flower of the Gramineae, has dealt more fully with the inflorescence, as such, in other plants than with the branching habit of the bamboo inflorescence and its significance (cf. Arber 1926–1929, 1934, and 1950). Holttum (1956*a*, *b,* 1958*a*) has made fruitful preliminary studies of the principal vegetative organs, the inflorescence, and the fruit (where available) in bamboos of seven genera represented in the flora of Malaya. In all of these genera the rhizome is pachymorph. In six of them (*Bambusa, Dendrocalamus, Dinochloa, Gigantochloa, Schizostachyum,* and *Thyrsostachys*) the branching of the inflorescence is indeterminate. In the seventh, *Racemobambos,* the inflorescence is determinate in its branching habit.

The prophyllum in inflorescence branching

Two-keeled prophylla are generally lacking on the branches of determinate inflorescences. In typical indeterminate inflorescences, on the other hand, each branch develops from a prophyllate bud. In gross dissections of such inflorescences, the distinctive features of the two-keeled prophyllum make it a marker useful in following the sequence of branching, especially where the branches of successive orders are extremely short (as in *Schizostachyum blumii;* see McClure 1934:546).

One-keeled prophylla appear in inflorescences that embrace structural features intermediate between the extremes of determinate and indeterminate branching, as in the transitional middle part of the inflorescence of *Thamnocalamus spathiflorus* for example, the prophylla are sometimes one-keeled and very narrow, and do not completely enclose the branch primordia on which they are inserted. In such cases, the branch promordium apparently does not have the true resting stage that is found in typical indeterminate inflorescences, where it is always completely enclosed by a two-keeled prophyllum.

In flowering axes having one-keeled prophylla, the counterpart of the other half of a two-keeled prophyllum usually appears on the approximately opposite side of the axis, at the second node (following a very short internode), and the second structure is usually one-keeled like the first. This pattern, and the independent origin (from distinct primordia) of the two parts of the divided palea in *Streptochaeta* (see Page 1951:28 and Figs. 1 and 2), support the theory that the vegetative prophyllum and its reproductive homologue, the palea, may have had their evolutionary origin in the fusion of two adjacent sheathing organs, made possible by the suppression of the intervening internode. The spontaneous suppression of culm internodes in *Arthrostylidium schomburgkii* is verified on p. 43.

Spikelet, floret, and flower

The spikelet is a basic structural unit of the bamboo inflorescence. Essentially, it represents a distinct and characteristic aggregation of flowers and structures intimately associated with them. The spikelet consists of a specialized axis, the rachilla, and its branches, the flowers. The rachilla is clothed in a series of imbricate sheathing appendages. Of these there are two major categories, called by American agrostologists glumes (or empty glumes), and lemmas (or flowering glumes). Empty glumes are characterized (1) by occupying the basal (physiologically transitional) portion of the spikelet axis; (2) by being "empty," that is, not subtending either a branch bud or a flower bud; and (3) by having, in determinate inflorescences generally, a smaller size and slightly different shape in comparison with the lemmas. The term lemma, or flowering glume, stands for a glume that subtends a flower. A lemma is referred to as sterile if its subtended flower is rudimentary or lacking. In some bamboos, the basis for a clear distinction between empty sterile lemmas and immediately adjacent empty glumes is not easy to establish.

The internode preceding the first sheathing appendage (usually an empty glume) of the spikelet is designated a pedicel. In paniculate and racemose determinate inflorescences (*Indocalamus sinicus; Aulonemia queko*) the pedicel usually is fairly long. In some determinate inflorescences (the spicate raceme of *Arthrostylidium cubense*) the pedicels are extremely short. In indeterminate inflorescences (made up of pseudospikelets) the spikelet has no other pedicel than the very short uppermost internode of the rachis. Being covered by the uppermost bract of the rachis, this pedicel is visible only upon dissection of the pseudospikelet.

Common agrostological usage designates as florets the units into which a spikelet breaks up when the rachilla segments disarticulate. But regardless of whether the rachilla segments are abscissile or not, a floret consists of (1) a lemma, (2) the rachilla segment immediately above the node which bears the lemma that embraces it, and (3) a flower, consisting of a branch of the rachilla subtended by the lemma and embraced by a prophyllum called the palea. Besides the palea, the floral axis bears lodicules (when these are present), stamens, or a pistil (or both). The floret thus includes structures from axes of two orders. The floret should be clearly differentiated from the flower proper in descriptive statements. The form and organization of the spikelet, floret, and flower of *Phyllostachys nidularia* are very effectively illustrated by Porterfield (see Figs. 52 and 53).

It appears that the incidence of two empty glumes at the base of the spikelet is generally accepted by agrostologists as a norm for the interpretation of spikelet structure in the nonbambusoid grasses. When only one empty glume is present in a spikelet, it is interpreted as being the second, or upper one, of two. Each of the sheathing structures above the second empty glume is conventionally called a lemma, whether it subtends a flower or not. In some bamboo genera with a determinate inflorescence (*Arundinaria* and *Sasa,* among others), the spikelet generally conforms to

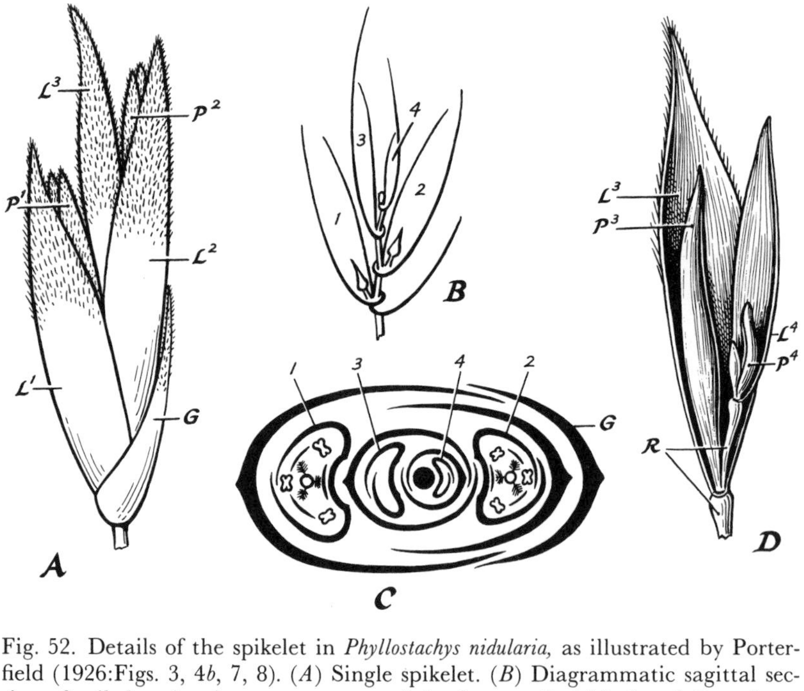

Fig. 52. Details of the spikelet in *Phyllostachys nidularia,* as illustrated by Porterfield (1926:Figs. 3, 4*b*, 7, 8). (*A*) Single spikelet. (*B*) Diagrammatic sagittal section of spikelet, showing arrangement of the florets, of which *1* and *2* are functional, *3* and *4* are rudimentary and sterile. (*C*) Plan of spikelet in cross section, showing florets *1–4*. (*D*) Longitudinal section of tip of spikelet, showing the rudimentary third and fourth florets. All much enlarged. *L,* lemma; *P,* palea; *G,* glume; *R,* rachilla segments.

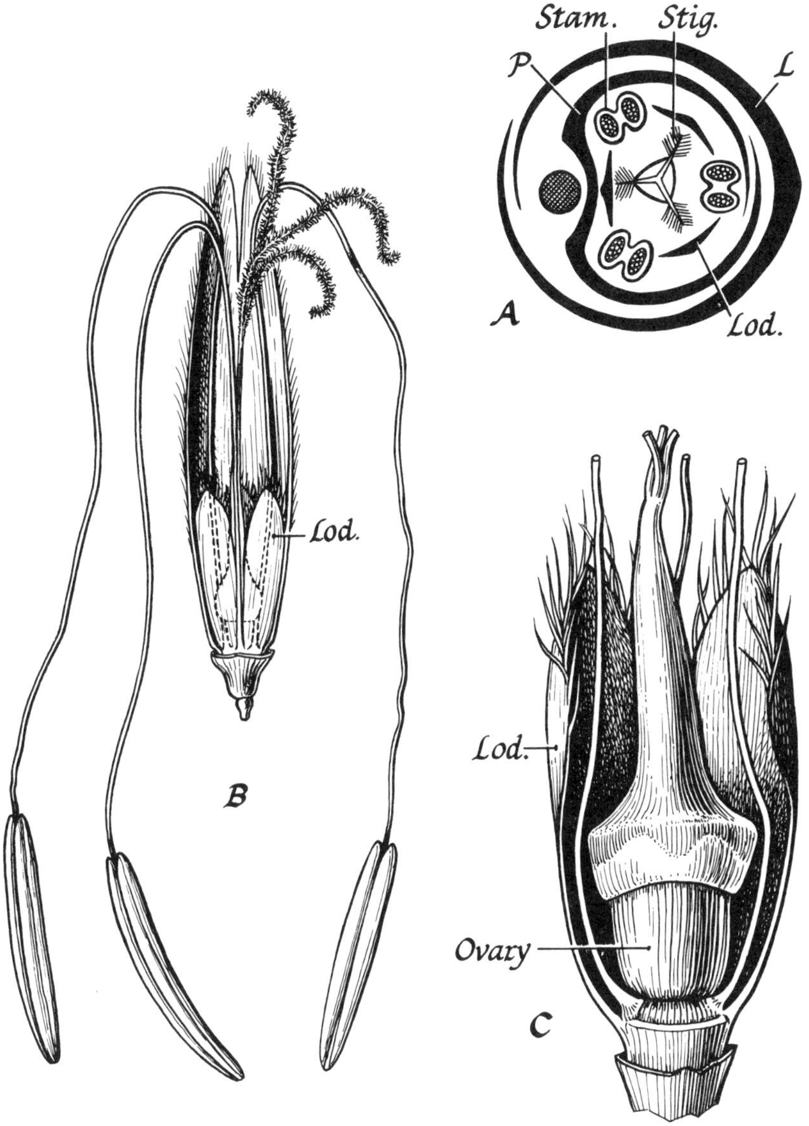

Fig. 53. The floret and flower of *Phyllostachys nidularia*, redrawn after Porterfield (1926:Figs. 4*a*, 5, 6). (*A*) Floret in cross section. (*B*) Floret (anterior aspect) with lemma removed (pubescent palea with notched apex seen at back). (*C*) Flower (posterior aspect) with dorsal lodicule removed and the anthers and stigmas cut off. All much enlarged. *L*, lemma; *P*, palea; *Lod.*, lodicule; *Stam.*, stamen; *Stig.*, stigma.

this "norm." That is, the empty glumes are regularly two, and the second one is followed by a flower-subtending lemma. In other genera, however, one to several glumes intermediate in size and form may intervene between the second empty glume and the first fertile glume, or lemma. These are empty, and they are conventionally referred to by agrostologists as sterile lemmas. However, this designation appears arbitrary in cases where there is no clear physical or physiological difference between structures called empty glumes and those called empty sterile lemmas. The numbers of "sterile lemmas" above the second one represent, is some species of *Arthrostylidium* and *Nastus,* respectively, the two extremes "one" and "several." Of more general occurrence is a situation in which the flowers subtended by lemmas at one or both extremes of the series in a given spikelet are rudimentary or reduced in size and in functional effectiveness.

Readers interested in pursuing the study of the gross floral anatomy of the bamboos will find a rich source of exemplary studies in the series of papers published by Agnes Arber in the *Annals of Botany,* 1926–1929, under the title "Studies in the Gramineae." Numbers 1, 2, 3, 4, and 8 in the series contain important textual and illustrative material on the floral anatomy of a wide selection of bamboo genera and species. Full bibliographic details are given in the list of references appended to the present work.

Arber observes (1934:108) that "the flowers of the bamboos are developed on a fuller plan that those of other grasses" (Fig. 54), and that they approach more nearly to the "complete monocotyledonous type." The flower of *Bambusa nutans* is described and figured (*ibid.*:133, Fig. 33) as having three "perianth members" (lodicules), three plus three stamens, and three styles. As Arber points out, the basic number of parts in any one, or all, of these categories may be augmented (as in *Ochlandra travancorica*) or reduced. In the evolution of the inflorescence, and in the reduction of floral parts, however, some of the bamboos have not fallen behind the herbaceous grasses as much as Arber suggests (*ibid.,* especially pp. 89–90). Throughout the genus *Chusquea,* for example, we find a determinate inflorescence typically bearing but a single functional flower in each spikelet. In this flower, the stamens are reduced to three, the styles are reduced to two, and the posterior lodicule is often very much smaller than the other two. This array of parts is very close to that commonly illustrated in diagrams of the structure of the "typical" flower of herbaceous grasses (Fig. 54).

Just as each vegetative branch primordium in the bamboo plant is enclosed in a prophyllum, the flower is embraced by the homologous palea. The palea is usually, but not always, strongly two-keeled. The prophyllate branch buds subtended by bracts on the rachis at the base of a pseudospikelet (in bamboos with indeterminate inflorescences) have often been mistaken for flower

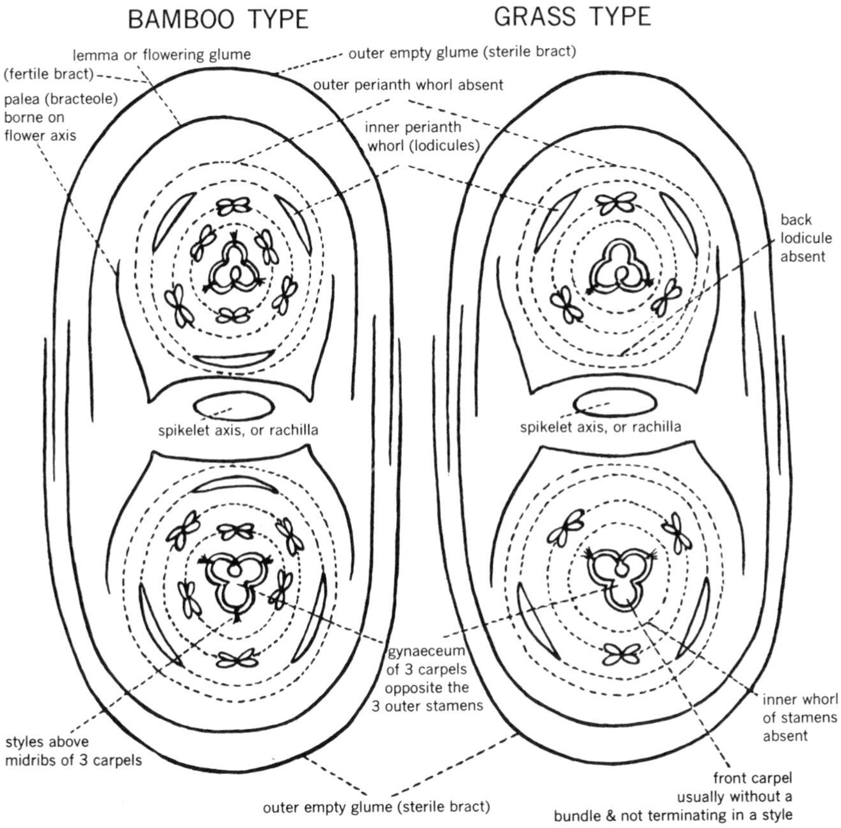

Fig. 54. Diagrammatic comparison of "typical" spikelet and floret structure in the bamboos (*left*) with that in the nonbambusoid grasses. Within the diversity of expression actually found in the spikelet and floret, particularly in the Bambusoideae, there are of course many features that do not conform to the pattern here shown as "typical." From Arber 1934:Fig. 58.

buds (cf. Munro 1868:Pl. 5, Figs. 4 and 5, and p. 153). In the vegetative axes of the plant, and in the spikelet itself, the insertion of the appendages is two-ranked, and distichous—alternating in a single plane. The axis of the flower, on the other hand, has its appendages inserted in whorls of three. The first whorl distal to the palea consists of lodicules—typically three, rarely two, sometimes lacking (in most known species of *Gigantochloa* and in some species of *Dendrocalamus* and *Schizostachyum*), occasionally more numerous (as in *Ochlandra travancorica*). The lodicules vary in size and texture, but are generally small, thin, and diaphanous, with venation irregular and often obscure. They are often noticeably unequal in size and shape. The posterior one that is addorsed to the palea is generally symmetrical and narrowly lanceolate. The anterior two, located near the margins of the palea, are in some genera broadened asymmetrically (subsemiovate) and paired. I have encountered no reference to this characteristic as a sign of zygomorphy in the bamboo flower. The appearance of zygomorphy is made even more noticeable in the flowers of a number of bamboo genera (*Chusquea* and others) by the occurrence of only two stigmas, these being matched with the paired lodicules.

In some bamboo species the anterior lodicules are thicker in substance than the posterior one and become swollen and turgid at anthesis. By pushing the lemma away from the palea, this causes the floret to open (*Bambusa multiplex*). The "forked" lodicules of *Bambusa arundinacea* (Arber 1934:Fig. 106, *F*), are so characterized because of their appearance in cross section. They are said (*ibid.*: 117) to have a "fish-tail-shape," and are examples of the form sometimes assumed by these structures after anthesis. As they lose their turgor and the floret closes, one edge of each lodicule is caught and pressed between the lemma and palea. What was a "belly" in the turgid state may become a thin flap in the dry state, as often seen in herbarium specimens. In some species of *Schizostachyum* with one-flowered spikelets, and elsewhere, lodicules are lacking, and the florets do not open at anthesis. Hackel (1881) discusses the lodicule complement of several bamboos and other Gramineae.

Next in order above the lodicules come the stamens, which appear typically in one or two whorls of three, according to the genus. However, in *Ochlandra,* the number of stamens may be very high and very variable (Beddome 1873:235 records 50–60; Gamble 1896:125 records up to 120 in a single flower of *Ochlandra travancorica*). Even when small, the number of stamens is sometimes irregular (*Pseudosasa japonica,* with 3, 4, or 5; *Dendrocalamus hamiltonii* with 6 or 7). A stamen consists of two parts: the anther and the filament. The filaments may be threadlike and free (*Phyllostachys*), flat, and free (or more or less coherent at first as in *Bambusa vulgaris*), or fused into a tube, in species of *Schizostachyum* (as *Neohouzeaua*), *Oxytenanthera,* and *Gigantochloa.* Normally the filaments elongate at anthesis and thrust the anthers out of the flower. Each anther consists initially of four locules (pollen-mother chambers) in two pairs (cf. Arber 1934:Figs. 201–210). The two pairs are joined through most of their length by the connective. The connective is sometimes prolonged above, into a smooth, or penicillate, appendage between the two apical lobes of the anther. The apical lobes of the anthers, usually subacute and rounded, are sometimes nipple-shaped (*Dinochloa scandens*), or divergently corniform and (exceptionally) twisted (Fig. 26, *12*). As the pollen matures, the two locules of each pair unite to form a sac. After the anther is exserted and pendent, the pollen spills out of each sac through an apical pore (Fig. 49) or, in some bamboos, by way of a long slit that follows the suture between the adjacent edges of each pair of locules.

The gynoecium, or pistil, is referred to by Arber (1934:163) as "simple and uniform" in members of the grass family. On p. 152 of the same work the author says, "When we turn to the gynaeceum we . . . receive no help from the ontogeny in deciding how many carpels are present . . . The mature structure, however, harmonizes perfectly with the descriptive convention that it consists of three carpels united edge to edge." And elsewhere (*ibid.*:120) Arber states that "the gynaeceum, though best interpreted as formed by three carpels, has an ovary with only a single cavity, with one ovule attached to its back wall." In the gynoecium, three more or less clearly distinct categories of component structures are externally discernible: ovary, style, and stigma. The ovary is the basal and basic structure. Its shape usually changes rapidly during the early stages of its growth. The diversity thus produced in the various specimens gives rise to discrepancies between different descriptions and different interpretations of the form characteristic of the ovary; and of the fruit, of a given species. The ovary has been described as stipitate (stalked) in some bamboos (*Phyllostachys;* see Siebold and Zuccarini 1843:746; Munro 1868:36), but this is an error due to misinterpretation of the dried form of an immature state of the ovary (often found in herbarium specimens) in which, owing to the thickening and induration of the apical part of the pericarp, this part retains its size and shape, while the soft lower part of the ovary shrinks to a slender stalklike form upon drying. Nakai interpreted this same phenomenon in *Bambusa* as a thickening of the base of the style, and used it incorrectly, along with thorniness, another character shown variably in some species, to separate this genus from Leleba (cf. McClure 1946*b*:106).

In most herbaceous grasses the ovary bears two styles separately inserted, while in the bamboos the ovary narrows at its apex into a usually single stylar column, or style (for alternative descriptions of the style, see p. 287f). In external aspect, the style commonly appears as an upward extension of the ovary. However, as Holttum has suggested, the relation of the style and the ovary is fully revealed only by a study of their inner structure. Holttum (1956*a*:Fig. 9) illustrates a completely hollow style in *Oxytenanthera* (species not designated); and in *Schizostachyum brachycladum* (*ibid.:* Fig. 10) a style that is hollow except for a central strand of tissue. Arber (1934:120) refers to this central strand as the "stylar core," and illustrates it (*ibid.:*Figs. 47 and 48) as it appears in *Ochlandra setigera, O. beddomei,* and *O. stridula.* The style may be greatly elongated and unbranched, with a single long stigma, as in *Dendrocalamus strictus.* Here the style is slender and fragile above; only the thicker, harder basal part persists as a short point on the fruit. In *Phyllostachys, Ochlandra,* and *Schizostachyum* the greatly elongated and more or less hardened style terminates in three short stigmas (four or five in some species of *Ochlandra*). Here, the style persists on the mature fruit as a long beak (Fig. 55). This character is often correlated with a tightly convolute and somewhat hardened condition in the lemma and palea. In such cases the floret is terete, or fusiform, with a minute apical opening through which the anthers and stigmas are exserted. In *Bambusa vulgaris,* the style is elongated and apically delicate (as in *Gigantochloa*), and may bear one, two, or three stigmas at its apex. In other species of *Bambusa* the style is commonly very much shorter and sometimes is divided almost to its base. In such extreme cases, as in *Bambusa multiplex,* for example, the three branches are sometimes interpreted as three styles. *Arthrostylidium cubense* exemplifies a condition common in its genus, in which the style is very short and bulbous, and bears two somewhat elongate branches, each terminating in a stigma of the form here designated by a new term, bottlebrush. Stigmas are variable in form as between genera as well as between species within a genus. Their branching often varies within a given species, and teratic manifestations of several kinds are common. Floral histogenesis in *Bambusa arundinacea* is effectively described and illustrated by Barnard (1957:2–7, Figs. 1–12) in what is apparently the pioneer contribution to our knowledge of this aspect of the Bambuseae.

Perfect flowers are the normal reproductive expression in most bamboos. However, of frequent occurrence are two states in which a secondary polygamous condition is produced:

(1) In many bamboos one or two so-called male flowers, in which the gynoecium is absent or rudimentary, appear in the transition zone just above the empty glumes at the base of the spikelet. These are followed by perfect flowers, as in *Bambusa multiplex,* of which (as *B. nana*) Munro (1868:90) says, "The lowermost 1 or 2 and the uppermost 2 or 3 flowers imperfect (either male or female), the intervening 3 to 6 flowers perfect." Of *Phyllostachys heteroclada,* Oliver (1894) says "the upper florets of each spikelet appear to be staminate."

(2) In the one-flowered spikelets that terminate the depauperate pseudospikelets that sometimes develop late in the life of an indeterminate inflorescence (*Schizostachyum lima*) the gynoecium may have become obsolete, while the androecium is either well developed or more or less rudimentary (see p. 100). Arber (1934:133) mentions other examples of secondary or fortuitous polygamy as occurring in *Gigantochloa* (as *Oxytenanthera*) *albociliata* and *Gigantochloa maxima.* In the determinate inflorescence of *Puelia* and *Atractocarpa,* the lowermost flower in a spikelet may be purely staminate and the uppermost purely pistillate.

The fruit

The bamboo fruit (Fig. 55) is indehiscent, and the single seed usually fills the pericarp completely. In external features the bamboo fruit embraces a wide range of forms: a caryopsis furnished

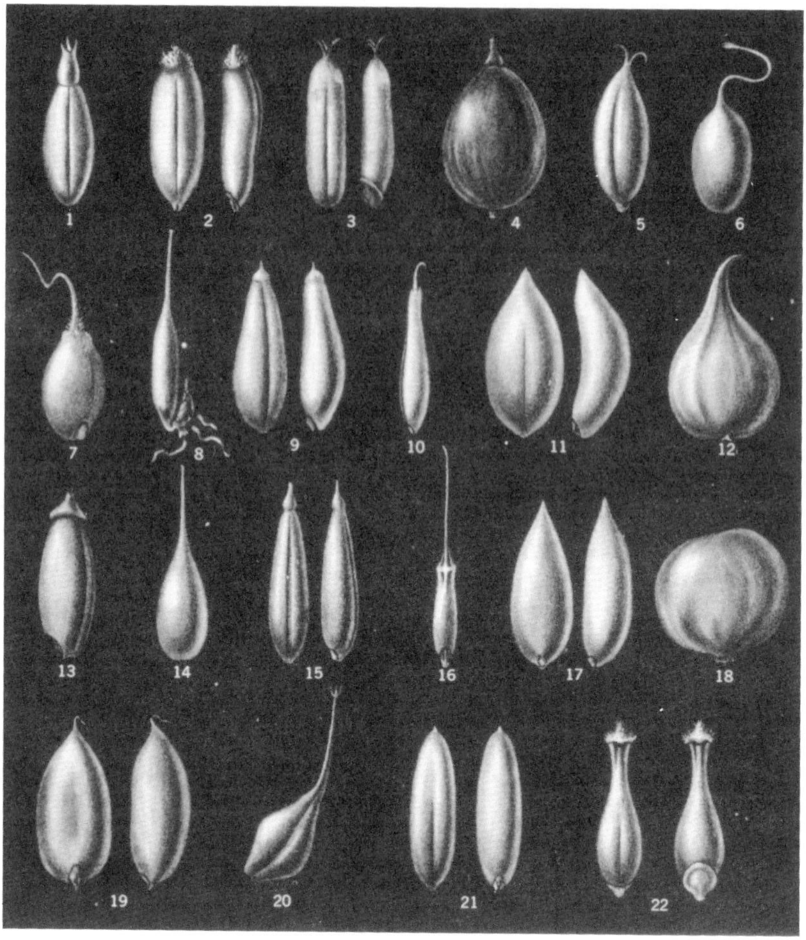

Fig. 55. Bamboo fruits, drawn by Elmer W. Smith.

1. Bambusa blumeana, dorsal aspect, showing the hilum or "sulcus." Redrawn, slightly modified, from Kurz 1876:Pl. II, Fig. 14*b*; ×2½.

2. B. multiplex, dorsal and lateral aspects, showing hilum and embryotegium, respectively. Redrawn from Muroi 1956:Pl. 49, Fig. 2 (as *Leleba multiplex*); ×1½.

3. B. longispiculata, dorsal and lateral aspects. Original; based on fruits from India, supplied by the U.S. Department of Agriculture under P.I. 117530; ×1¼.

4. Cephalostachyum sp.; persistent axis of flower shown at base. Original; based on fruits from Burma, supplied by the U.S. Department of Agriculture (without P.I. number); × 1½.

5. Chimonobambusa marmorea, dorsal aspect. Original; based on fruits collected by Dr. Jisaburo Ohwi (s.n.) near Osawa, Saitama prefecture, Japan, May 15, 1952; × 2½.

6. Dendrocalamus asper. Redrawn from Kurz 1876:Pl. II, Fig. 15*b* (as *Bambusa aspera* in legend, *Gigantochloa aspera* in the text, p. 221); ×4.

with a thin pericarp and shaped like a grain of wheat but larger and with the short, two- or three-pronged stylar column (or the base of it) persistent at the apex (*Arundinaria* and arundinarioid genera); a similar fruit with the pericarp thickened at the apex only (*Bambusa, Dendrocalamus, Nastus*); a long-beaked structure with either a moderately thickened hard pericarp (*Cephalostachyum* and *Schizostachyum*), a thick, hard pericarp (*Ochlandra*), or a very thick, tough pericarp (*Melocanna baccifera*); a subspherical chestnutlike

---

7. D. strictus. Original; based on fruits collected by Dr. I. D. Clement in March 1957, from plants cultivated at the Atkins Institution of Harvard University, Soledad, Cuba; $\times 2\frac{1}{4}$.

8. Dendrochloa distans. Original; based on fruits from India, supplied by the U.S. Department of Agriculture, under P.I. 117532, in an early stage of germination; $\times 1\frac{1}{4}$.

9. Elytrostachys clavigera, dorsal and lateral aspects. Original; based on fruits from El Recreo, Nicaragua, collected by McClure (No. 21478); $\times 1\frac{1}{2}$.

10. Gigantochloa nigro-ciliata. Redrawn from Kurz 1876:Pl. II, Fig. 16*a, b;* $\times 1\frac{3}{4}$.

11. Guadua aculeata, dorsal and lateral aspects. Original; based on fruits collected from plants cultivated at Chocolá, Guatemala, by McClure (No. 21591); $\times 3$.

12. Melocanna baccifera. Original; based on fruits supplied by Denis Koester from plants cultivated at Rosario, Alta Verapaz, Guatemala, under P.I. 164567; $\times \frac{1}{3}$.

13. Nastus elegantissimus. The fruits not seen. Copied from an unpublished sketch by J.S. Gamble (as *Oreiostachys pullei* Gamble) based on one of ten fruits collected by K. A. R. Bosscha (s.n.) in Java (Koorders 1908:130); courtesy of the Keeper of the Herbarium at the Royal Botanical Gardens, Kew; $\times 2\frac{1}{2}$.

14. Ochlandra travancorica. Original; based on fruits supplied by the U.S. Department of Agriculture under P.I. 190905; $\times 1$.

15. Oxytenanthera abyssinica, dorsal and ventral aspects. Original; based on fruits from Eritrea, supplied by the U.S. Department of Agriculture, under P.I. 22776; $\times 1\frac{1}{2}$.

16. Phyllostachys pubescens, ventral aspect. Redrawn from Muroi 1956:Pl. 49, Fig. 1 (as *Ph. heterocycla* var. *pubescens*); $\times \frac{3}{4}$.

17. Pseudosasa japonica, ventral and lateral aspects. Redrawn from Muroi 1956:Pl. 51, Fig. 12; $\times 2\frac{1}{4}$.

18. Melocalamus compactiflorus (Kurz) Bentham. Original; based on fruits collected in South Vietnam by McClure in December 1953; $\times \frac{1}{2}$.

19. Sasa nebulosa, dorsal and ventral aspects. Redrawn from Muroi 1956:Pl. 51, Fig. 11; $\times 3$.

20. Schizostachyum gracile (teste Holttum), dorsal aspect. Redrawn from Kurz 1876:Pl. II, Fig. 2 (as *S. chilianthum*); ca. $\times 1\frac{1}{2}$. See also Fig. 63.

21. Sinobambusa tootsik, dorsal and ventral aspects. Redrawn from Muroi 1956:Pl. 49, Fig. 3; ca. $\times 2\frac{1}{2}$.

22. Bambusa copelandii, dorsal and ventral aspects, the latter showing the prominent embryotegium at the base. Redrawn from Raizada 1948:Pl. 1, Fig. *C* (as *Sinocalamus copelandii*); $\times 2$.

fruit with a very short or obsolete beak and a thickish leathery pericarp (*Melocalamus compactiflorus*), and numerous variants of these main types. See also Kurz 1876:266–7, and Holttum 1956*a*: Figs. 1–14, 1958:Fig. 10, *C–E*.

In spite of the striking diversity of external features manifested by the fruits of different bamboos, and the usefulness of these features for the recognition of some genera, the inadequacy of the gross morphology of the bamboo fruit alone for the purposes of comprehensive classification is evident in the weak and inconclusive role it has played, even when a conscientious attempt has been made to use it in traditional systems. Recent anatomical and systematic studies published by Holttum (1956*a*, 1958*a*) have greatly improved our perspectives on the form and gross anatomy of the fruits of the bamboo genera of Malaya. Holttum has brought into focus fundamental similarities and differences that apparently are of dependable value for characterizing some genera, and for uniting related genera into groups of major phylogenetic importance. Further studies of this nature are greatly needed. The vast majority of preserved flowering specimens of the known species of bamboo do not present mature fruits. For this reason, published descriptions and illustrations cover only very incompletely the fruits of the known genera. A sustained effort to bring together documented studies of all of the existing examples will be very rewarding, and will gradually illuminate an important aspect of the nature and relationships of the different kinds of bamboos. Usui (1957*a*) published a study of the anatomy of the embryo in *Sasa nipponica* and *Arundinaria* (as *Pleioblastus*) *chino.* The characteristics of the embryo of the fruits of bamboos of different genera are also being studied by Reeder (1961, 1962) particularly in comparison with the embryo of the fruits of other genera of the Gramineae (cf. also Kennedy 1899). Such studies may reveal features useful, in combination with other characters, for showing hitherto neglected affinities, or sharpening the differentiation of the bamboo genera from each other, and from related genera of the nonbambusoid Gramineae—particularly if they embrace the formative stages of the ontogeny of the embryo (cf. M. V. Brown 1960:218) as well as its dormant stage.

Brandis (1907:87) states that in the Indian species of the genera *Dinochloa, Melocalamus, Melocanna,* and *Ochlandra* the ripe seed has no endosperm. This statement is verified by Stapf (1904:408) for at least one species, in an exhaustive study of the developmental anatomy of the fruit of *Melocanna baccifera* (as *M. bambusoides*). No comparable description of the fruit of any other bamboo has been encountered. Holttum is of the opinion (expressed in personal conversation) that in the genera mentioned by Brandis as having no endosperm the embryo has no resting stage, and the endosperm is absorbed through the scutellum as fast as it is formed.

3 Vegetative Phase: The Seedling

Published accounts of the course of development of the bamboo seedling are so fragmentary in nature, and are limited to such a small array of entities, that scarcely any basis exists for perspectives of either particular or general application. Brandis (1899) pointed out that in his day the remarkable process of the development of the seedling into a mature clump had not been sufficiently studied and, as Arber observes (1934:65), this unfortunately still holds true.

Between the time of germination of the embryo and the acquisition of its full complement of vegetative structures, the bamboo plant may be called a seedling. Leaving aside detailed characteristics of the embryo as not pertinent to the present account, we may describe the initial stage of the seedling (Figs. 57–66) as consisting essentially of a root (the primary root) and a shoot (the primary culm).

First to emerge, the primary root (developed from the radicle) is a slender, unsegmented axis, cylindrical or nearly so, with a subapical body of meristem that produces new cells both proximally and distally. The tissue that results from the proximal increase causes the elongation of the root, and that resulting from the distal increase makes the root cap. No experimental evidence has been found as to whether the primary root in any bamboo is actuated in its initial orientation by positive geotropism, or partly or wholly (if only temporarily) by negative phototropism. In the zone of active elongation just proximal to the apical meristem, many of the epidermal cells produce root hairs that establish effective contact with the ambient medium. Although the primary root develops a system of vascularized lateral branches, it is essentially limited in its growth, and is soon overshadowed and superseded by roots of an "adventitious" origin, whose primordia emerge just above each of several of the lower sheath nodes of the primary culm (see p. 79). According to Shibata (1900:444ff), as the proximal part of each adventitious root matures (in bamboos of *Phyllostachys* and *Arundinaria* at least) the cells of the epidermis and some parts of the underlying tissues die progressively and eventually disintegrate, leaving the endodermis as the protective outer layer. In at least four species of *Bambusa,* on the other hand, the root epidermis is described by Shibata as persistent.

The primary culm (developed from the plumule) is a segmented axis of clearly negative geotropic reaction, bearing a foliar appendage at each sheath node. The course of development, and the mechanism of growth, in a young bamboo culm shoot taken from a plant of mature stature (exemplified by *Phyllostachys nigra;* see Fig. 56) are described and illustrated by Porterfield (1930*a*). The apex of the growing point is protected by many layers of overlapping sheathing appendages (culm sheaths), which are the first lateral organs to be differentiated.

The segmented axes of a bamboo plant elongate principally during a "grand period of growth" (Porterfield 1928). This elongation is effected by means of intercalary growth, a process described in great detail by Porterfield (1930*b*) as it occurs in the bamboo culm (see also Holttum 1955:403). In intercalary growth the immature axis increases in length by the elongation of cells in zones of secondary meristem each located just above a node (Fig. 56). As defined by Jackson (1949:199), an intercalary vegetative zone of growth lies between zones of mature tissue. In the elongating segmented axes of a bamboo plant the locus of each zone of intercalary growth is just above the locus of insertion of a sheath. There is empirical evidence that the sheath may be the origin of substances that control, or at least influence, the process of intercalary growth and possibly also the initiation of root and branch primordia. When Chinese gardeners wish to dwarf a bamboo, they remove each culm sheath prematurely, beginning with the lowest, before the elongation taking place above its node is completed. Upon the removal of a sheath, the elongation above its node ceases. The initiation of branch buds and root primordia on any segmented axis always takes place within a zone of intercalary growth, before the tissues lose their meristematic potential, and while the subtending sheath is still living (see also pp. 12 and

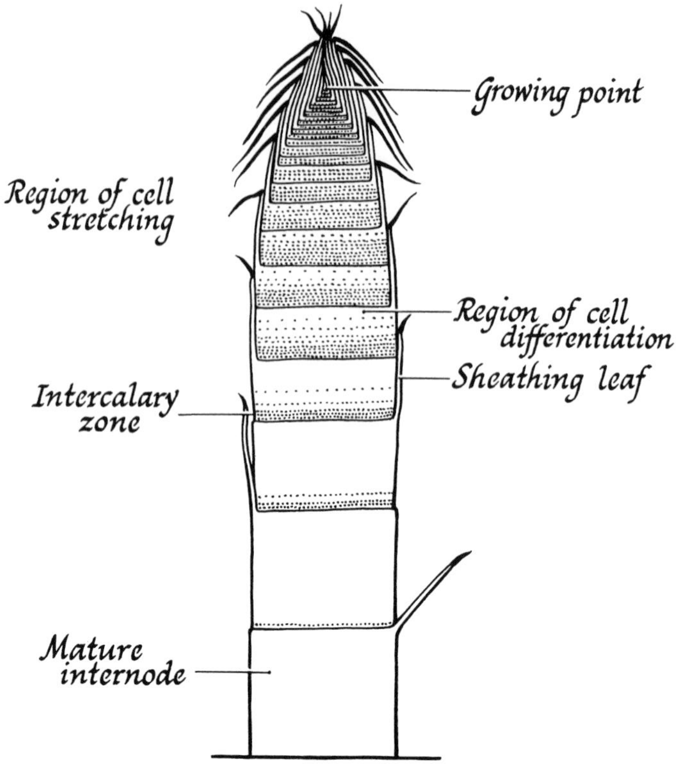

Fig. 56. Diagrammatic longitudinal section of a young bamboo culm shoot, showing stages of development and zones of intercalary growth. Stippled areas represent regions of active differentiation and growth of tissues; unstippled areas, mature tissues. In the bamboos, all segmented axes elongate by intercalary growth. Redrawn from Porterfield 1930*b*:Fig. 1.

61). In the culms of many bamboos (all known species of *Phyllostachys* and *Shibataea,* for example), each zone of intercalary growth is marked at branch-bearing nodes by a transverse thickening or "supranodal ridge," that is, the culm node, *sensu stricto* (Fig. 27). At nodes that do not bear buds or branches, this ridge is usually much less conspicuous, and in some bamboos it may be absent, It is lacking throughout or inconspicuous (even at branched nodes) in the culms of *Melocanna baccifera,* and in those of most known species of *Schizostachyum.*

In the seedlings of some bamboos (species of *Arundinaria, Bambusa,* and *Dendrocalamus*) the primary shoot is generally rather short, reaching a few inches or at most a foot or so in height. Elsewhere (as in *Melocanna baccifera,* for example) it may reach the phenomenal height of 18 ft, as observed by Dr. Ernest Imle on a visit to the Federal Experiment Station in Puerto Rico (personal communication). In some bamboos the primary shoot (or culm) produces other culms or rhizomes only from buds at its lowermost nodes (Figs. 58 and 62), or it may also ramify freely at upper nodes as well, to form leafy branches. The primary culm in some seedlings of *Melocanna baccifera* grown under artificial (fluorescent) light and controlled temperature (25°C) in quartz-pebble culture irrigated with a standard nutrient solution, produced lateral branches from buds at mid-culm nodes before producing secondary culms from their basal buds. More commonly secondary culms may develop by tillering before leafy branches appear. Occasionally, a secondary shoot may emerge when the primary one is only a few inches long and still lacks foliage leaves (Fig. 62). The emergence of a rhizome proper is usually delayed until several erect culms have emerged by tillering, as in those species of *Arundinaria* whose seedlings have been studied (Figs. 64 and 65). The first branch to develop from a basal bud of the primary culm of a seedling is usually a culm that arises by tillering, without an intervening rhizome. However, it sometimes happens in a seedling of *Melocanna baccifera* that the first basal branch to develop may be a rhizome (Fig. 61).

After the successful establishment of a rhizome axis a seedling bamboo plant may be said to be adolescent, since as a rule it has by this time developed all of the structures that will characterize it as vegetatively mature. However, as a plant increases in stature, successive generations of its culms will manifest a gradual change in the shape, dimensions, vesture, and texture of constituent parts. The attainment of mature stature may require from 3 to 20 years or more, depending on the genetic constitution of the plant and the nature of its environment. The leaf blades of the seedling may be smaller than those of the mature plant (*Arundinaria tecta*) or they may be much larger (*Melocanna baccifera*). The culm sheaths in small plants may be provided with auricles or oral setae or both. In rare cases these become progressively smaller in subsequent orders of culms and, as the plant develops stature, disappear altogether (*Phyllostachys viridis*). More common is the case where the auricles and oral setae are lacking in the culm sheaths in small plants but become well developed in the sheaths of subsequent orders of culms, as the plant increases in stature (*Phyllostachys bambusoides*).

The seedling in bamboos with pachymorph rhizomes

Bambusa arundinacea Retz. On the postgermination history of seedlings of *Bambusa arundinacea* we possess but few observations. Arber's account (1934:65), largely a condensation from Brandis (1899:4–5), follows:

> It is said that in the first stages of their existence the young plants are very delicate and, except under the influence of plenty of moisture, they are unable to resist the scorching effect of the sun's rays; on the other hand, excess of water about their roots causes them to die off rapidly. Moreover, they are incapable of competing with the minor grasses, by which they are easily and speedily choked and destroyed. Brandis's account of the stages in their development is that in March, 1882, he found large patches of young seedlings from seed which had been produced in 1881 and had germinated during the rains of that year. The youngest plant consisted of one shoot, about 6 in. long, bearing two or three leaves at the tip and, below these, a sheath with a small imperfect blade. Near the ground the shoot bore a short, membranous-pointed sheath, at the base of which were two rootlets, about 3 in. long. At a later stage, several conical side shoots made their appearance, just below the surface of the ground; they were bent, first downwards, then upwards, and were covered with numerous membranous, white sheaths. These side shoots, which would ramify later, were the beginnings of the rhizome. They are destined to turn upward at the tip, thus forming leaf-bearing stems, each rooting from the bend. Besides these underground side shoots, with short internodes, others arose which had moderately long internodes, and rooted at the nodes, sending up leaf-bearing stems from these points also. In this manner it came about that seedlings, not quite a year old, had an underground rhizome of complicated build, pushing numerous rootlets into the soil, and bearing a number of shoots, of which the first to be formed were short-lived. The other bamboos which Brandis examined showed a general similarity to *Bambusa arundinacea.*

A remarkably rapid rate of growth in seedlings of *Bambusa arundinacea* is recorded by White in connection with his account of the longevity of bamboo seeds stored under controlled conditions (see p. 203f).

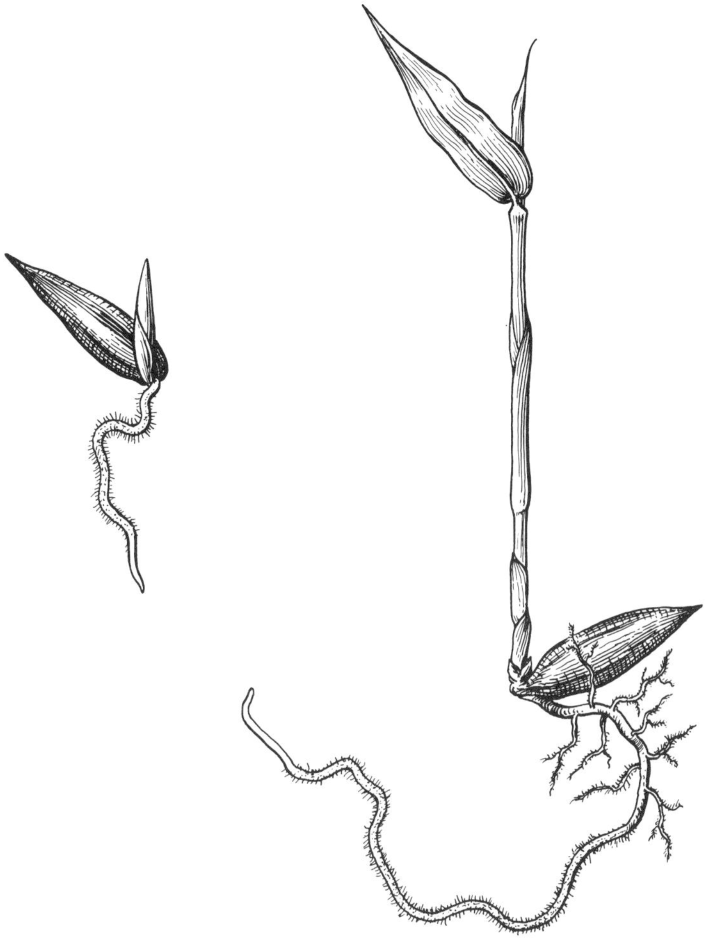

Fig. 57. *Bambusa multiplex.* Seedlings in two stages of development: (*left*) fruit with radicle and plumule just beginning rapid elongation; (*right*) fruit with attached young plant showing branching root, and culm shoot with first foliage leaf. Redrawn from Muroi 1956:Fig. 47 (as *Leleba multiplex*).

Takenouchi (1932:Fig. 127) presents illustrations of two stages in the germination of seeds of *Bambusa arundinacea,* without textual discussion.

Bambusa multiplex (Lour.) Raeusch. (Fig. 57). Muroi (1956:Fig. 47) illustrates two early stages in the development of the seedling of *Bambusa* (as *Leleba*) *multiplex.* Muroi's notes, in Japanese, have not been translated.

Chimonobambusa falcata (Nees) Nakai (Fig. 58). Troup 1921:Fig. 373) illustrates the early stages of the development of a seedling of *Chimonobambusa* (as *Arundinaria*) *falcata* to the age of 4 months. Of this species (*ibid.*:985) he says only this: "A tufted growth may commence at an early age." At 4 months this seedling had four stems, with a fifth already initiated as a small shoot. This shoot barely shows, basally, the beginning of the horizontal growth that will characterize the successive rhizome axes more conspicuously as the plant develops more culms.

Dendrocalamus sikkimensis Gamble (Fig. 59). Arber (1934:Fig. 22) presents sketches illustrating an early stage in the development of a seedling of *Dendrocalamus sikkimensis.*

Dendrocalamus strictus (Roxb.) Nees (Fig. 60). Troup (1921:985) describes the development of the seedling in *Dendrocalamus strictus* in the following words:

> The plumule emerges in the form of a pointed conical bud with sheathing scale-like leaves, which rapidly develops into a thin, wiry stem bearing single foliage leaves arising alternately at the nodes, the bases of the leaves sheathing the stem. Meanwhile, fibrous roots develop from the base of the young shoot. The tufted form of the young plant commences to show itself at an early stage. This is effected by the production on the rhizome of successive pointed buds, from which are developed short rhizomes which curve upwards and form aerial shoots. The buds and rhizomes, and the shoots arising from them, become successively larger and larger [Fig. 74]. The earlier shoots are thin, wiry, and grass-like; but subsequently a time comes when woody culms are produced, which bear some resemblance to the adult culms in form, in the shape of the sheaths, and in other particulars.

The most dramatic feature brought out by Troup is the manner in which the strong positive geotropism of the neck of the pachymorph rhizome operates in young seedlings to carry the successive rhizome axes deeper and deeper into the soil. Troup observes: "In this species the new rhizomes of seedlings take a decided bend downwards before curving upwards to form aerial shoots; thus successive shoots, besides being larger than the preceding ones, arise from rhizomes deeper in the ground." This behavior stands in striking contrast with the gradualness of the manner in which leptomorph rhizomes usually penetrate the soil

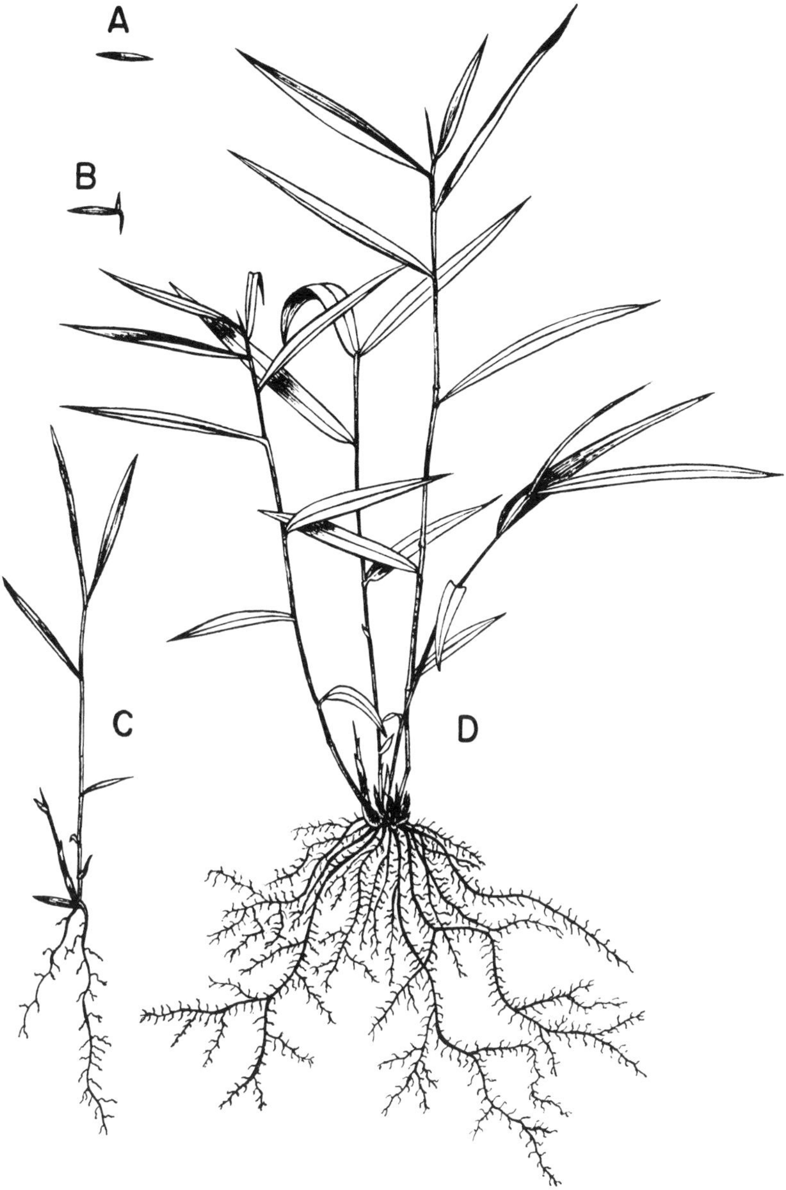

Fig. 58. *Chimonobambusa falcata.* Fruit, and stages in the development of seedlings: (*a*) fruit, in dormant state; (*b*) early stage of germination; (*c*) seedling 1 month old, with fruit still attached, showing leafy culm and a second shoot arising, by tillering, from a bud at the base of the first culm; (*d*) well-rooted seedling 4 months old, with four leafy culms and a fifth already visible as a small shoot. Redrawn from Troup 1921:Fig. 373 (as *Arundinaria falcata*).

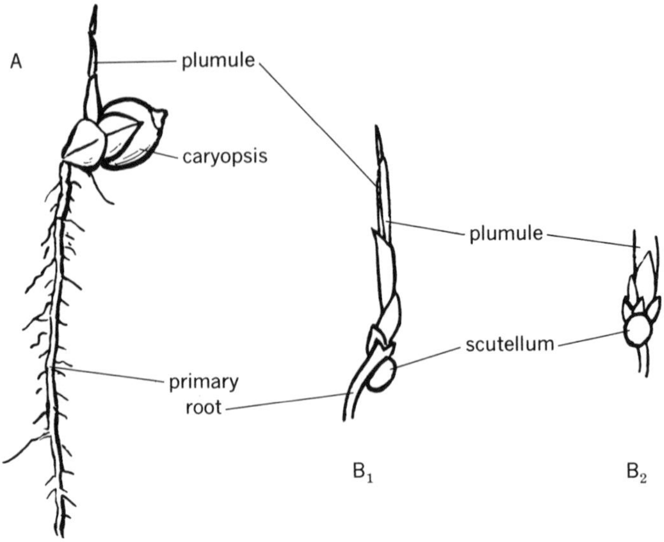

Fig. 59. *Dendrocalamus sikkimensis.* (*A*) Seedling, natural size, the root incomplete. The fruit or "seed"—"caryopsis"—is still enclosed in a glume and the lemma and palea. As Arber indicates in her caption to the figure, the vigorous first root has burst through the glume. (B_1, B_2) Two views of another seedling, with the glume, lemma, and palea removed to reveal the plumule, scutellum, and base of primary root. Redrawn from Arber 1934:Fig. 22.

to greater depths (see discussion of rhizome neck behavior, p. 17f). Troup adds:

> The rate of development of the clump depends very largely, even in the same species, on the conditions under which it has been grown [p. 985] . . . In vigorous nursery-grown seedlings of Dendrocalamus strictus as many as ten shoots have been counted at the end of the first season . . . Under favorable conditions the shoots of bamboo seedlings may die back for some years in succession before the plant finally established itself; this is particularly common in the case of *Dendrocalamus strictus* in dry situations. Even under more or less favorable conditions, the first shoot of the seedling may die off at the end of the first season [p. 987] . . . A clump [of bamboo] may be said to have attained maturity[1] when it commences to produce full-sized culms [p. 990].

The age of the plant when this point is reached will vary, however, since the manifestations of maturity are as much a function

[1]Clear thinking requires that we discriminate between vegetative maturity (here intended) and sexual maturity in bamboos.

F.A.M.

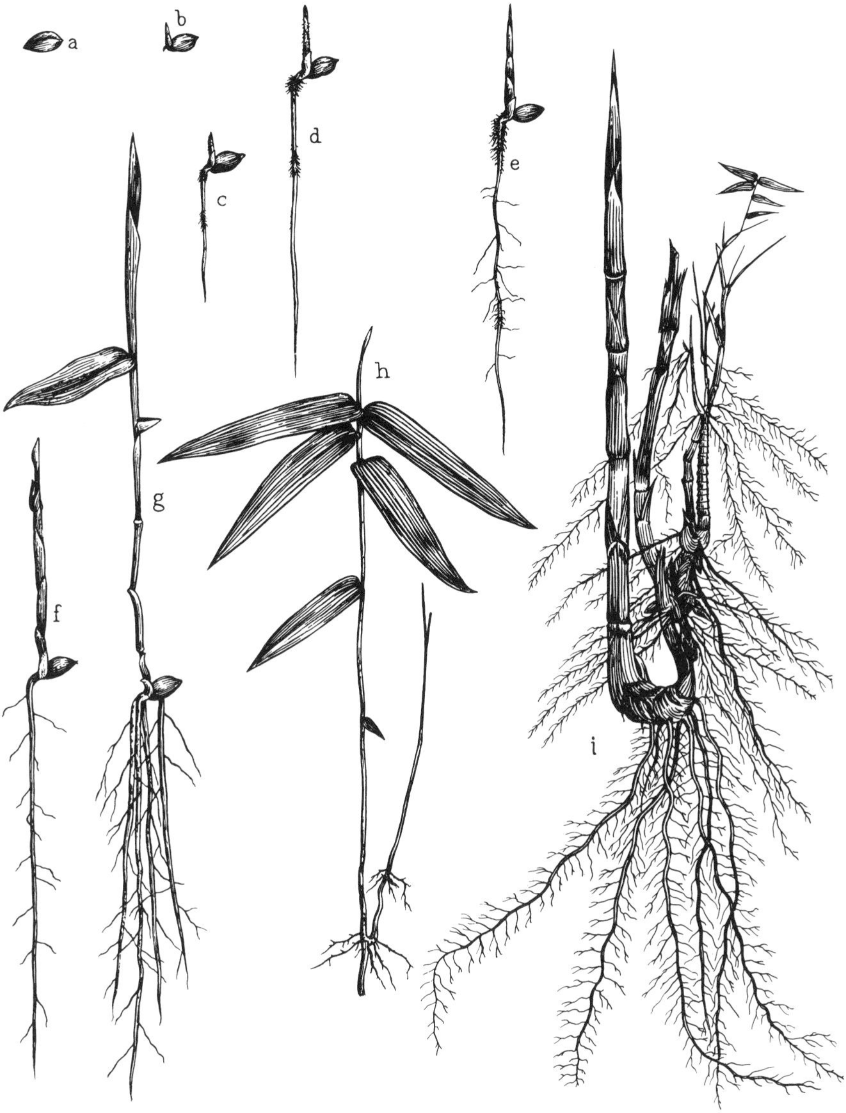

Fig. 60. *Dendrocalamus strictus:* (*a–g*) fruit in dormant state, and stages in the development of seedlings "under more or less favorable conditions"; (*h*) seedling at 1 year; a second culm has arisen from the base of the primary one, the latter now dead; (*i*) nursery seedling 14 months old. Note that the neck of each successive rhizome axis carries its rhizome deeper into the earth. The rhizome then takes a horizontal course, only to turn upward when a change in the physiological state of the growing point induces a changed pattern in the tissues subsequently produced. The axis then takes on the form of a developing culm, with negative geotropism and other distinctive physiological behavior strongly developed. Redrawn from Troup 1921:987 and Figs. 374 and 375.

of the site where the plant grows, and of forces that impinge upon the clump from without, as of the innate potentialities of the plant. Deogun (1937:Pl. III) illustrates later stages in the developing clump (see Fig. 74).

Melocanna baccifera (Roxb.) Kurz (Figs. 61 and 62). According to Kurz (1876:266), the seed of *Melocanna baccifera* often germinates before the fruit falls to the ground. It is a curious fact that, in the long period since this bamboo was first made known to science by Roxburgh (1819), the only published notes on the course of development of the seedling are the following observations by Troup (1921:1012):

> Germination commences with the first heavy showers of the rainy season, roots and shoots being produced from the thick end of the fruit; roots often begin to appear before the fruit falls. The seedlings, unlike those of most bamboos, make vigorous growth from the commencement. By the end of the first season each fruit will usually have produced about five shoots, of which the latest may be as much as 10 ft. high; these shoots are crowded together in a clump. During the second season more shoots are produced, the clump expands somewhat, and the largest culms reach a height of about 20 ft. By the fifth season the culms attain almost their maximum height, but are still thin and crowded together, and it is not until later that they become spaced out with the gradual extension of the rhizomes. [Troup (Fig. 397) shows a plant seven years old from seed; see Fig. 82.]
>
> This bamboo spreads to a remarkable extent by its long vigorous rhizomes. [Actually, the neck is the only markedly elongated part of the rhizome in this bamboo.] At the last general fruiting in Arakan it was also observed to spread, owing to the rolling of the heavy fruits down the hill-sides, to places where it did not exist before, and was found springing up on savannahs and in beds of streams. It does not thrive well under shade, but springs up readily in gaps.
>
> As an instance of the great vigour and vitality of the rhizomes, it may be mentioned that Mr. W. D. Turner of Hurbanswala, Dehra Dun, obtained seeds from Assam in 1912, of which six germinated successfully and produced strong plants, which grew and spread rapidly on moist fertile ground. In 1917 he was able to dig up no fewer than 400 offsets for transplanting elsewhere, after which a fairly large grove of bamboos still remained in the parent crop.

For additional observations on the seedling stage of *Melocanna baccifera* see p. 125.

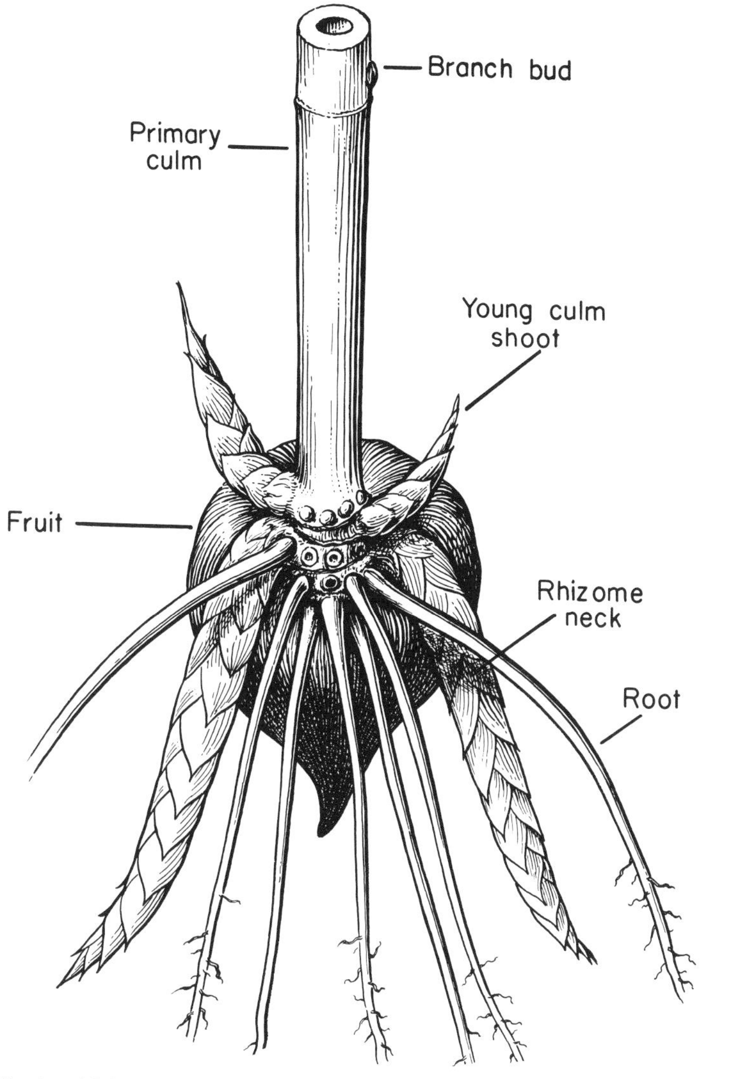

Fig. 61. *Melocanna baccifera.* Fruit with basal portion of attached seedling, in the second year of its growth at the Federal Experiment Station, Mayagüez, Puerto Rico, under P.I. 164567. The primary culm had reached a height of 5.4 m (18 ft) according to Dr. Ernest Imle, who supplied the annotated specimen. The development of this plant is unusual in the great size attained by the primary culm and in the precocious emergence of long-necked rhizomes. The rhizome neck is much elongated in this species, and its development (shown here in strongly geotropic orientation) preceded that of the rhizome proper, which is terminal to it. Four of the roots of the culm have been cut off (circles with dark centers); four dormant root primordia are shown between the two young culm shoots. Fruit, 6 cm in diameter.

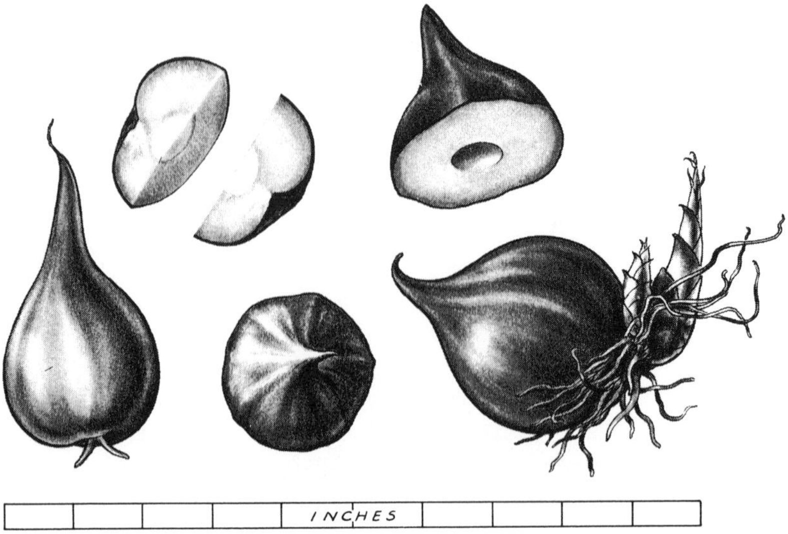

Fig. 62. *Melocanna baccifera.* Fruits collected by Denis Koester from plants cultivated at Rosario, Alta Verapaz, Guatemala, under P.I. 164567: (*above*) sectioned fruit, showing the thickness of the pericarp to be as great as the diameter of the seed cavity; (*lower right*) fruit bearing two shoots. The precocious emergence of additional shoots from the primary one, before the development of foliage leaves (rare elsewhere) occurs frequently in this species. This apparently is related to the abundance of food stored in the scutellum and pericarp of these fruits as compared to that found in the small thin-walled caryopses of most other bamboos. Drawing by Elmer W. Smith, from McClure photo. See Stapf 1904.

Schizostachyum acutiflorum Munro. Takenouchi (1932:Figs. 127 and 128) presents illustrations of three stages in the germination of seeds of *Schizostachyum acutiflorum,* and two stages of the further development of the seedling, without textual discussion.

Schizostachyum gracile (Munro) Holttum (teste Holttum) (Fig. 63). Kurz (1876:Pl. II, Figs. 6–12) illustrates the early stages of the germination of the seed and the growth of the seedling of *Schizostachyum gracile* (as *S. chilianthum*) to the 35th day. On pp. 267–268 of the same work, Kurz supplements the captions to the figures by the following brief account of the development of the seedling during its first year. It is at some points difficult to follow with confidence Kurz's terminology and the sequence of events as he describes them in this account:

Here the lower blunt end of the cotyledon [coleorhiza, Fig. 63, *6*] protrudes through the pericarp about the fourth day after sowing, and

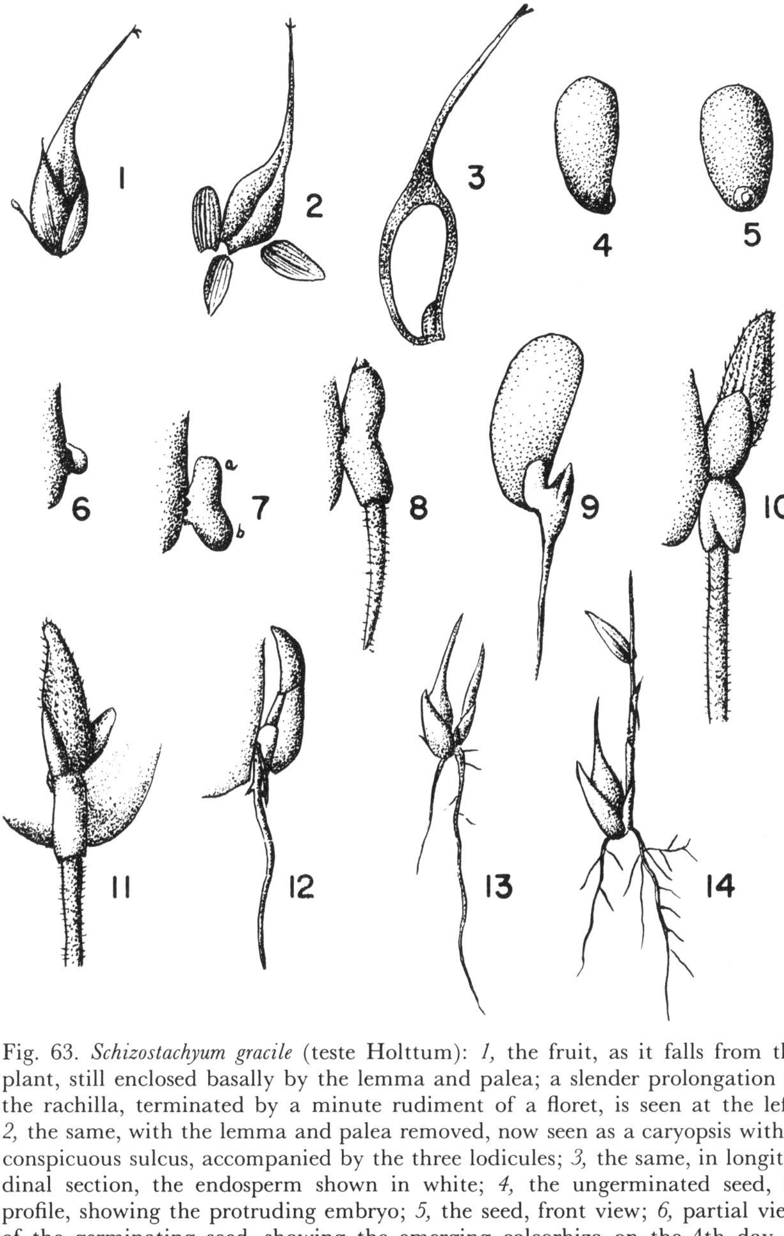

Fig. 63. *Schizostachyum gracile* (teste Holttum): *1,* the fruit, as it falls from the plant, still enclosed basally by the lemma and palea; a slender prolongation of the rachilla, terminated by a minute rudiment of a floret, is seen at the left; *2,* the same, with the lemma and palea removed, now seen as a caryopsis with a conspicuous sulcus, accompanied by the three lodicules; *3,* the same, in longitudinal section, the endosperm shown in white; *4,* the ungerminated seed, in profile, showing the protruding embryo; *5,* the seed, front view; *6,* partial view of the germinating seed, showing the emerging coleorhiza on the 4th day of germination; *7,* the same, on the 5th day, showing the coleoptile (*a*) and coleorhiza (*b*); *8,* the same, on the 6th day; *9,* the same, on a smaller scale, including the seed, with the first sheath of the primary culm exposed by the removal of a portion of the coleoptile; *10* and *11,* two views of the same on the 9th day; *12,* the same, on a smaller scale, on the 11th day; *13,* the whole plant, with the fruit still attached, on the 15th day; *14,* the same, on the 35th day, showing the first leaf blade. Redrawn from Kurz 1876:Pl. 2 (as *Schizostachyum chilianthum*).

is followed the next day by its upper part [coleoptile, Fig 63, *7*]. Already the following day the primary rootlet, which is hairy, forces its way downwards to a considerable length, while the upper part has enlarged and separated into two equally large lobes which are separated from the downward growing part by a more or less distinct constriction. These two lobes enclose in their axil the plumule which is stiff, hairy and striped and quickly protrudes from between them, as can be seen in Fig. 9 (3 days later) [Fig. 63, *10, 11*]. The subsequent stages of development of the young plant are represented in Figs. 10–11 [Fig. 63, *12, 13*], as observed on the eleventh and fifteenth day after sowing. At the latter stage the growth of the plantlet becomes considerably slower, and although still connected with the seed, the cotyledon [endosperm?] was entirely absorbed already before the fifteenth day of [after] sowing, and thus the young plant is left to itself for further nourishment from the soil alone. On the thirtieth day after sowing, the halm-sheaths and a leaf are fully developed (see Fig. 12) [Fig. 63, *14*], but instead of seeing the growth now accelerated, it becomes considerably slower, so much so, that after a lapse of a year the plants reached only 2–2½ feet in height.

The seedling in bamboos with leptomorph rhizomes

Arundinaria nikkoensis Nakai (Fig. 64). Hisauchi (1949) describes and illustrates a few details of the development of the external features of two seedlings of *Arundinaria nikkoensis.* The following interpretation of Hisauchi's account is based on a translation, by David Ray, of the Japanese text.

Each germinating seed produced first of all a single primary root and a single upright, leafy culm. In the course of the first year the primary root sent out many short branches, and the culm sent out a single leafy branch from each of two successive nodes near the ground. Early in the second year, an underground bud situated at the uppermost of several nodes crowded near the base of the culm gave rise to an upright branch which is interpreted as a culm (culm no. 2). Then, later in the same year, a bud still lower on the first culm (near its very base) developed into a horizontal axis (a rhizome) which, after growing for a considerable distance laterally, turned up at the tip to form an unbranched culm, whose leaves were larger than those of the aerial branches of the primary culm [Fig. 64, *B*]. From the proximal bud of the horizontal axis (rhizome) a horizontal axis of the second order arose. Again, from the proximal bud of the second culm another horizontal axis arose. This latter gave rise, from its proximal bud, to still another horizontal axis of the second order.

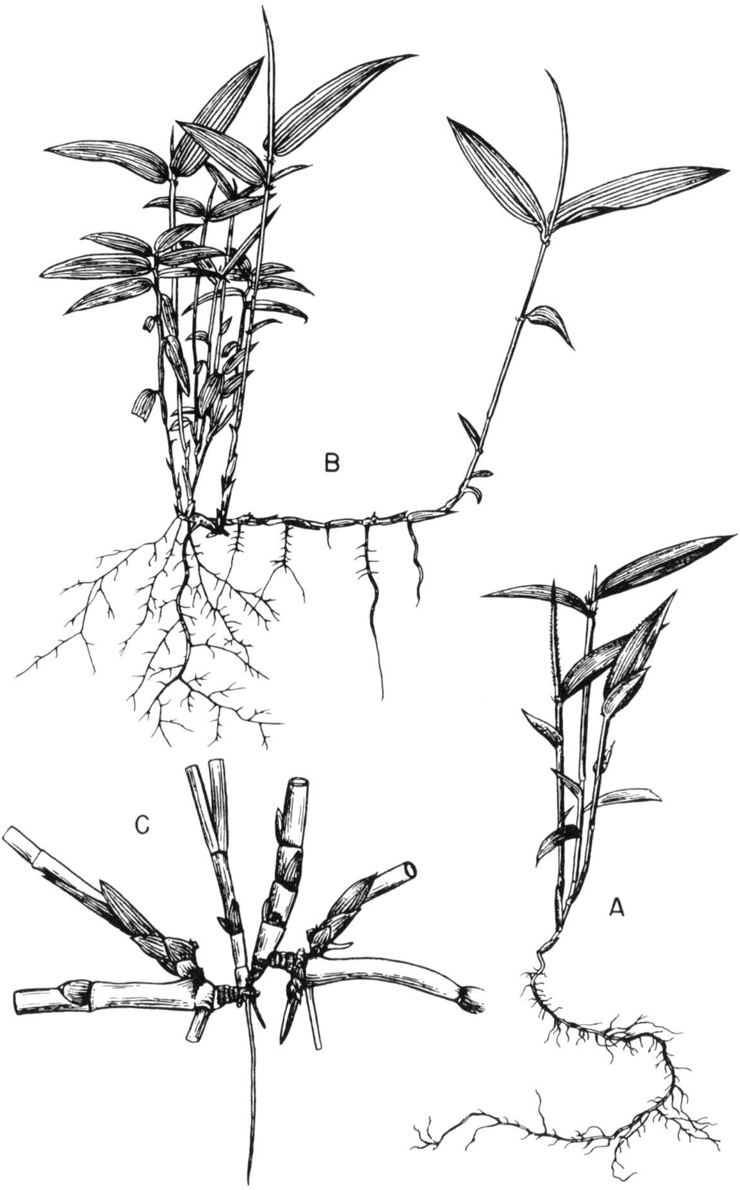

Fig. 64. *Arundinaria nikkoensis:* (*A*) 1-year-old seedling; (*B*) 2-year-old seedling, showing two leptomorph rhizome axes; (*C*) subterranean part of a 2-year-old seedling, enlarged to show details and relation of the segmented axes of different orders and categories that have developed. The horizontal or downward-pointing segmented axes bearing sheaths are rhizomes. The unsegmented downward-pointing axes are roots. Rearranged from Hisauchi 1949:Figs. 1–3.

The subsequent formation of culms from these rhizomes is not recorded by Hisauchi, who concludes his notes with the observation:

> It must be confessed that, even after such a long period of observation, the rhizome's manner of branching is yet to be explained.

Arundinaria simonii (Carr.) A. & C. Rivière (Fig. 65). Muroi (1956:Fig. 48) illustrates seedlings of *Arundinaria* (as *Pleioblastus*) *simonii* at three stages of their early development. At 150 days from germination, two rhizome axes were already making strong growth, and showing oblique geotropic orientation. Here, the rhizome itself performs the depth-regulating function that in bamboos with pachymorph rhizomes, is effected by the rhizome neck (see p. 17).

Arundinaria tecta (Walt.) Muhl. [det. F.A.M.]. Hughes (1951:118) says:

> The first rhizome on a seedling [as *Arundinaria* without specific identification] was observed in August, 1949, on a plant grown from seed [sown] in 1947. This initial root stock [rhizome] elongated for a distance of five inches horizontally through a layer of mulch, developed three nodes each having numerous roots . . . and then turned directly upward into an aerial leafy stem.

It has been established, by personal observation, that in this species the transformation of the terminal bud of a rhizome axis into a culm, as illustrated in the young seedling (*ibid.*:Fig. 3) is a character that is retained by the plant thereafter as the normal form of growth in the rhizome. See Fig. 10, *8* and p. 34.

Sasa nipponica (Makino) Makino et Shibata (Fig. 66). Usui (1957*a*) illustrates the basal portion of a seedling of *Sasa nipponica,* still attached to the fruit.

Recapitulation of significant events in the ontogeny of a bamboo plant

Differentiation between tissues emerging from the distal and the proximal facies of the apical meristem of the radicle.

Emergence and elongation of the unsegmented radicle to form a primary root.

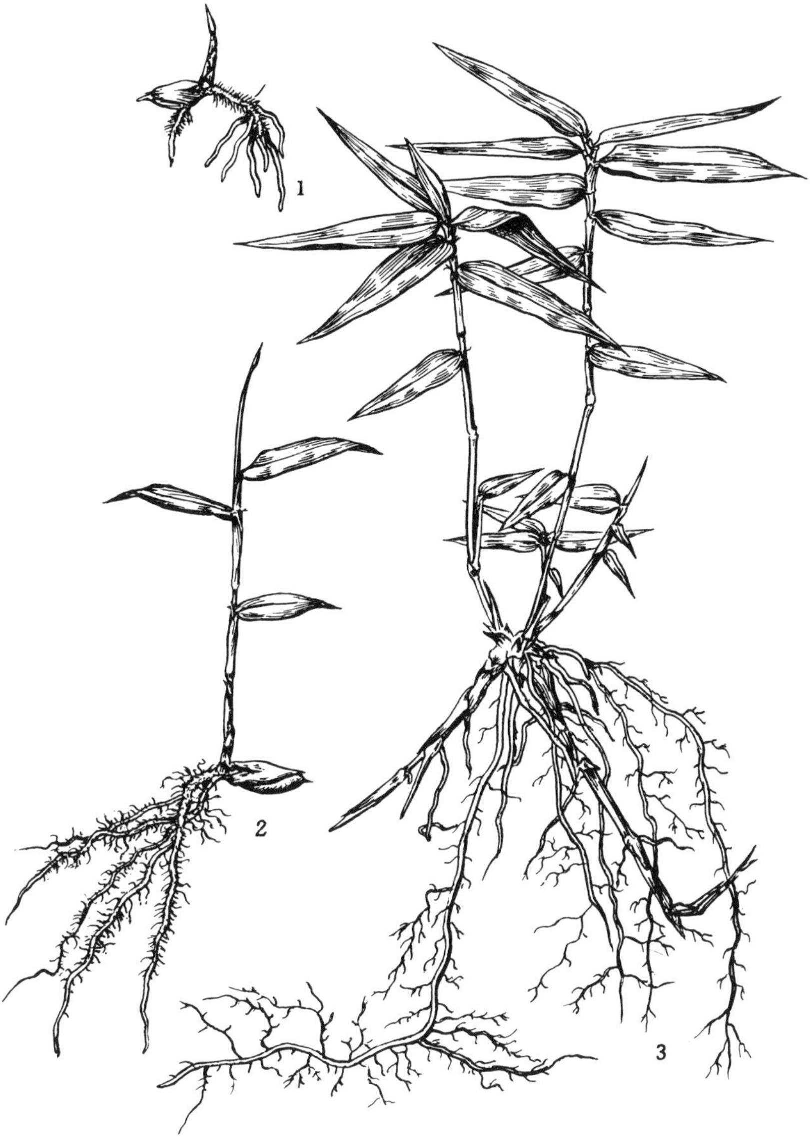

Fig. 65. *Arundinaria simonii.* Seedlings: *1,* at 10 days; *2,* at 30 days; *3,* at 150 days from germination, showing at this stage two leptomorph rhizome axes descending obliquely. From Muroi 1956:Fig. 48 (as *Pleioblastus simonii*).

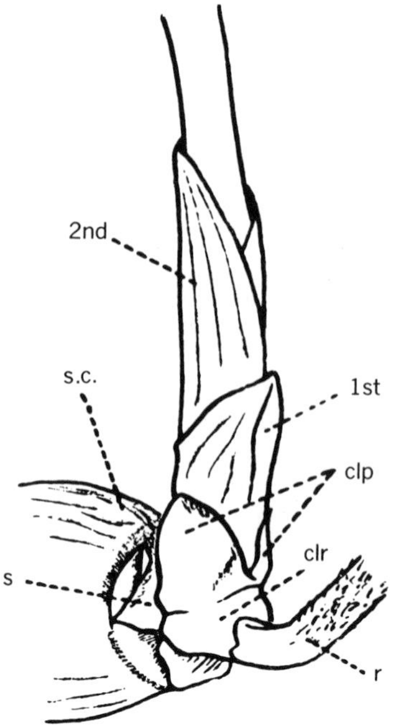

Fig. 66. *Sasa nipponica.* Basal portion of very young seedling, with adjacent part of the fruit. The author's legend: "*2nd.*, second leaf; *1st.*, first leaf of the plumule; *clp.*, coleoptile; *clr.*, coleorhiza; *r.*, primary root; *s.c.*, seed coat; *s.*, scutellum." From Usui 1957*a*: Fig. 4.

Development of root hairs on the "piliferous zone" of the primary root.

Development of branch roots on the newly matured part of the primary root.

Initiation of sheath primordia from apical meristem in the plumule and in the primary shoot into which it develops.

Initiation of segmentation (differentiation of nodes and internodes) following the initiation of sheath primordia from apical meristem in all axes except the roots.

Initiation of branch-bud primordia axillary to developing sheaths.

Emergence and elongation of the segmented plumule to form the primary shoot.

Development of adventitious roots from the zone of intercalary growth at basal nodes of the young culm.

Differential development of successive internodes of the primary shoot in respect to their shape and dimensions, the character of the foliar appendages, and the incidence of axillary buds.

Development of plain sheaths (without appendages) at the lowest nodes (the neck) of the primary culm.

Differentiation of a ligule at the apex of sheaths proper inserted at levels above basal ones on the primary culm.

Differentiation of (sessile) sheath blades from the sheath proper at all nodes (except several basal and several apical ones) of each aerial axis.

Differentiation of foliage leaves with petiolate blades on sheaths at several nodes at the apex of each aerial axis (culm or branch).

Differentiation (in most species) of auricles or oral setae, or both, at the apex of sheaths proper (except those at basal nodes) on all aerial axes.

Development of abscission in certain foliar appendages, including petiolate leaf blades, and in twigs bearing senile leaf blades.

Development or suppression of (usually solitary) vegetative buds in the axils of sheaths on all segmented axes.

Duplication of the primary culm from its basal buds, by tillering, and the differentiation of an important new organ, the culm neck.

Differentiation of leafy aerial branches from mid-culm nodes.

Differentiation of a pachymorph or a leptomorph rhizome axis from a basal bud of a culm.

Duplication of rhizome axes from (always solitary) buds at nodes of the rhizome.

Development of culm axes from lateral buds, or the growing tip, of a rhizome.

Development of adventitious roots at nodes of a rhizome.

Development of waxy exudate as epidermal vesture, particularly on exposed parts of segmented axes.

Development of pubescence as epidermal vesture here and there, particularly on sheaths or their blades, or both.

Differentiation of reproductive axes, of either determinate or indeterminate branching, marked by:

Changes in interval and degree of development of segmentation of the axes, and changes in the dimensions, form, and texture of the axes and of their foliar appendages, from those characteristic of the vegetative state to those characteristic of the flowering state (inflorescence).

Differentiation of pseudospikelet and spikelet as structural units.

Differentiation of determinate inflorescence branching.

Differentiation of bracts, glumes, lemmas, paleas, lodicules, androecium, and gynoecium.

Anthesis, pollination, and fertilization.

Maturing of fruit, or withering of pistil containing unfertilized ovule.

Withering and drying of the floral parts.

Development of abscission at the nodes of the rachilla axis, or at the apex of a rachis, which releases the part(s) of the inflorescence containing the ripe fruit.

The principal objective here is the provision of an aid to the visualization of the more important stages or advances in the ontogeny of the individual plant. Since many of these changes occur simultaneously, or overlap each other, it is not possible to list them in a linear sequence that corresponds to the chronological order of their appearance. In view of the dichotomies that develop sooner or later as between the respective ontogenies of plants of the great groups of genera and species (dichotomies resulting from diversities in the manner of branching of the rhizome and of the inflorescence, and in the respective timing of initiation and termination of growth in various axes, for example) an attempt to make an exhaustive list of foci of attention (characters) would run the account into a degree of complexity beyond the limits established for the present work. The repeated occurrence of certain of the events (such as the development of a culm from a lateral bud, or the growing tip, of a rhizome), combined with the progressive increase in the size of the culms that emerge from the successive orders of branches of the rhizome system, results in the progression of the plant toward its mature stature.

As indicated elsewhere, gametic reproduction in a bamboo plant usually is consummated only after a relatively long period of vegetative development, but may, in certain well-authenticated cases, erupt within a few months or a year or so of the germination of the seed (see p. 276). We are still in ignorance of both the nature and the sequence of the physiological events that precede or accompany the flowering of the plant, as well as the factors that determine the length of the period of vegetative development; and whether they are innate in the plant, or dependent upon the fulfillment of a series of specific stimuli from without, or both. See Seifriz 1920 and Wardlaw 1961.

To talk of the habits of Bamboos, and of the management of Bamboo plants,
has little meaning and is of no practical use. Each species
has its own peculiarities and its own requirements . . . and without
a reliable guide, the study of Bamboos . . . would be hopeless.

—BRANDIS (1899:3)

Part II Elite Bamboos and Propagation Methods

4 Selected Species

It has been intimated that, from the points of view of both pure and applied science, more comprehensive studies of the bamboos are urgently needed. The successful and fruitful execution of a program of bamboo studies, however, calls for the establishment of a source of documented materials appropriate to the needs of each pertinent discipline. Dried herbarium specimens may yield material suitable for certain studies such as leaf anatomy and the morphology of the reproductive structures. Fiber-dimension determinations relating to papermaking require selected and comparable portions of the culm. Certain studies in anatomy and cytology require material preserved in liquid. If arrangements are made to have such material prepared by the general collector, he should be provided with suitable vials and preservatives and be given very full written instructions concerning the selection, processing, documentation, and labeling of the material, and how to pack it for shipment. The documentation of the material should embrace an adequate system of cross references, connecting the study material with the herbarium vouchers, special notes, and photographs. Above all, the collector's field number corresponding to the herbarium specimens should be associated with all of the study material taken from the same plant (cf. p. 6).

Stern and Chambers (1960) have recently stated the reasons, and described the procedure, for the use of herbarium specimens suitable for identification by a taxonomist, as vouchers for the documentation of wood samples for anatomical studies. Actually the principles and requirements described apply to all plant materials intended for study from the point of view of any botanical discipline.

For work in physiology, genetics, breeding, propagation and silviculture, living plants are essential. The development and disciplined application of a routine for the testing and screening of the many bamboos living plants of which are already available in this country is long overdue. There are in our collections bamboos of evident superiority for various purposes, some of which were introduced a full half-century ago, but which have not yet been subjected to any properly monitored basic investigations. Moreover, there are almost certainly in existence elsewhere other bamboos that for special purposes may prove superior to any of the numerous kinds immediately available in the United States. Some of these could readily be introduced from countries still accessible and cordial to the free exchange of plants and information. Others are to be had only in areas where, for political reasons, the accomplishment of such objectives is rapidly becoming more difficult.

The importance of establishing, maintaining, and studying living collections of bamboos as a means of advancing the knowledge necessary to their taxonomic disposition and their economic exploitation has been stressed by McClure (1935) and Holttum (1958). There exists an urgent need for the development and discriminating application of criteria more suitable than those used in the past to select and introduce bamboos for agriculture, certain domestic industries, and the agencies for soil and water conservation. Basic studies in the various biological sciences can be fully revealing and fruitful only when living plants of the species most suitable for the particular objectives are available. Following systematic exploratory studies carried out collaboratively over a 10-year period in laboratory and field, along lines adopted under my direction, a progressive domestic paper company has recently developed a plantation of over 3000 acres of selected species of bamboos as a first step in establishing a permanent and adequate source of paper pulp of proved superiority.

The bamboos treated in the following pages were selected for discussion because they possess features of outstanding scientific interest and industrial promise. Living plants of all of them are available in this country. Yet none of them has ever been made the object of comprehensive or sustained studies here. The importance of a competently directed program of studies in relation to the general problem of bringing new crop plants into any agricultural economy is stressed by Jones and Wolff (1960:57) in the following words:

> Results will not be quickly or easily achieved. None of our present crops, we can be reasonably sure, attained an established niche in agriculture without at least decades of directed endeavor. Many others have failed to attain crop status perhaps more because of a lack of sustained research interest than lack of economic potential in the species involved. A carefully planned program, based from the beginning on the best possible information and with provisions for frequent refinements of evaluations of promising species, as new information is brought to light, offers the best formula for success.

On studying the accounts of the bamboos included here, the inquiring mind probably will notice at once the almost total omission of data on yields. It is characteristic of the backward state of existing knowledge of this group of plants that data on annual yields from a given site are hardly available for any bamboo—even those that have assumed supreme importance in local economies elsewhere in the world. Some very limited yield data are summarized below under *B. vulgaris*, *G. verticillata* and *M. baccifera.* Reliable figures of this sort on other species might afford a key to the economic promise of commercial ventures. Therefore, in the list of suggested studies offered in the following pages, first place may well be given to a project for determining, on a long-term basis, the pattern of the per-acre-per-year yields of plantings of biometrically optimum size, located on ecological sites representative of areas likely to be available in a given land-use program. Different harvesting regimes, and other devices for discovering how to get long-term sustained yields, should be incorporated.

In the selection of bamboos for individual treatment here, no attempt has been made to give special emphasis to the needs of the domestic situation in the United States. Of the seven species discussed, only one (*Arundinaria amabilis*) is considered to be eminently eligible for major attention in this economy. The aim has been, rather, to demonstrate and recommend a rational approach to the selection, out of the hundreds of available kinds, of those bamboos that stand out in ways that clearly relate to the needs and the conditions of a given economy.

The brief description of each species is intended solely to convey a general idea of the size, habit, and over-all appearance of the plant in the sterile condition in which it would usually be encountered. It is not intended as a means of specific identification.

Arundinaria amabilis McClure (Figs. 67–70 and 92)

Because of the superior technical properties of its culms, *Arundinaria amabilis* held a pre-eminent position among bamboos in world trade over a period of about 50 years, beginning late in the last century. During this period, this bamboo supplied the preferred material for split-and-glued fishing rods in England and America, and for hop poles in Germany. Its culms are still in demand in the United States for rug poles, and for fine handicraft productions, including fishing rods. While this report was being prepared, an inquiry was received from a leading manufacturer of fishing rods, who desired to place an order for $10,000 worth of select culms of this bamboo.

The following properties are responsible for the esteem in which the culms of *Arundinaria amabilis* are held: natural straightness, slight taper, and freedom from branches in the commercial cuts, stiffness and resiliency (slowness to take a set when held under strong flexure), lack of prominence in the nodes, and high density, toughness, and strength of the wood (McClure 1944:38–40, 51). The high commercial value of *Arundinaria amabilis* was sustained by the maintenance, at the source, of a rigorously executed regime of selection, processing, storage, and special packing of the culms for shipment. Until the supply was cut off by political events on the Chinese mainland, this bamboo was without a significant competitor on the world market.

The Plant (Vegetative Characters). Rhizome leptomorph; clump open; culms distant, strictly erect to tip, straight or nearly so; branches typically 3 at each mid-culm node, slender, stiff, wiry, the central one dominant, appressed; the incidence of branches and branch buds at culm nodes retreating from the base of the culms as the plant develops, branches lacking in the lower ½ to ⅔ of the height of culms in plants of mature stature; leaf blades dark green above, paler (glaucescent) below, oblong-lanceo-

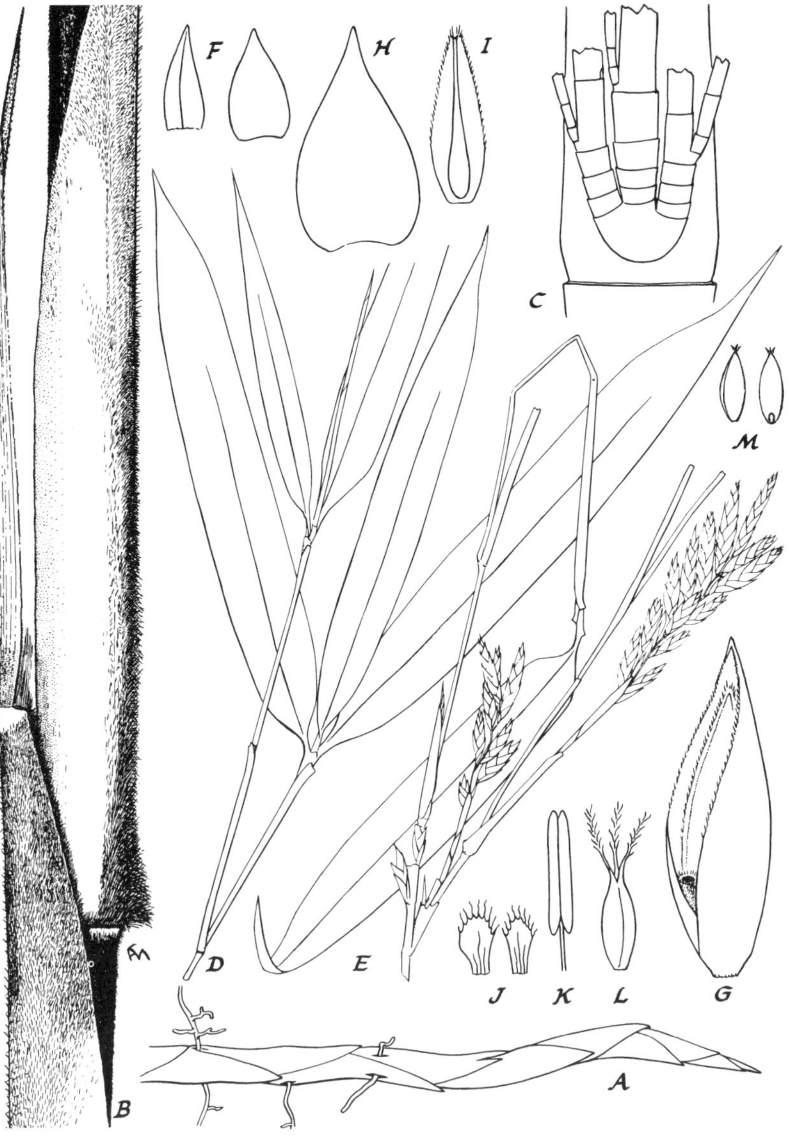

Fig. 67. *Arundinaria amabilis:* (*A*) tip of the leptomorph rhizome; (*B*) mid-culm node of young culm shoot showing tip and base of culm sheaths; (*C*) base of initial branch complement at mid-culm node; (*D*) leafy sterile twigs; (*E*) flowering twigs associated on same branch with leafy twig; (*F–M*) reproductive parts; (*F*) empty glumes I (*left*) and II, from the base of a spikelet; (*G*) floret; (*H*) lemma, abaxial aspect; (*I*) palea, adaxial aspect; (*J*) lodicules (one of the anterior pair *at left,* and the smaller, posterior one); (*K*) stamen; (*L*) gynoecium; (*M*) fruit, lateral aspect (*left*), and dorsal aspect showing embryotegium. Based on specimens from the type plant, LUBG 1880, in the Lingnan University Bamboo Garden.

late, subcoriaceous in texture, durable, the larger ones generally pendent when mature, with transverse venation conspicuous on both surfaces.

In a fully established plant, as seen under plantation management at Au Tsai in Wai-tsap District, Kwangsi Province, China (Fig. 68), the clump habit is open, the culms reach a height approaching 40 ft and a diameter that may exceed 2 in. The lower half of such culms is only slightly tapered, if at all, and branches or branch buds are lacking entirely (McClure 1931*a*).

The chromosome number of *Arundinaria amabilis* apparently has not been determined.

Infraspecific Variation. No variants have been described.

Flowering and Fruiting. Flowering in *Arundinaria amabilis* is cyclic and gregarious within a given population of plants with a common origin; the length of the flowering cycle is unrecorded. As observed in a planting established at Canton, China, under LUBG 1880, the flowering period lasted for 10 years (1929–1939); its termination was accompanied by a gradual recovery of vegetative vigor. Individual flowering culms eventually died after losing their leaves, but the rhizomes remained viable. Flowering was active throughout the period, but fruit production was meager and only a few hundred mature fruits were discovered, in the course of much tedious searching.

Distribution and Ecology. As far as published records are concerned, *Arundinaria amabilis* is known only in cultivation. The only known area of commercial production is the Kwang-ning District of Kwangtung Province, and the adjacent part of Wai-Tsap District of Kwangsi, China. The persistence of the name tonkin cane in the American trade suggests a possible primordial origin in Tonkin Province of Indochina (northern Vietnaam), whence the first supplies of this bamboo to reach the Western market may have come.

At Canton, China, plants grown in unfertilized "lateritic" soil derived from granite, with about 75 in. (1880 mm) of rain falling in the six warmest months of the year, and a temperature pattern similar to that of southern Florida, were slow in establishing themselves. During the first 10 years they developed a wide-

Fig. 68. *Arundinaria amabilis.* A typical cluster of characteristic culms in a plant of medium stature (20 ft tall) in the midst of a commercial plantation at Au Tsai, Kwangsi province, southern China. The superior technical properties of the bamboos known in commercial circles as tsingli or tonkin cane have been recognized in Europe and in the United States for three-quarters of a century. Product of the plant known to science as *Arundinaria amabilis* (McClure 1931*a*), these canes were prepared for the Western market (McClure 1931*b*), by disciplined procedures characteristic of the sophisticated craftmanship of China—the old China, long respected for its mature cultural traditions and its distinguished artistic productions. McClure photo, 1928.

Fig. 69. *Arundinaria amabilis,* vicinity of Au Tsai, Kwangsi province, China: (*above*) bundles of freshly harvested culms being assembled into rafts for transport to the scouring beach; (*below*) the beach where the culms are cleansed at once by scouring with sand.

Fig. 70. *Arundinaria amabilis,* village scenes at Au Tsai, Kwangsi province, southern China, the principal center of production of culms of this bamboo for export: (*above*) the stack of culms at the right represents culls or current overproduction; the walls of the granary are woven from culled culms; the roof is thatched with the bark of China fir (*Cunninghamia lanceolata*); (*below*) sturdy fences woven with culled culms guard the sunning yards where hourglass-shaped bundles of culms selected for export are being dried in the sun.

spreading system of rhizomes whose course was marked by widely spaced slender culms from 0.3 to 2 m in height. In the eleventh year the number of new culms initiated jumped to double what it had been previously, and the maximum height of new culms increased to about 4 m.

After repeated failures, living plants of *Arundinaria amabilis* were successfully introduced into the United States in 1936, accessioned by the U.S. Department of Agriculture under P.I. 110509, and established in cultivation at the U.S.D.A. Plant Introduction Garden at Savannah, Georgia. These plants have sustained, without damage, temperature minima that frequently reach 17°F, and culms 30 ft tall were measured in 1959.

This species has been established in cultivation at the McIlhenny Jungle Gardens, Avery Island, Louisiana; at San Andres, El Salvador; at the Imperial College of Agriculture, Saint Augustine, Trinidad, B.W.I.; and at the Federal Experiment Station, Mayagüez, Puerto Rico.

Propagation. From the single plant available at Mayagüez in 1948 (Fig. 92) about 200 plants were propagated by means of clump divisions and rhizome cuttings. These were used to initiate three experimental field plantings in Puerto Rico: at Mayagüez (sea level) and at two field stations, Maricao (elev. 2000 ft) and Toro Negro (elev. 3000–3500 ft). Owing to more adequate rainfall and soil fertility prevailing at the latter two stations, the plants established themselves promptly, and have shown excellent growth.

Arundinaria amabilis has proved very refractory to propagation by means of rhizome cuttings without attached culms. The best results have been achieved by the use of clump divisions with one or more culms attached to a relatively long section of rhizome. The larger the propagule, the more quickly it will establish itself. Protection from intense insolation and wind is important. The practical upper limit of propagule size is set by decreasing convenience in handling, and by the limited availability of material. However, Dr. John Creech (in an unpublished study) succeeded, by the use of controlled temperature and moisture conditions (in a propagating box maintaining high atmospheric humidity and with sand as the medium) in securing rooted plants from cuttings consisting of culm segments each bearing a full branch complement, with the branches cut back to a length of 3 or 4 in. The material used embraced an age range of about 1–5 years. As is the rule in the vegetative propagation of bamboos, the resulting plants consist of spontaneously rooting new shoots arising from pre-existing buds. Success depends upon keeping the cuttings under conditions that will awaken dormant buds and cause the new axes to root.

In 1954, I saw at the Bogor (formerly Buitenzorg) Botanic Garden living plants of *Arundinaria amabilis* that had been propagated from seeds I sent there from Canton, China, in 1937. The retarded state of the development of these plants suggests that the climate prevailing at Bogor may be unsuitable for this species.

Suggested Studies. *Arundinaria amabilis* should be given a place in any field trials that may be set up in mild temperate or cool tropical regions. Studies to develop efficient methods for its propagation on a large scale are in order, especially since it has proved relatively refractory in the conventional large-scale procedures. Successful development of a superior procedure for its propagation would increase the likelihood that large-scale, or at least widespread, cultivation of this bamboo might develop eventually. This species should also be included in studies aimed at the artificial combination of desirable technical and ecological characters by controlled hybridization.

Bambusa vulgaris Schrad. ex Wendland (Figs. 5, 7, 71, 72, and 99)

Bambusa vulgaris is a bamboo remarkable for a number of reasons. Among these are the high strength of its culms and their adaptability to a wide variety of uses, the readiness with which the plant responds to propagation by vegetative means, the rarity and restricted extent of the incidence of flowering in plantations, and the prompt recovery of plants after severe harvesting—even clear-cutting of the clump. However, the susceptibility of the harvested culms to invasion by the powder-post beetle (*Dinoderus spp.*) was rated as the highest of all among a dozen important economic bamboos studied by Plank at Mayagüez (1950:8). This susceptibility limits its value for many conventional purposes, but its reputation is redeemed by pulping studies that rate *Bambusa vulgaris*

Fig. 71. *Bambusa vulgaris: 1,* leafy twig; *2,* flowering branch bearing tufts of pseudospikelets; *3,* section of a young culm shoot; *4,* culm sheath, abaxial aspect; *5,* young branch or small culm (green-striped yellow form); *6,* top of leaf sheath and base of leaf blade; *7,* spikelet; *8,* empty glume; *9,* lemma; *10,* palea; *11,* lodicule; *12,* stamen; *13,* gynoecium. Redrawn from Gamble 1896:Pl. 40.

very high among nearly 100 species selected from those available in the Western Hemisphere.[1] According to the records of C. T. B. Ezard, the General Manager, in February 1946, of the now defunct Trinidad Paper and Pulp Co., Ltd., well-established plants of *Bambusa vulgaris* produced, at Saint Augustine, Trinidad, over 4 tons of pure, dry cellulose pulp per acre per year on a 3-year cutting cycle. Indications are that this could have been substantially increased by the use of a longer cutting cycle (McClure 1948:735).

After trying many devices (including posts of galvanized iron pipe, and stays of galvanized iron wire, as well as supports made from the culms of several species of bamboo) for the prevention of lodging in their banana plantations, the United Fruit Company selected *Bambusa vulgaris* as the most satisfactory source of banana props in its Central American plantations.

The Plant (Vegetative Characters). Rhizome pachymorph; clump caespitose, rather open; culms erect or suberect, generally more or less curved, commonly 20–50 ft, rarely to 60 ft tall, and up to 4 in. in basal diameter, commonly bearing a ring of roots at several lower nodes, and sometimes up to well above the middle of the culm; branches unarmed, several at each node, the middle one of each complement strongly dominant; in large plants the branch bud may remain dormant at several of the lower culm nodes; leaf blades narrowly to broadly lanceolate, concolorous or nearly so on the two surfaces, the transverse venation sometimes visible, especially on the abaxial surface.

According to Hubbard and Vaughan (1940:30), the somatic chromosome number of *Bambusa vulgaris* is 72.

Infraspecific Variation. Kurz (1876:339–340) lists as varieties of *Bambusa vulgaris* the following color forms distinguished by the Malays: "The natural species, bamboo hower hedyoo [aur hijou, in current Malay, teste Holttum], also called bamboo hower gullies and bamboo ampel, with uniformly green culms and branchlets; bamboo hower kenneng, also called bamboo koonieng or yellow bamboo, with culms uniformly yellow, or rarely with an occasional one green with yellow stripes; bamboo hower seh-ah, also called

[1]Unpublished results of tests by technicians of Champion Papers, Inc.

Fig. 72. *Bambusa vulgaris. 1.* When a representative clump of 60-ft culms (from the grove in the background) growing on alluvial land at the Raheen Estate, Jamaica, was clear-cut, the culms racked up exactly a cord, and weighed 2576 lb in the fresh, green state (data courtesy of Champion Papers, Inc.). *2.* Base of a well-developed clump showing that in large plants the branch buds may remain dormant at several of the lower nodes in culms of mature stature. *3.* A close-up view featuring the sheaths that clothe the internodes of the young culms while they are in a growing condition. As each internode reaches its full development, an absciss layer forms at the base of the sheath that clothes it, and the sheath gradually dries up and falls away. *4.* The type form of the species, with green culms, gives rise by spontaneous mutation to several color forms in which yellow occurs alone or in combination with green stripes. These forms are generally rather unstable, and the color pattern often reverts to green in new culms of the same clump. In the commonest form (the one with typically green-striped yellow culms), individual culms may be almost pure yellow, with only a suggestion of green (*left*) or an occasional one may be half green and half yellow (*right*). Specimens photographed in El Salvador. *5* and 6. *B.* vulgaris, cultivar "Wamin," cv. nov., is characterized by a shortening and basal inflation of the internodes, particularly those of the lower part of the culm. The habit photo (*5*) was taken at the National Botanic Garden, Calcutta; the near view (*6*) represents a plant in the Royal Botanic Garden, Port-of-Spain, Trinidad.

1

2

3

4

5

6

bamboo kooda, with culms green beautifully striped with yellow; bamboo tootool, or blotched bamboo, with culms green at first then becoming blotched with black on aging." The first three apparently are unstable, and reciprocal reversion takes place spontaneously between them. According to personal observation, the green-striped yellow form predominates on the Central American mainland, while the green form predominates in frost-free areas of continental United States and on the islands of the Caribbean area, at least in Cuba, Puerto Rico, Jamaica, and Trinidad.

Bambusa vulgaris, cultivar 'Wamin' cv. nov. (Fig. 72, *5, 6*) is distinguished from the type form of the species principally by the greatly shortened and basally inflated internodes in the lower part of the culms. This bamboo was first described by Brandis (1906:685) under the Burmese vernacular name Wamin, without assigning it to a genus, though he alluded to the resemblance of the culm sheaths to those of *Bambusa vulgaris.* Camus (1913:135) listed it under *Bambusa* as "B ? Wamin Brandis," and simply repeated Brandis's English description in French translation. This provisional name is invalid. The geographical origin of this apparently quite stable form (referred to by Brandis as possibly a deformity) is still a matter of conjecture. It may be assumed to have originated in cultivation, however, since the native home of *Bambusa vulgaris,* from which it clearly is derived, is still unknown.

Flowering and Fruiting. The flowering and fruiting habits of *Bambusa vulgaris* are treated on p. 82.

Distribution and Ecology. As far as published records are concerned, *Bambusa vulgaris* is known only in cultivation, although according to Trimen (1900:314) Kurz regarded it as indigenous in Java and Thwaites treated it as a native of Ceylon. The consensus of modern published opinion is that the native home of *Bambusa vulgaris* is unknown. It is quite pantropic in its distribution today, but its penetration into the temperate zones is limited by its vulnerability to cold (its culms may be killed to the ground at 32°F). The green form has become naturalized in large areas on the island of Jamaica in the wake of a sort of migratory agriculture. Stakes freshly cut from culms of the living plant and used to support yam vines take root and, where neglected, produce extensive groves.

Bambusa vulgaris thrives under a wide range of soil and moisture conditions. The Inter-American Institute of Agricultural Sciences, at Turrialba, Costa Rica, has a clump growing well on an island in a small lake, where the water table is within 1 ft of the surface of the soil. On the other hand, in El Salvador, the green-striped yellow form thrives in areas where the dry season is so severe that the plants become completely defoliated. In continental United States, the plant is common only in frost-free or nearly frost-free parts of Florida and California, but it abounds in the populated parts of Puerto Rico where, since its introduction early in the 19th century, it has been an important source of building material for temporary use. In the Spanish colonial period, strips split from the mature culms were effectively used, and proved very durable, in a special lath-and-plaster construction of house walls. Unless given special protection, however, it is soon destroyed by xylophagous insects.

Propagation. The propagating of *Bambusa vulgaris* is discussed under various headings in Chapter 5, and illustrated in Figs. 94 and 95.

Suggested Studies. Because of its vegetative vigor, ease of vegetative propagation, wide ecological tolerance, the ability of established clumps to withstand severe harvesting, and the high rank given the culms as a source of paper pulp, *Bambusa vulgaris* should be included in any program of study aimed at the development of a tropical source of cellulose fiber. According to personal experience in Guatemala, it should be possible, under favorable ecological conditions, to begin harvesting this bamboo, by clear-cutting of the clumps, within 5 or 6 years after the establishment of a planting. Optimum planting distances and optimum patterns and cycles of harvesting have to be determined for each distinct set of field conditions.

Hybridization studies and a search for the causes of self-sterility in the flowers of the common strain, which is the typical form of the species, could be fruitful. One or more flowering plants can usually be found here and there in the wide range of its distribution, at the beginning of the rainy season.

Additional References. Wendland (1810) contains the original description and the earliest known illustration of the species; A. and

C. Rivière 1879:191–203, 328–331; Gamble 1896:43–45, Pl. 40; Moebius 1898; Arber 1934:64, 66, 67 (Fig. 23), 68; McClure 1955:60, Fig. 11.

Dendrocalamus strictus (Roxb.) Nees
(Figs. 55, *7*, 60, 73, 74, and 93, *1–3*)

Dendrocalamus strictus has assumed great importance in the economy of India, chiefly as the foundation of that country's paper industry. It is India's principal accessible source of paper pulp of acceptable quality. Its pulp is used wholly for blending with inferior pulps from other domestic botanical sources. By the papermaker's standards, pulp from *Dendrocalamus strictus* is inferior in some respects (fiber length,[2] for example) to pulp from some other bamboos, but this species owes its dominance in India's paper industry to the abundance of natural stands of it within reach of transportation. Another reason for giving special attention to this species here is its possession of a property apparently rare among the bamboos, namely, a high degree of drought resistance (Deogun 1937:83). On this account, it could become an important source of germ plasm to impart drought resistance to synthetic bamboos through hybridization. Again, the scope and volume of published data on the silviculture and management of *Dendrocalamus strictus* in India has set it off, in this respect, from all other bamboos. Familiarity with the long history of the efforts of the Indian Forest Service toward the economic conquest of this bamboo should, in spite of their inconclusive nature (Deogun 1937:vii, 147) provide a useful background for the design of similarly oriented studies of other species, in other settings.

In evaluating data published under the specific name of this (or any) bamboo, it should be remembered that distinct (though unrecognized and unnamed) strains or clones from which the data were drawn may differ in important characteristics (see p. 170). It is well to keep this consideration in mind also in selecting living material for silvicultural studies related to culm production for particular purposes. Deogun refers frequently to a strong tendency toward congestion in the clumps of *Dendrocalamus strictus*. Congestion in this species is due primarily to a shortness of the rhizome

[2]An average fiber length of 1.32 mm was found in culm material studied by technicians of Champion Papers, Inc.

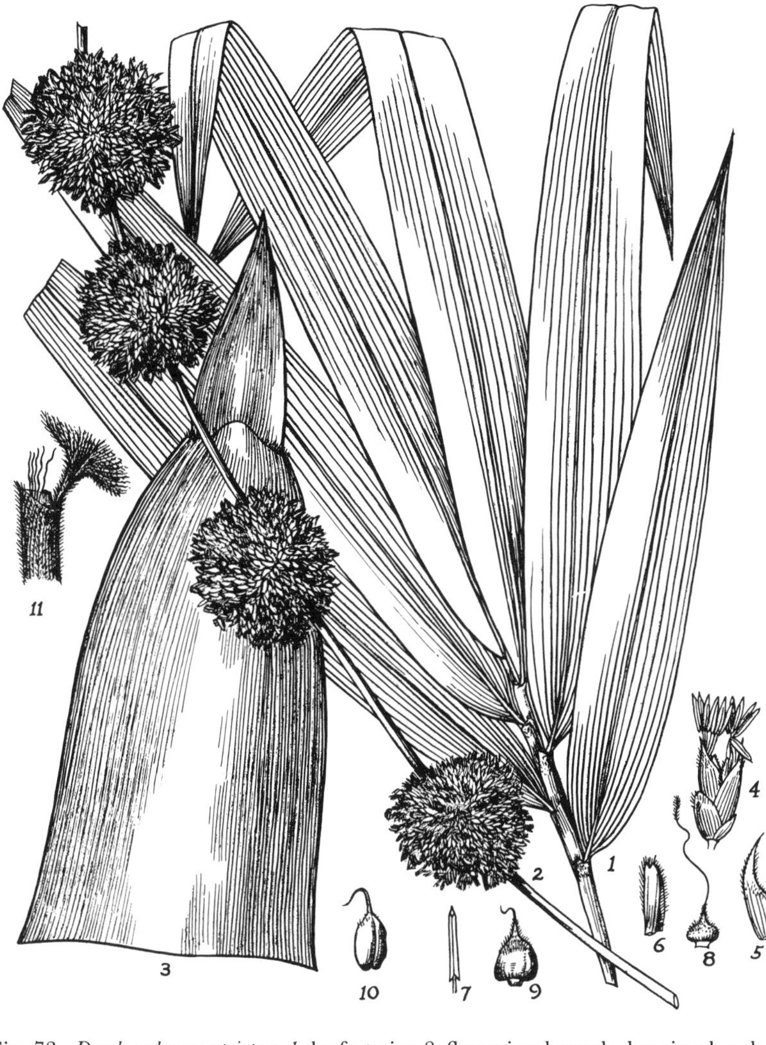

Fig. 73. *Dendrocalamus strictus: 1,* leafy twig; *2,* flowering branch, bearing heads of pseudospikelets; *3,* culm sheath; *4,* spikelet; *5,* lemma; *6,* palea; *7,* stamen; *8,* gynoecium; *9, 10,* fruits in different stages of development; *11,* top of leaf sheath and base of leaf blade. Redrawn form Gamble 1896:Pl. 68.

Fig. 74. *Dendrocalamus strictus.* A young clump in the process of formation, showing the successive development of culms. The oldest one is on the left; the youngest on the right. From Deogun 1937:Pl. III.

axes, but it may be complicated by an excessive development of long branches in the lower part of the culms. Congestion in a clump renders the extraction of culms very difficult. Deogun's opinion that some strains of *Dendrocalamus strictus* are more or less completely free from the tendency to congestion is confirmed by my own observations. Deogun (1937:Pls. IV, *a, b,* and V) illustrates clumps of *Dendrocalamus strictus* before and after thinning to relieve congestion.

The Plant (Vegetative Characters). Rhizome pachymorph; clump more or less densely caespitose; culms glaucous at first, up to about 60 ft tall and 5 in. in diameter, of variable habit, broadly arched to erect, strongly curved to fairly straight; branches several at each culm node, unequal, the central one strongly dominant, of variable length and habit, buds and branches present at all nodes or (in full-grown plants of some strains such as P.I. 254921 lacking in the lower ½ to ⅔ of the culm; leaf blades of variable size, small in some strains to large in others (apparently of intermediate size in the type form), usually pale green above, paler below, typically pubescent on both surfaces, the obscure transverse venation, visible only in transmitted light, is described by Gamble (1896:78) as consisting of "pellucid dots."

Richharia and Kotwal (1940:1033) give the somatic chromosome number of *Dendrocalamus strictus* from root-tip counts as 72; Parthasarathy (1946:234) gives the somatic number as 70, and the "*n*" number at metaphase as 35, both numbers based on examinations of root-tip sections.

Infraspecific Variation. *Dendrocalamus strictus* is a polymorphic species. Plants in seedling populations vary, apparently both phenotypically and genotypically. As observed at the Federal Experiment Station, Mayagüez, Puerto Rico, and elsewhere (without benefit of comparison on uniform habitats), seedling populations are seen to vary widely in ultimate stature and habit of culms, thickness of culm wall, texture and pubescence of the culm sheaths, branching habit, size of foliage leaves, and disposition toward congestion of culms in the clump. Two named varieties have been based on extremes of variation in the pubescence of the flower-subtending lemmas: var. *prainiana* Gamble, with lemmas nearly glabrous, and var. *sericeus* Gamble, with the lemmas silky pubescent. On the basis of the variegation that appears as silvery white lines on the leaf blades and dark-green and yellowish stripes on the culm sheaths, A. and C. Rivière (1879:681) described a third variant, "*Bambusa stricta* var. *argentea*" (a name not yet transferred to *Dendrocalamus*). Numerous seedling strains, only one of which (P.I. 254923) has been formally distinguished and documented, are under cultivation at the Federal Experiment Station in Puerto Rico. One of these strains (undesignated) is established at the Atkins Institution, Soledad, Cuba, where it began flowering in 1956 (Clement 1956) and fruited abundantly in 1957 and 1958.

Deogun (1937:79–80) has the following to say about variation in *Dendrocalamus strictus* as observed in the field:

> Growth forms.—Different conditions of soil, aspect and drainage conditions appear to have resulted in the differentiation of recognisable growth forms which may possibly be more or less inherited through the seed, though this needs further examination.
>
> Three main forms have been found to be recognisable in the course of the present study.
>
> 1. The common type.
>
> This type can further be subdivided into three minor variations;
>
> (*a*) Culms with moderately thick walls. This is the ordinary form producing medium-sized culms and is met with everywhere.
>
> (*b*) Culms hollow with relatively thin walls. Generally found in depressions, on cooler aspects, and where conditions are more favourable. It attains quite a big size.
>
> (*c*) Culms solid or nearly so. Generally found growing on ridges and on hotter aspects. On the whole this variation does not attain a big size.
>
> 2. Large type.
>
> This type of bamboo is met with in forests in the United Provinces, in Bihar and Orissa, and to a limited extent elsewhere, where growth conditions are optimum.
>
> This type has practically no side branches to a great height and seldom shows signs of congestion. It attains a very big size with long, straight, and smooth internodes.
>
> 3. Dwarf type.
>
> This [type] is of a small size and only exceptionally forms clumps. Typical examples are met with in Balaghat division of the Central Provinces, where it is known as karka, and to a limited extent elsewhere . . . This is the poorest form of *D. strictus.*

It is not clear which of these is the typical form of the species nor which variations are phenotypic and which genotypic. Elsewhere (*ibid.*:163) Deogun describes in the following words what he apparently considers to be a distinct form, with stable characteristics:

> The bamboo in certain localities, e.g., Nauri, Lansdowne division in United Provinces, has long internodes and is practically free from side branches to a great height. It is of superior quality and congestion is never seen. Even clumps growing on a camping ground, [which are] always rubbed by bullocks and buffaloes, showed no signs of congestion. The clumps are quite big, but not congested.

Flowering and Fruiting. Deogun (*ibid.*:107–109) assembled the existing records of the gregarious flowerings of *Dendrocalamus strictus* by provinces and localities, and summarized his discussion of its flowering cycle in the following words (p. 111):

> The general conclusion is that the cycle for *Dendrocalamus strictus* in one locality is more or less constant but differs in localities remote from one another and with appreciable climatic and soil differences. More than two generations ago Kurz noted the [flowering] cycle [of *Dendrocalamus strictus*] as between 25 and 35 years (*Ind. For.*, 1876, pp. 256–257), and after the lapse of so many years with a number of flowerings, we are still unable to be more precise. We can only say that there may be a bigger variation, viz., 20–40 years or so.

Sporadic flowering in parts of a continuous population, and even in parts of individual clumps, is of frequent occurrence in this species (Mathauda 1952). Moreover, a number of cases of the precocious flowering of seedlings of *Dendrocalamus strictus* are on record. See p. 274 for further details.

Distribution and Ecology. According to Troup (1921:1006),

> This is the best known, commonest, and most widely distributed of all Indian bamboos, occurring in deciduous forests throughout the greater part of India, except in northern and eastern Bengal and Assam; it is common in the drier types of mixed forest throughout Burma. It is found typically on hilly country, ascending to 3500 ft and occurring gregariously sometimes almost to the exclusion of tree growth, but usually forming an under-story to, or a mixture with, deciduous trees. It is not uncommon in certain types of sal forest on hilly country. It is abundant in many parts of the Siwalik tract and outer Himalaya from the Punjab eastward to Nepal, occurring most plentifully between the

Ganges and Ramganga rivers. It is also common in most of the hilly parts of the Indian Peninsula, except in very moist regions. In Burma it is the typical bamboo of the drier types of upper mixed deciduous forest with or without teak; it also extends into dry dipterocarp forest.

This is the hardiest of all Indian bamboos, thriving in regions which suffer periodically from excessive drought. Within its habitat it is frost-hardy. In the abnormal frost of 1905 in northern India it withstood the effects of the frost better than almost any of the tree species; only isolated clumps suffered, and in these only the younger and more tender culms were affected. In the abnormal drought of 1899–1900 in the Indian Peninsula it escaped damage, although other species of bamboo and many tree species suffered severely.

Deogun (1937:79) describes *Dendrocalamus strictus* as generally deciduous except on moist sites, adding the following notes (pp. 82ff) on its ecological tolerance, without reference, however, to possible infraspecific genotypes or ecotypes that may be involved:

A minimum of about 30 inches and a maximum of 200 inches average rainfall, and a maximum shade temperature of 116°F, and a minimum of 22°F, are recorded from bamboo-bearing localities. Atmospheric humidity is said to be one of the determining factors in the distribution of the species. Thus, in Orissa [according to] J. W. Nicholson (*Ind. For.* 1922, p. 425) it flourishes in regions where the relative humidity of the air is low, as in the interior tracts which are beyond the influence of sea breezes, but as the humidity rises, it becomes scarcer . . . In general it may be said that it can stand drought better than any other species of bamboo.

It is a common experience that frost and drought, although of no consequence to grown-up bamboo, may be very harmful to seedlings and young plants, and this has been confirmed in the course of experiments on artificial regeneration of bamboo in Dehra Dun Division, United Provinces.

It appears that, within its climatic habitat, *Dendrocalamus strictus* grows on practically all types of soils, provided there is good drainage. It does not grow on water-logged or heavy soils, such as pure clay or clay mixture with lime. Well-drained localities with sandy loam overlying boulders are the best. . .

Dendrocalamus strictus prefers hilly ground and is generally more predominant, and of better quality on cooler aspects.

Deogun (*ibid.*:91) quotes from the observations of M. D. Chaturvedi (1928) on the effects of light and shade on *Dendrocalamus strictus* as shown by an experiment set up by P. C. Kanjilal:

> A certain amount of overhead cover is necessary during the earliest stages of development of bamboo seedlings before they actually form clumps. Natural regeneration of bamboo is conspicuously absent from areas entirely exposed to the sun . . . A judicious amount of shade minimizes the effect of frost . . . The largest number of bamboos is obtained when the light falling on bamboo crops is at the maximum. But . . . a certain amount of overhead cover improves the quality of the bamboo at the expense of its quantity.

Figures from three experimental plots in Lansdowne Division, United Provinces, quoted by Deogun (1937:92) to document the unfavorable effects of shade on number of culms produced do not take into account the competition of the trees for other requirements of the bamboos such as water and nutrients that come from the soil. The effects of wind are described as unfavorable in proportion to its intensity. Deogun (p. 163) says that "in areas exposed to strong winds the individual culms produce more side branches and so become congested."

Dendrocalamus strictus has been introduced into the United States, on several occasions, by means of seeds from India. It is commonly found in cultivation in Florida and southern California but is rare elsewhere, except in Puerto Rico, where numerous seedling strains are rather widely disseminated.

Propagation (Figs. 60 and 93, *1–3*). A summary of experience in the vegetative propagation of *Dendrocalamus strictus* by various methods is given in Chapter 5.

Suggested Studies. Search for elite strains.

Additional References. Arber (1926:458, Fig. 8, *A*, *B*; 461) illustrates the anatomy of the florets, by cross sections.

Bhargava (1946:Pt. I, Introduction) describes the use of this bamboo for paper pulp in India.

Blatter (1929–30:459ff) discusses possible "causes" of flowering in bamboo.

Brandis (1874:569, Pl. 70) presents a description in English, and the first good illustration of the inflorescence.

Church (1889:283) gives principal constituents of the fruits.

Gamble (1896:68, Pls. 68–69) presents the first illustration of the culm sheath; includes a good, brief treatise on distribution and uses.

Gupta (1952:547) presents a treatise on the flowering habits of this species; of interest for comparison with Mathauda's paper on the same subject.

Kadambi (1949) deals with the ecology and silviculture of bamboo forests which have been set apart for exploitation of *Dendrocalamus strictus* for papermaking.

Mathauda (1952:86) presents a brief but provocative discussion of recorded data and observations on the highly divergent flowering behavior manifested by different plants (strains?) of this species.

Mooney (1933, 1938) makes interesting observations on the local distribution of *Dendrocalamus strictus* in Orissa and in western Singhbhum, India, as affected by pedological factors.

Plank (1950) reports results of comparative studies of this and eleven other bamboos in respect to factors influencing attack and control of the bamboo powder-post beetle (*Dinoderus minutus*).

Rebsch (1910): "The chief object of the present article is to draw attention and invite discussion on the management and working of bamboo forests in a locality where the demand is great and the working consequently intense. The past history of these forests is reviewed, and the gradual evolution of the present working method, and changes in system of management are discussed." *Exp. Sta. Rec. 23:*644 (1910).

Rivière (1879:675) presents the first description giving detailed attention to vegetative characters of the plant (as *Bambusa stricta*).

Trotter (1922) completes the record begun by Troup (1921) of the development of plants of *Dendrocalamus strictus* from natural seedlings (from seeds that germinated in 1911) to exploitable age.

Gigantochloa verticillata (Willd.) Munro, at least in part[3] (Figs. 75 and 76). Syn.: *Bambusa verticillata* Willdenow, Linn. *Sp. Pl.* 2:245 (1799).

Among the twenty-odd introduced species that constitute the

[3]That part which pertains to plants accessioned by the U.S. Department of Agriculture under P.I. 79568.

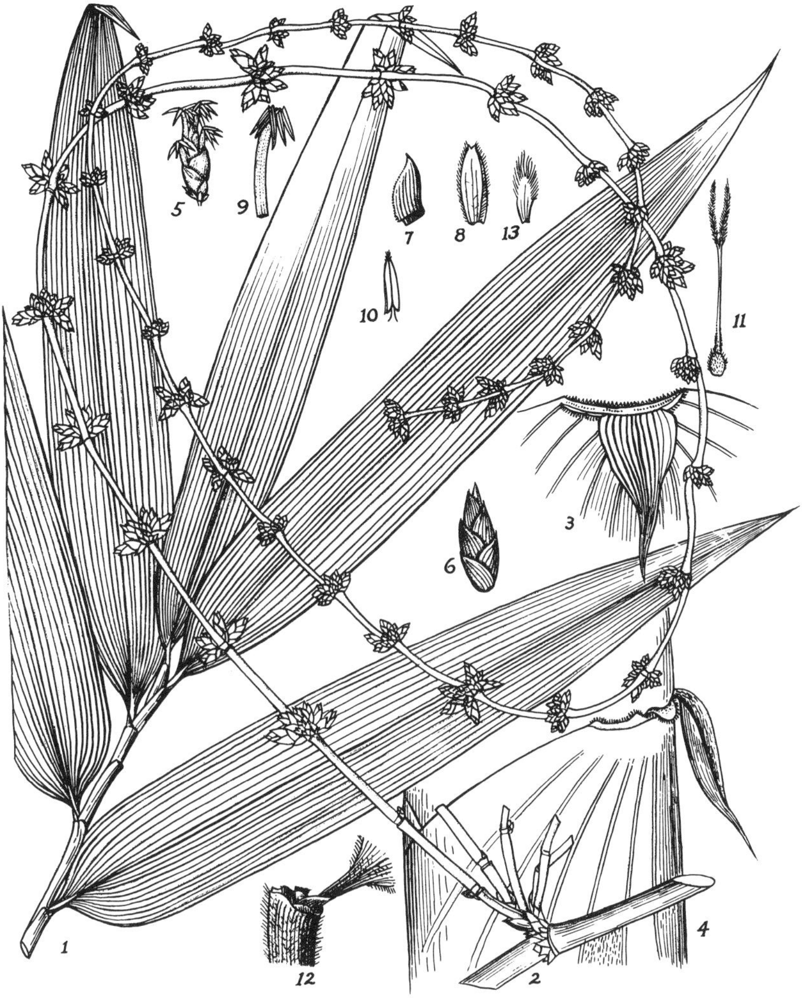

Fig. 75. *Gigantochloa verticillata: 1,* leafy twig; *2,* flowering branch bearing tufts of pseudospikelets; *3,* top of culm sheath, dorsal aspect; *4,* top of culm sheath, lateral aspect; *5,* pseudospikelet; *6,* spikelet; *7,* lemma; *8,* palea; *9,* androecium, showing monadelphous stamens with stamen tube; *10,* stamen; *11,* gynoecium; *12,* top of leaf sheath and base of leaf blade; *13,* lodicule. Redrawn from Gamble 1896:Pl. 52.

1

2

3

Fig. 76. *Gigantochloa verticillata.* (*Top*) A plant of mature stature, with culms up to 65 ft tall and 5 in. in diameter, that died from an unknown cause. The habit of culms and clump are more clearly shown here than in a plant in full leaf. Taken at the Agricultural Experiment Station of the Tela Railroad Company, Lancetilla, Honduras. (*Middle*) A young, vigorous clump with culms up to 35 ft tall, at the Escuela Agricola Panamericana, Zamorano, Honduras. (*Bottom*) The lower nodes of culms of mature stature are characteristically ringed with a verticil of rather prominent roots and root primordia. Plant cultivated at Rosario, Alta Verapaz, Guatemala, under P.I. 79568.

principal known bamboo flora of Java, *Gigantochloa verticillata* ranks third in importance (after *G. apus* and *Dendrocalamus asper*) as a source of building material and edible shoots. Since, according to Heyne (1950:300), the culms are durable only as long as they are not exposed to the weather, the standing that this species has acquired in Indonesia is due not to the durability of its culms, but rather to their versatility and their good natural shape, straightness, and workability. Heyne says that in Java, Bali, and elsewhere the posts, walls, ceilings, and rafters of houses are made from the culms of this bamboo, when the more durable species are not available.

Gigantochloa verticillata is included here, however, primarily because of its remarkable qualities and performance in relation to papermaking, brought to light only recently. During the course of a survey of the pulping properties of bamboos available in the Western Hemisphere, this bamboo stood in general performance among the top half-dozen of about a hundred species tested comparatively in the laboratories of Champion Papers, Inc. But it is in the field (at Rosario, in the Polochic Valley, Guatemala) that it made its most spectacular showing. One plant, clear-cut after growing in the nursery 3 years from the time it was transplanted from the propagating bed (that is, 4 years from the time the single-culm offset from which it originated was prepared) showed a total increment of substance in the harvested culms corresponding to a yield of more than 4 tons of oven-dry cellulose per acre per year from plants spaced at 7 × 7 m.[4]

The Plant. Rhizome pachymorph; clump caespitose; culms closely spaced, exceeding 15 m (50 ft) in height and 10 cm (4 in.) in diameter, erect, nodding apically; the nodes perceptibly flared and fringed with hairs at the sheath scar, the lower ones ringed by a rather prominent verticil of roots or root primordia just above the sheath scar; internodes plain green, remarkably cylindrical, at first appressed pubescent with dark hairs, later glabrescent; the wood (culm walls) tapering upward gradually, commonly from a basal thickness of about 3/4 in; branches more or less completely lacking in the lower 1/2 to 2/3 of the height of the culm in plants of mature stature; leaf blades large, oblong-lanceolate, glabrous

[4]Data courtesy Champion Papers, Inc.

on the upper surface, pubescent at first and later glabrescent on the lower surface; transverse veinlets distant, irregularly spaced, oblique, weakly manifest, on the lower surface only, as a rule.

The chromosome number of *Gigantochloa verticillata* apparently has not been determined.

Infraspecific Variation. *Gigantochloa verticillata* apparently is a highly variable species. Ochse (1931:Fig. 204) following Gamble (1896:61) illustrates as the typical form Awi gombong, which is characterized by green and yellow striping of the lower culm internodes. Ochse (pp. 323ff) ascribes to it as minor variants three bamboos under the vernacular names Awi andong, Awi lèah, and Andong kèkès (Figs. 205–207). Heyne (1950:300) lists and weakly characterizes these and several other variants. However, the taxonomic disposition of the numerous forms of *Gigantochloa* under cultivation in Java and adjacent areas is in need of critical study.

Flowering and Fruiting. Although the flowers of *Gigantochloa verticillata* have been described, the fruit has not. Evidently plants of this species flower rarely and sporadically, while fruit production, as in many bamboos known only in cultivation, is sparse or nil.

Distribution and Ecology. Gamble (1896:62) says of *Gigantochloa verticillata:*

> Wild, or more usually cultivated, in the Malay Peninsula and throughout the Malay archipelago, probably extending northwards to Tenasserim; cultivated in the Calcutta Botanic Garden.

However, it appears that its precise origin has not yet been identified and, according to Holttum (personal communication),

> Gamble never saw specimens of this species from the Malay Peninsula. Specimens [from this area] reported by Roxburgh [as this bamboo] were another species.

In fact, the typical form of the species may no longer exist in a primeval state. Holttum (1958:4) says:

> It is fairly clear that the main center of distribution of *Gigantochloa* lies north of Malaya, probably in Lower Burma, where no intensive study of bamboos has been made. So that the *Gigantochloas* of Java, and some planted in Malaya, probably came from the Burma region, and

may be interesting records of the migration of men; but in the absence of a good knowledge of the bamboos of Burma, those records cannot be interpreted. It is strange that the most distinctive and most useful *Gigantochloas* of Java seem hardly known in Malaya (*G. apus* and *G. maxima*) . . . I have seen no native plants of that genus in southern Malaya. The first appears in the neighborhood of Tampin (at the southern end of the Main Range) and as one travels northwards the variety is quite bewildering, and I believe that hybrid swarms exist.

Holttum intimates that early migrants probably carried with them bamboos of mixed heredity. The multiplicity of closely related forms of bamboos found today in cultivation in Java and elsewhere in Malaysia certainly suggests the occurrence of introgression, or the segregation that takes place in the F_2 generation following the union of heterozygous gametes.

The U.S. Department of Agriculture accessioned, under P.I. 79568, plants of the plain green form of *Gigantochloa verticillata* collected by Fairchild and Dorsett at the Sibolangit Botanic Garden, Sumatra, in May 1926. This bamboo has become established in cultivation in the Western Hemisphere, at the U.S.D.A. Plant Introduction Garden, Coconut Grove, Florida; at Lancetilla, Honduras, in the arboretum founded for the United Fruit Company by Dr. Wilson Popenoe; and at Chocolá and Rosario in Guatemala. The account given above, relating to the outstanding performance of this species in relation to papermaking, both in the laboratory and in the field, is based on plants derived from P.I. 79568.

Propagation. No mention of any experience in propagating this bamboo by any means other than conventional clump division has been encountered in the literature.

Suggested Studies. It would be worth while to assemble all available variants of *Gigantochloa verticillata* and other outstanding species of *Gigantochloa,* to pave the way for comparative studies of the performance of their cellulose pulp in the paper laboratory, and comparative trials of living plants on selected ecological sites and under diverse patterns of harvesting. In view of the generally excellent form and workability of the culms of many bamboos of this genus in relation to the requirements for structural and handicraft material prevailing in certain local economies, a search for strains resistant to rot fungi and xylophagous insects is indicated. Studies on the propagation of each bamboo of economic promise are always in order.

Additional References. Backer 1928:275–276); Holttum (1956*b*; 1958:114).

Guadua angustifolia Kunth (Figs. 6 and 77–80). Syn.: *Bambusa guadua* Humboldt & Bonpland

Among the bamboos native to the Western Hemisphere, *Guadua angustifolia* is outstanding in the stature, mechanical properties (strength and workability), and durability of its culms, and in the importance their many uses have given this species in the local economy wherever it is available.

On the occasion of describing the species and giving it a scientific name, Humboldt and Bonpland (1808:65–66) made the following observations (here given in free translation from the original French text):

"In America the bamboos offer the same benefits as they do in India. *Bambusa guadua* is used alone for building entire houses [Fig. 78]. The walls are made of the oldest and largest culms; the first layer of the roof (the foundation of rafters) is made with smaller ones, while the second layer is thatched with young branches still bearing leaves. In using the culms of this plant instead of the hard timber of the tall trees that often surround it, the native Americans avail themselves of the following advantages: (1) the ease with which they can be cut and transported over long distances; (2) the slight amount of labor required to prepare them, since they are used either whole or split in two; (3) the durability of their wood, which may compare favorably with that of the best timber; finally (4) open construction of their houses, and the protection from the burning rays of the sun afforded by the wide, thick roof, maintains a cool and agreeable temperature during even the hottest part of the day."

The diversity of uses and the general importance of *Guadua angustifolia* appear to have increased rather than diminished in modern times. In addition to supplying almost all of the building materials for rural dwellings, and other uses mentioned by Humboldt and Bonpland, whole culms, boards, and lath from *Guadua angustifolia* are used today in large structures such as apartment houses and factories in Cali, Guayaquil, and other large cities. The

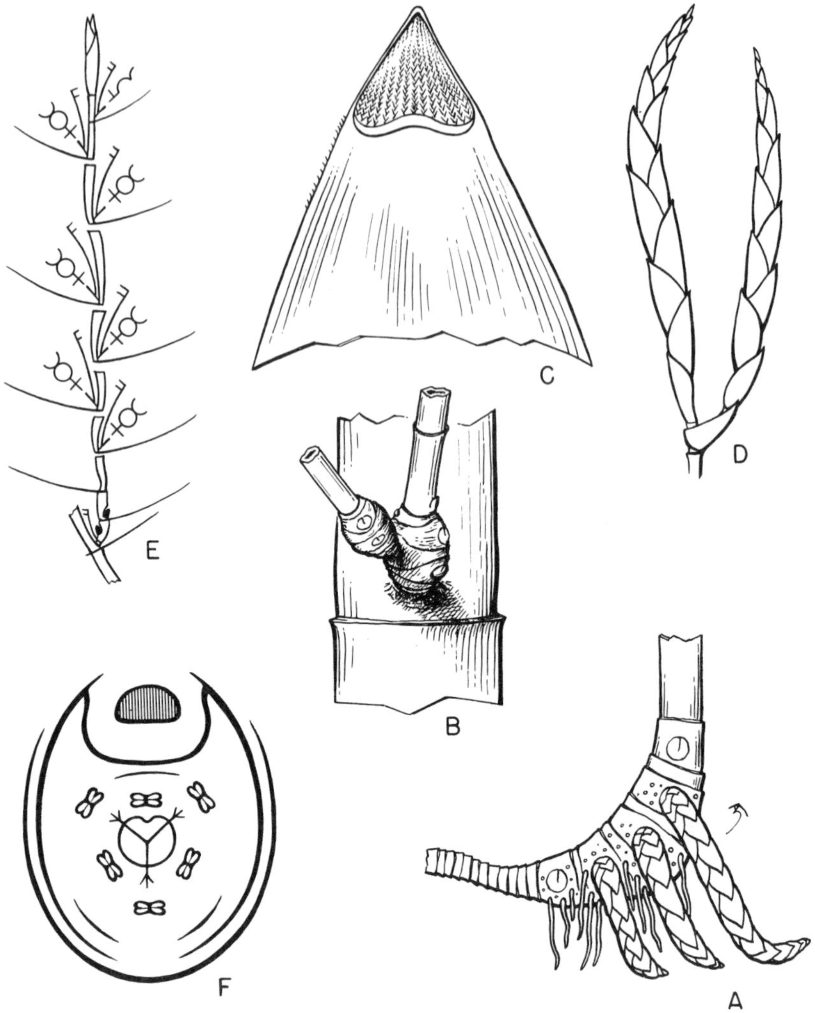

Fig. 77. *Guadua angustifolia:* (*A*) rhizome, with base of its culm, and the necks of new rhizome branches; (*B*) mid-culm branch complement; (*C*) apex of sheath from mid-culm node, in adaxial aspect, showing the ligule at the base of the sheath blade; (*D*) young inflorescence, consisting of two pseudospikelets, the one on the right developed from a bud at the base of the other; (*E*) diagram showing the relation of the constituent parts of a primary pseudospikelet; (*F*) diagrammatic cross section of a floret.

scraps of waste from the large-scale fabrication of bamboo boards and lath are used by the bakers of Cali as fuel to heat their ovens. The scaffolding that supports the operations of building, painting, and repair of the highest structures is erected with the tall, slender, strong, and resilient culms of *Guadua angustifolia*. The rafts by which many of the culms are transported to the markets by river are also utilized for shipping farm produce and for carrying passengers (Fig. 79). Although not ranking with the best, the cellulose manufactured from the culms of *Guadua angustifolia* makes very good paper. Large quantities of the culms are imported by Peru, particularly for use in building purposes in the city of Lima. These

Fig. 78. *Guadua angustifolia*. (*Left*) Culm material of this species "is used alone for building entire houses. The walls are made of the oldest and largest culms; the first layer of the roof (the foundation of rafters) is made with smaller ones." In these words, Humboldt and Bonpland (1808:65) described the place of *Guadua angustifolia* in the economy of New Grenada more than 150 years ago. That these words remain true today is due to the superior technical properties and the incomparable durability of its culms. (*Right*) In Colombia and Ecuador boards like these, fashioned by opening out large culms of *Guadua angustifolia* with an axe, replace in a hundred ways the sawn boards of timber woods in conventional applications. McClure 1953.

Fig. 79. *Guadua angustifolia.* (*Upper*) The vast forests of pure or nearly pure stands of this bamboo "several leagues in extent" mentioned by Humboldt and Bonpland as existing in Quindíu, New Grenada (now the Department of Caldas, Colombia) have in recent decades been severely reduced in extent by clearing to establish cattle farms. Very thorny branches form an effective barrier around the base of the culms in this typical form of the species. (*Lower*) Rafts, by which a large share of the culms of this bamboo are transported to the markets by river, are utilized for shipping farm produce and for carrying passengers as well.

have their origin chiefly in the littoral of Ecuador, where this species is possibly even more important than it is in the economy of the Department of Valle, Colombia. Everywhere in rural areas where they are available, one sees the culms of *Guadua angustifolia,* either whole or as unsplit lengths, serving as electric power-line poles, as fence posts, and in the construction of sturdy gates and durable woven fences, corrals, bridges, and water pipes. In Colombia and Ecuador, boards fashioned by opening large culms out flat replace in a hundred ways the sawn boards of timber woods in conventional use elsewhere (Fig. 78). These boards serve in place of concrete for surfacing the floors of houses, and the sunning floors where seeds of coffee and cacao are dried. Seasoned culms of *Guadua angustifolia* apparently have a relatively high resistance to both rot fungi and wood-eating insects. It has been observed repeatedly that ordinary hardwoods used in conjunction with this bamboo have had to be replaced because of insect damage, while the bamboo still remains serviceable. The original untreated siding, consisting of boards of this bamboo, in a 40-year-old plantation house at Pichilingue in the Department of Los Rios, Ecuador, was still in a serviceable condition in 1945, long after the hardwood floors had been replaced because of insect damage (McClure 1953:37).

The Plant (Vegetative Characters; Figs. 6, 77, 79, and 80). Rhizome pachymorph, very thick, the neck somewhat elongate; clump open-caespitose; culms commonly to 60 ft, sometimes approaching 100 ft, in height, commonly 4–6 in., exceptionally 8 in., in diameter, erect, broadly arched above; internodes hollow, usually perceptibly sulcate above the point of attachment of a branch complement, lower ones very short, the wood up to 1 in. thick at the base of the culm; culm sheaths deciduous in the upper part of the culm, usually more or less persistent at the lower nodes, densely and minutely tomentose on the back, especially toward the base, with small, brown, persistent hairs, and more or less densely strewn with longer, stiffer, coarser, sharp, antrorse-spreading, persistent or easily detachable hairs; auricles and oral setae usually lacking entirely in the lower sheaths; ligule very variable, usually more or less strongly convex, sometimes truncate or humped; culm sheath blade roughly triangular, about as broad at the base as the

Fig. 80. *Guadua angustifolia.* A fine clump of a thornless or nearly thornless strain of this species observed at finca Santa Julia, 24 mi south of Vinces, Ecuador. The culms reach an estimated height of 80–90 ft, and a basal diameter of about 5 in. In this plant the culms are relatively free of the long thorny branches that make a formidable barrier around the base of the clump in the typical form of the species (Fig. 79). U.S.D.A. photo by McClure.

apex of the sheath proper, persistent, appressed to the culm. Branches (in large culms, suppressed throughout the lower ½ or ⅔ of the height except the basal 6–10 nodes) solitary and very thorny at the basal nodes, usually 1 + 1 or 1 + 2 above the middle of the culm and progressively more fasciculate above. Leaf blades extremely variable in size and shape, those on young growth ovate-lanceolate to oblong-lanceolate, up to 7 × 2 in., those on old wood oblong- to linear-lanceolate, up to 8 in. long and ½ in. broad, commonly glabrous or nearly so on the abaxial surface, sparsely strewn with coarse white bristles or more rarely glabrous on the adaxial surface, sometimes glabrous on both surfaces; transverse ridges between the veins often visible here and there on the lower surface.

Infraspecific Variation. A number of interesting variants of *Guadua angustifolia* have come to my attention. The Milagro strain, mentioned on p. 275, is characterized by annual flowering. Another strain found at Pichilingue, Ecuador in 1945 under the vernacular name "caña mansa" (McClure 1955:152) differs from the typical form in the following respects: (1) weaker development of the branches and leaves at the base of the culms; (2) fewer, shorter, and blunter spines on the lower branches; (3) the generally somewhat broader, shorter leaf blades; (4) stronger tendency of the mid-culm branches to root spontaneously. To these may be added the difference between a reputed 3-year life of service given by boards made from culms of "caña mansa" when used to make drying floors, and the reputed 5 year service given by boards made from culms of the typical form of the species.

A third strain was observed at finca Santa Julia, 24 miles south of Vinces, Ecuador, in 1945 (Fig. 80). This plant is distinguished most strikingly by the almost complete absence of the long thorny branches that constitute such a formidable barrier around the base of culms of the common form of the species (Fig. 79, *1*). While stationed in Cuba, Dr. Frank Venning secured from Ecuador propagules alleged to have been taken from this nearly thornless strain. These propagules were presented to the Horticulture Department of the Estación Experimental Agronomica at Santiago de las Vegas, near Havana. Ing. Julian Acuña Galé, Chief of the Department of Economic Botany at the Station, reports (in letters dated July 14 and August 11, 1960) that the two young plants, which are growing well in Matanzas clay soil, show some small thorns on basal branches of the culms. It remains to be seen whether these plants will grow up to be as free from thorny basal branches as those observed at finca Santa Julia.

Distribution and Ecology. Native to northeastern South America, and extending into Panama in a cultivated state at least, the typical form of *Guadua angustifolia* is especially common in well-watered, fertile regions at elevations below 5000 ft, particularly in Colombia (where it is known by the vernacular name "guadua") and Ecuador (where it is known as "caña brava"). Parodi (1936: 235) reports the spontaneous occurrence of this species in Venezuela, Brazil, Paraguay, and as far south as northern Argentina.

Through the agency of the U.S. Department of Agriculture, the typical form of *Guadua angustifolia* has been established in cultivation in Florida, but the plants suffer from frost injury at temperatures below 27 or 26°F and are cut to the ground at temperatures 2 or 3 degrees lower, while at 17°F they may be killed outright (Young 1946*b*:360).

The natural habitat of *Guadua angustifolia* corresponds to superior farm land. The vast forests of pure or nearly pure stands several leagues in extent mentioned by Humboldt and Bonpland as existing in Quindiu, now the Department of Caldas, Colombia, have in recent decades been severely reduced in extent by clearing to establish cattle farms (Fig. 79). The rhizome appears to be very persistent. According to ranchers interviewed personally, repeated cutting and burning in at least three successive years is required to eradicate the plant.

Flowering and Fruiting. The flowering and fruiting habits of *Guadua angustifolia* are treated on pp. 274f.

Propagation. When new plantings of *Guadua angustifolia* are started, in the area of its natural occurrence, single-culm clump divisions are used by tradition. In trials of whole culms (aged 1 and 2 years, and 3 years or more) as cuttings, set up in collaboration with Dr. Alberto Machado in December 1949 at Chinchina, Colombia, it was found that only culms aged 3 years or more gave good yields of rooted plants (see Triana 1950). For results of similar trials at the Federal Experiment Station in Puerto Rico, see Table 5 and text, p. 234.

In the variant form "caña mansa" described above, the primary branch at mid-culm nodes, and to some extent the secondary branches, bear root primordia on their swollen, rhizomelike base. In trials in Guatemala in 1950, I found that this bamboo could be propagated readily by means of cuttings consisting of intact primary branches alone, or whole branch complements, from mid-culm nodes.

Suggested Studies. Representatives of the available variants of *Guadua angustifolia* should be included in any silviculture studies that may be carried out in tropical regions to discover plants of superior capacity to store the sun's energy in a form of special value in the human economy. Judging by personal observations made in its native habitat, it seems likely that *Guadua angustifolia* may embrace forms outstanding among the bamboos in their tolerance of poor drainage, and in the natural durability of their culms.

Comparative studies of the chemistry and physiology of sterile plants of the typical form of the species with flowering plants of the annually flowering strain should yield information of value in relation to the problem of inducing flowering in *Guadua angustifolia* and perhaps in other bamboos by artificial means.

Comparative studies of the pulping qualities of bamboos growing in the Western Hemisphere should include all available strains of *Guadua angustifolia.*

Additional References. McClure (1944:65–66, Tables 18–25) gives distribution and availability, source and documentation of test material, with results, and comments relating to performance tests of ski poles made from *Guadua angustifolia;* McClure (1945*c*, 1953) gives a picture of *Guadua angustifolia* as it appears in the present-day economy of Ecuador and Colombia.

Melocanna baccifera (Roxb.) Kurz (Figs. 55, *12,* 61, 62, 81–85, and 89, *1* and 2). Syn.: *Bambusa baccifera* Roxb.; *Melocanna bambusoides* Trin.

Melocanna baccifera is most widely known in the Western Hemi-

Fig. 81. *Melocanna baccifera: 1,* leafy flowering branch, showing some of the flowers with stigmas exserted; *2,* leafless flowering branch, showing some flowers in anthesis; *3,* culm sheath (abaxial aspect); *4,* upper part of a pseudospikelet (the prophyllum is lacking); *5,* floret, showing androecium and gynoecium; *6,* lodicule; *7,* stamen; *8,* fruit, much reduced; *9,* fruit in longitudinal section. Redrawn from Gamble 1896:Pl. 109.

Fig. 82. *Melocanna baccifera.* A plant 7 years old, "artificially raised from seed" at Dehra Dun, India. From Troup 1921:Fig. 397 (as *M. bambusoides*).

sphere as the bamboo with a large, thick-walled fruit commonly referred to as "like a pear" in size and shape (Figs. 55, *12* and 62; cf. Stapf 1904:Pl. 45, Fig. 3). In form and structure, however, the fruit actually is more like the fruit of some species of *Sechium.* In its native home (East Bengal, Pakistan), *Melocanna baccifera* occurs in vast, often pure or nearly pure stands whose yield of culms constitutes the principal material available for economical housing. In this area it is also the principal source of withes for making basketry and matting. In pulping studies made in the laboratories of Champion Papers, Inc., the culms of *Melocanna baccifera* were rated among the top half-dozen of nearly 100 selected species of bamboo available in the Western Hemisphere. For a good many years, earlier in this century, the *Annual Report* of the Forest Research Institute at Dehra Dun, India, was published on excellent paper made from *Melocanna baccifera.* In 1954, a paper mill with a capacity of 100 tons per day was put into operation at Karnaphuli, near Chittagong, East Bengal, to utilize local natural stands of bamboos, of whose yield *Melocanna baccifera* provides 85 to 100 percent (McClure 1956*a*:182); see Fig. 83. The same report states that

> The existing sources of supply (in areas of which cutting rights are held by the Pakistan Industrial Development Corporation) comprise about 696 square miles of bamboo (either in pure stand or admixed with forest trees) situated in the drainage basin of the Karnaphuli river above the mill. The potential annual yield of air-dry bamboo [from this area] on a 3-year, selective cutting cycle, is estimated at 351,360 tons. Of this about 85 percent is the muli [*Melocanna baccifera*].

With regard to the density of pure stands of *Melocanna baccifera,* Troup (1921:1011) records that enumerations carried out in Arakan in East Bengal, on two separate plots of 1 acre each, gave respective totals of 10,575 culms weighing 27,404 lb and 6,855 culms weighing 33,248 lb. The moisture content of the culms when weighed is not indicated.

In preliminary small-scale field trials, carried out in the Polochic Valley, Guatemala, two plants of *Melocanna baccifera* increased during the first 4 years of their growth from single-culm offsets at an average rate of 24,714 lb of air-dry culms (equivalent to about 6 tons of air-dry cellulose) per acre per year, calculated at a 7×7-m spacing (unpublished data developed in collaboration

Fig. 83. *Melocanna baccifera.* (*Above*) A harvesting station on the Karnaphuli River, East Bengal, Pakistan. Bundles of culms intended for paper pulp are rafted to the Karnaphuli mill site from various parts of this bamboo-producing area, over distances ranging from about 5 to 100 mi. (*Below*) From the same area, millions of culms selected for structural purposes are delivered annually to Chittagong, East Bengal, Pakistan, by raft.

between the U.S. Department of Agriculture and Champion Papers, Inc.). The rate of recovery of *Melocanna baccifera* after clear-cutting was also phenomenal, especially in young plants. However, no long-range data relating to sustained yields under controlled harvesting rates have been developed.

The Plant (Vegetative Characters) (Figs. 61, 81, 82, and 89, *1, 2*). Rhizome pachymorph, the slender neck strongly elongate (up to 3 ft long in mature plants); young clumps usually densely caespitose at first, becoming more open on the periphery as they develop; culms distant, erect or ascending, up to 70 ft tall and 3 in. in basal diameter (more commonly 50 ft tall and 1.5–2 in. in diameter), with nodding tip, the internodes cylindrical, glabrous, to 20 in. long at mid-culm, hollow, thick-walled for 3 or 4 ft at the base, thin-walled above, the nodes marked by a thin sheath scar only, not at all pulvinate or swollen above the sheath scar, glabrous, those in the lower $\frac{1}{2}$–$\frac{2}{3}$ of the culm usually without buds or branches in plants of mature stature; branches in each complement numerous, slender, subequal, the point of attachment of the branch complement small; leaf blades oblong-lanceolate, glabrous on the upper surface, glaucous and pubescent on the lower surface when young, at length glabrescent except at the tip, the inner edge densely fringed with slender hairs, the outer edge sparsely spinulose with short, sharp, thick-walled teeth; transverse veinlets (called "pellucid glands" in the literature) sometimes apparent on both surfaces, especially in young leaf blades, obscure in older ones. In shape and texture, the culm sheath is quite distinctive (see Fig. 81).

No reference to the chromosome number of *Melocanna baccifera* has been found in the literature.

Infraspecific Variation. No infraspecific taxa have been described, but fruits of three slightly different shapes that have come to my attention are presumed to be those of *Melocanna baccifera.* Field observations indicate that in the vegetative condition the plant is remarkably uniform in its morphological and physiological expression.

Natural Distribution and Ecology. Gamble (1896:119) gives the natural distribution of *Melocanna baccifera* as "throughout Eastern

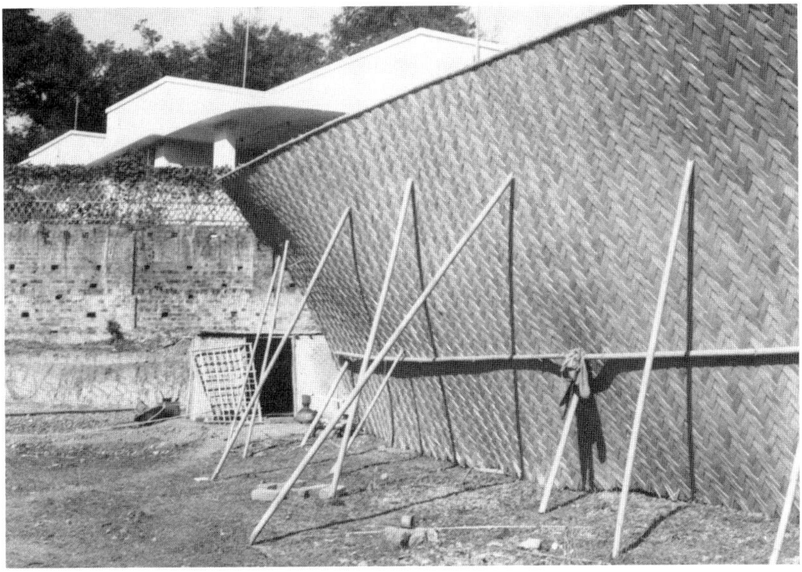

Fig. 84. *Melocanna baccifera.* (*Above*) A prefabricated wall for a house is not a modern innovation in East Bengal. This traditional version is a 10 × 30-ft unit, woven from flexible strips—flattened-out whole culms—of this bamboo, which is called *muli* in Bengalese. (*Below*) No special decoration is needed, when the natural pattern of units of construction has its own aesthetic merit.

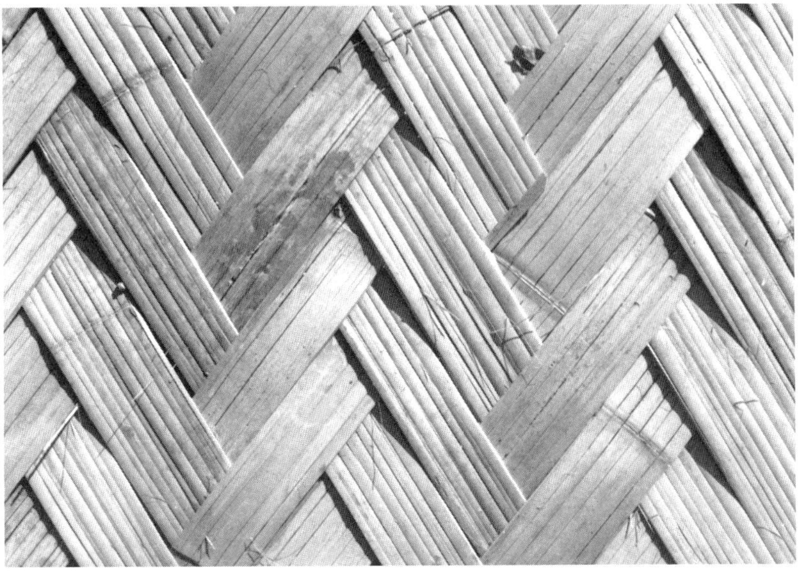

Fig. 85. *Melocanna baccifera.* (*Above*) This bamboo has culms so versatile for structural purposes that no other material is needed—even for nails or hinges or shingles—to make a beautiful and comfortable home, such as these in Chittagong, East Bengal, Pakistan. (*Below*) At the International Housing Exhibition held at New Delhi, India, in January 1954, the delegation from West Bengal presented this prize-winning home, which has many attractive features. Excepting the framework of durable hardwoods, it was constructed of two locally abundant and inexpensive materials: plastic earth and the culms of the *muli* bamboo.

Bengal and Burma from the Garo and Khasia Hills to Chittagong and Aracan, and again in Tenasserim. In parts of the above region, and certainly in Chittagong, this is the most common species, and the one most universally used for building purposes." He adds, "Major Lewis says that white ants [termites] do not touch it."

The plant apparently is in its natural home in the Hill Tracts of East Bengal. I have observed that it thrives there almost equally on the well-watered sandy clay loam alluvial soils and on the well-drained residual soils consisting of almost pure sand, even to the summits of the low sand stone hills. It springs up in practically pure stands after repeated clearings for agricultural purposes have destroyed the natural forest. The rhizomes are remarkably tenacious of life, and survive the burning *in situ* of the felled culms, leaves, and branches. The annual rainfall in the Hill Tracts of East Bengal where this bamboo thrives ranges from 78 to 120 in., with a long dry season each year from November to March. Average minimum temperatures reach a low of 50°F in January and a high of 77°F in July and August. Average maximum temperatures reach a high of 97°F in April and a low of 75°F in December. Nursery trials made in Florida with plants of *Melocanna baccifera* from fruits introduced by the U.S. Department of Agriculture early in the 20th century were unsuccessful.

In 1948, I secured propagules from a large clump of *Melocanna baccifera* near the central kiosk at the old Castleton Garden in Jamaica. Fifty of these were sent to Guatemala. Six others were sent to the U.S. Department of Agriculture, where they were accessioned under P.I. 164567. Upon their release from quarantine at Glenn Dale in 1949, the surviving plants were sent to the Federal Experiment Station in Puerto Rico. In the somewhat unfavorable soil and climate at Mayagüez the development of the plant has been less than spectacular. The material taken to Guatemala throve in the rich volcanic soil at Chocolá, where the maximum annual rainfall approaches 200 in. Plants for the field studies on rate of increase carried out in the Polochic Valley (see p. 190) were drawn from this source.

Flowering and Fruiting. The plants at the old Castleton Garden and those established in Guatemala and in Puerto Rico all began to flower in late 1957, produced a large crop of fruits in 1958, and continued to flower and bear fruit in 1959. The plant is known to be monoperiodic, however, and all of the flowering and fruiting stands have doubtless died. Fruits produced in Puerto Rico and Guatemala have been widely disseminated to centers of agricultural research in tropical parts of the Western Hemisphere.

Plants of *Melocanna baccifera* established in Honolulu, from an independent introduction of unrecorded history, flowered and fruited in 1948 and 1949 and died afterward. This lot of plants must represent a hereditary line distinct from that of the Castleton Garden plant. However, no difference has come to light, other than the date of initiation of flowering, which in this species heralds the approach of the end of a life cycle.

On the basis of other recorded occurrences of gregarious flowering and monoperiodic fruiting in *Melocanna baccifera,* its life cycle has been variously estimated at about 30 years (Gamble 1896:120), 30–35 years (Kurz 1876:257), and about 45 years (Troup 1921:984). Blatter 1929–1930:916) says: "The data at present available are not sufficient to justify any definite conclusions." The plant discovered at Castleton Garden, Jamaica, was received there, the head gardener asserted, "around the turn of the century, as a little seedling I could hold in my hand." Its flowering and fruiting in 1957–1959 gives it a reproductive cycle of about 60 years.

Propagation (Figs. 61, 62, and 89, *1, 2*). When fruits are locally available, they afford the best means of propagating *Melocanna baccifera.* However, since the seed usually germinates very promptly upon maturing (Fig. 62), even while the fruit is still in place on the branch, procurement of fruits from a distance presents special problems. The development of the plant from seed is described on p. 132. The vegetative propagation of *Melocanna baccifera* is treated briefly on pp. 216f, and illustrated in Fig. 89, *1, 2.*

Suggested Studies. Because of its great vegetative vigor, and the rapidity with which the plant establishes itself from seeds or single-culm clump divisions (offsets) giving high early yields, and in view of the versatility of the culms in industrial and handicraft applications, *Melocanna baccifera* should be included in any study of the development of a tropical source of building material, handicraft materials, or cellulose. Unpublished comparative studies indicate that its cellulose fiber ranks close to that of *Bambusa vulgaris* by the standards of the paper technologist.

Additional References. Roxburgh (1819:col. 37–38, Pl. 213) presents the original description and illustration of the species, as *Bambusa baccifera* Roxb. Trinius (1821:43) first described the genus *Melocanna* on the basis of *Bambusa baccifera* Roxb. Stapf (1904:401–425, Pls. 45–47) gives a detailed description and illustration of the morphology and anatomy of the fruit. Troup (1921:1011–1013, Figs. 396–397). Kurz (1875) published the correct binomial.

Ochlandra travancorica (Bedd.) ex Gamble (Figs. 55, *14* and 86). Syn.: *Beesha travancorica* Beddome

Since the individual plant itself is unimpressive, Beddome's reference to it (1873:235) as "a magnificent species" and Gamble's characterization of it (1896:125) as "magnificent and most interesting" probably allude to the bizarre and noteworthy features of the flowers and fruits, and perhaps to the vast extent of its natural stands.

Ochlandra travancorica first attracted the attention of the paper world by the remarkable maximum length of 9 mm to which, according to Bhargava (1946:App. 1, p. 20), individual cellulose fibers of its culm tissues attain, and by the vastness and exuberance of the natural stands that occur in South Travancore and South Tinnevelly mountains. Its culms were found by the Technological Laboratory of the Indian Central Cotton Committee to yield, by the sulfate process, a pulp containing 92.66 percent of alpha cellulose, with an ash content as low as 0.2 percent, characteristics indicating that this pulp has possibilities for rayon manufacture. Some years ago, it was announced in an anonymous article published in *Fibres* (8:82–84, March 1947) that a mill operated by Travancore Rayons, Ltd., and equipped to produce 5 tons of rayon and $1\frac{1}{4}$ tons of transparent paper daily, had recently been established in Travancore to utilize the natural stands of *Ochlandra travancorica* found there. No recent notice has come to light, either of this mill, or of the one mentioned before the turn of the century by Bourdillon, then Conservator of Forests in the Travancore State. Bourdillon is quoted by Gamble (1896:126) as saying of *Ochlandra travancorica:* "It makes a splendid paper, and we have a

Fig. 86. *Ochlandra travancorica: 1,* leafy flowering branch; *2,* culm sheath; *3,* pseudospikelet, showing one keel of the lowermost bract (prophyllum); *4,* the one-flowered spikelet, with two bracts still attached, and the flower in anthesis; *5, 6,* bracts; *7,* lemma; *8,* palea; *9,* lodicules; *10,* stamen; *11,* gynoecium; *12, 13,* stigmas, much enlarged; *14,* fruit, still surrounded by the lemma, palea, and bracts. Redrawn from Gamble 1896:Pl. 111.

mill which uses it almost exclusively. The fibre has been pronounced superior to 'Esparto.' Our only difficulty in connection with it is the great cost of the chemicals required."

The Plant. Rhizome unknown (judged by descriptions of the clump habit, the rhizome should be pachymorph); the clump (according to Beddome 1873:235) close and impenetrable; culms commonly 6–8 ft, but in favorable habitats to 20 ft tall, and 1–2 in. in diameter, straggling (even climbing) except where self-supporting by virtue of the denseness of the clump habit; branching habit unknown; internodes gray-green, rough; nodes somewhat swollen and marked with the base of fallen sheaths; 1.5–2 ft or more in length (Gamble), sometimes even to 5 ft and very thin walled (Bourdillon); leaf blades broadly oblong-lanceolate, 6–18 in. long by 2–4.5 in. broad, glabrous or slightly rough on both surfaces; rounded (often unequally) at the base into a thick, broad, somewhat concave petiole, the apex long-acuminate, twisted, bearing stiff hairs, and often scabrous; transverse veinlets submerged but often showing on the abaxial surface as "oblique pellucid glands." (Except as noted, the characters for the foregoing description were drawn from Gamble 1896:125).

Darlington and Wylie (1956:458) cite "EKJ" (Janaki-Ammal) as authority for a $2n$ chromosome count of "about 72" for *Ochlandra travancorica.*

Infraspecific Variation. *Ochlandra travancorica* var. *hirsuta,* described and illustrated by Gamble (1896:126, Pl. 111) from specimens collected by Beddome in the Travancore Hills, is differentiated from the typical variety by thicker leaves, with their margins more cartilaginous, their sheaths strewn with appressed hairs emerging from bulbous bases; and spikelets thickly clothed with light-brown velvety pubescence.

Flowering and Fruiting. Of the incidence of flowering in *Ochlandra travancorica,* Brandis (1906:685) says: "Beddome collected it in flower about 1868. I found it in flower on the Tinnevelli ghats in February, 1882, and in Travancore the species was expected to flower in 1905. Believed to die down after flowering." Gamble (1896:126) quotes Beddome, the discoverer of the species, as saying that "it flowers almost every 7 years and dies down." Fischer (1934:1863) notes that it "flowers at long intervals and dies down."

Ochlandra travancorica apparently is unique among the bamboos in the degree of duplication of its essential floral parts, especially the stamens, of which Gamble reports that as many as 120 have been found in a single flower. Other known species of *Ochlandra* show duplication of the floral parts in a more moderate degree. In common with other members of the genus the fruit in *Ochlandra travancorica* is characterized by (1) a very thick pericarp that is at first fleshy, later induresecent, and (2) a long stiff basally conical persistent beak. Gamble describes the fruit proper in *Ochlandra travancorica* as reaching a length of 2 in., with a beak of equal length (Fig. 55, *14*). When mature, the fruit retains the rachilla, the very persistent glumes, lemma, lodicules, and palea—the whole of the one-flowered spikelet (less the stamens)—firmly attached to its base.

Distribution and Ecology. Beddome (1873:235) says:

> This magnificent species is most abundant on the South Travancore and South Tinnevelly mountains, [at] 3000–5500 feet elevation, where it covers many miles of the mountains, often to the entire exclusion of all other vegetation; in open mountain tracts it generally only grows to 6–8 feet in height, but most close and impenetrable, elephants even not attempting to get through it; inside sholas [small ravines] and their outskirts it grows to 15 feet high and is much more straggling. It is called Irul by all the natives, and by Europeans the Elephant grass.

Gamble records (p. 125) that *Ochlandra travancorica* is planted in Madras and at Peradeniya in Ceylon.

Attempts to introduce it into the United States by means of seeds have been unsuccessful, but seedling plants sent by Dr. Boshi Sen from Almora, United Provinces, India, in 1951 and accessioned by the U.S. Department of Agriculture under P.I. 198012 have been established in cultivation at the Federal Experiment Station in Puerto Rico.

Propagation. No recorded experience in the propagation of *Ochlandra travancorica* has been encountered.

Suggested Studies. The culms of *Ochlandra travancorica* and those of its congeners should be studied in the paper-technology laboratory, to determine their papermaking characteristics for comparison with results already obtained from other bamboos available in the Western Hemisphere.

The high maximum fiber length of 9 mm reported by Bhargava (*vide supra*) for *Ochlandra travancorica* certainly suggests further fiber studies. However, this recorded maximum, and the minimum of 1.0 mm, standing alone, carry little weight with the discerning paper technologist. It is only the proportion of fibers falling within each length class that has significance.

Also needed are field studies to determine whether *Ochlandra travancorica* is a desirable silvicultural subject, in relation to the high- and sustained-yield requirements of a paper mill. The reputedly short flowering cycle of *Ochlandra travancorica,* and the subsequent death of the flowered culms, may, if true, be a disadvantage to its cultivation as a source of cellulose pulp. However, the plant apparently fruits freely, and there is the possibility that the development of a seedling progeny would be sufficiently rapid to restore the stand to a productive state by the time the dying flowered culms had been harvested—if the stands maintained are of sufficient extent to provide several years' supply from a single progressive-harvesting cycle. Flowered culms of *Dendrocalamus strictus* are usable, and give a slightly enhanced cellulose yield of undiminished quality, after standing in the field 4 years (Deogun 1937:115).

Additional Reference. Gamble (1896:121–128) is the most comprehensive available treatise on the known species of this genus.

5 Propagation

Bamboo heads the list of eleven new crops selected in 1957 by the President's bipartisan Commission as meriting special attention. In the words of the U.S. President's Commission on increased industrial use of agricultural products (1957:IV-4): "Bamboo has been the subject of little research in this country. Major efforts will have to be made in every phase of the program if large acreage production and use is to be developed rapidly."

The economic success of large-scale exploitation will depend to an important extent on the cultivation of elite bamboos, selected for outstanding quality and high productivity. Such traits will be found associated only in an occasional individual clone. When it is realized that the establishment of plantings to supply a modern paper mill of a size suitable for economical operation may call for the prompt production of at least a million rooted propagules from a single elite plant, the importance of developing the most efficient means of propagation becomes clear. The production of bamboo plants of uniform stature and of a size suitable for laboratory studies will also require the development of special techniques of propagation.

New bamboo plants may be produced by means of seeds, vegetative fractions, or layers. Experience has shown that each of these methods has certain advantages, and each may in certain circumstances be subject to limitations, for the propagation of any particular bamboo.

Propagation by means of seeds

The availability of bamboo seeds is conditioned on unpredictable events and circumstances. The time of fructification of any given stand cannot be foretold with certainty, and there is no assurance either that the event will be discovered or that the availability of seeds will become known to those most interested. It is at present, therefore, usually not possible to secure, on demand, seeds of a particular bamboo that may be desired for study or exploitation.

The genetic constitution of bamboo seeds from open-pollinated wild plants is also unpredictable. It may be heterogeneous, and often appears to be so, in the light of variation shown in some seedling populations. Seeds of *Bambusa longispiculata* (P.I. 93573), *Dendrocalamus membranaceous* (P.I. 74229), and *D. strictus* (P.I. 77061) from natural sources were introduced by the U.S. Department of Agriculture. The seedling populations were grown to maturity at the Federal Experiment Station in Puerto Rico. The plants of each population showed great diversity in vigor, habit of growth, and ultimate stature. Probably present also were variations in features of interest from the technical point of view. Natural seedling populations of bamboos showing genotypic diversity may afford an opportunity for selecting, as clones, individual superior plants.

Several forms of bamboo "seeds" (one-seeded fruits) are shown in Fig. 55. Seedling stages of several bamboos are described on pp. 126ff and illustrated in Figs. 57–66.

Duration of viability in bamboo seeds

Nothing approaching the duration of viability normal to the common cereals has been recorded for the seeds of any bamboo. Published information on the experimental storage of bamboo seeds to preserve their viability is, however, very meager. The two items of experience reported below are purely exploratory.

White (1947*a*) summarizes the results of his studies of seeds of *Bambusa arundinacea* as follows:

> In these tests the most practical method of preserving viability of bamboo seed was storage over calcium chloride at room temperature. Storage over hydrated lime or over charcoal was also a good method if refrigerated. Drying of the seed to a moisture content of about 12% definitely increased longevity when stored over hydrated lime under refrigeration. In other cases there was little or no advantage gained by drying. When no drying agent was used, exposed seed retained viability longer than seed sealed airtight.

Deogun (1937:117) reports briefly the results of experiments on the storage of the seeds of *Dendrocalamus strictus* carried out at the Forest Research Institute, Dehra Dun. Seeds showing an initial germination capacity of 56 percent were stored dry in sealed tins. Deogun does not give either the moisture content of the seeds when sealed up or the temperature at which they were stored. Germination after 1, 2, and 3 years was 54, 43, and 5 percent, respectively. No seeds germinated after storage in gunny sacks for 1 year, under conditions not otherwise described.

The viability of corn seed has been successfully extended by refrigeration at the Plant Industry Station, Beltsville, Maryland, and elsewhere. Methods for extending the period of viability in bamboo seeds should be extensively tested as material becomes available. The establishment of a successful "germ-plasm bank" of fully documented bamboo seeds would do much to encourage, and make possible, the initiation of a wide range of much-needed descriptive and experimental studies of valuable bamboos.

Planting the seeds

Hughes (1951:119 and personal conversation) reports that seeds of a species of *Arundinaria* native to North Carolina (probably *A. tecta*) germinated when gathered in the late dough stage, if planted and watered immediately. Fully formed seeds of *Chimonobambusa marmorea* will also germinate promptly while still green and soft, when planted in a moist medium (personal experience). White (1948:13) relates his experience with another species in the following words:

> Seed of *Bambusa arundinacea* were sown in soil at a depth of 1/4 inch and about an inch apart in rows 3 or 4 inches apart. Germination occurred in about one week and the seedlings grew quite rapidly. When the plants were 6 to 8 inches high they were transplanted to individual containers holding about one gallon of soil. Transplanting to the field was done after the plants were 2 1/2 to 3 feet in height. Growing the plants from seed is undoubtedly the most economical and convenient method of propagating large numbers of plants.

The Rivières (1879:463) advise that, in the absence of a knowledge of the viability of bamboo seeds acquired from any source, it is best to plant them at once. At the same time, they stress the need of studies to ascertain the best season for planting—should seeds of any bamboo become available in sufficient quantities. The history of the conditions under which the seeds were stored, from the time of collection until they are sown, should of course be a part of the documentation.

The potential practical advantages of propagating bamboos by means of seeds of known genetic constitution should not be neglected. The artificial induction of flowering and fruiting in this group is therefore a problem that challenges the creative imagination and merits the active collaboration of persons suitably qualified and equipped for the task. Until this challenge is successfully met, however, the handicaps inherent in the "natural" situation will continue to prevail.

Dendrocalamus strictus—the principal bamboo of economic interest in India—is the one species concerning which somewhat extensive notes on establishing plants from seeds have been published. Deogun (1937:118–119) gives general directions for the preparation of sites and treatment of the soil, both for direct sowing (to produce plants *in situ,* without transplanting) and for nursery practice. Details pertaining to quantity of seed required, shading, irrigation, weeding, and thinning are also given, but without reference to locality. Troup (1921:994) gives a more circumstantial, illustrated account of personal experience in the same matters, acquired at the Forest Research Institute at Dehra Dun.

Requirements relating to scientific and commercial objectives

Two specifications, whose importance in relation to the requirements of scientific or commercial objectives must be recognized, are: uniformity of genetic constitution, and uniformity of size, in the plants used.

The chief reason why such objectives call for vegetatively propagated material is the requirement for uniformity of genetic constitution. In contrast with seedling progenies, which are generally heterogeneous with respect to one or more characters, the vegetative progeny from a given clone may be assumed, in the absence of evidence to the contrary, to be identical with the original plant in all respects, including genetic constitution. Such evidence to the contrary has come to my personal attention in only two cases. Both involved the appearance of minor spontaneous mutations consisting of distinctive color patterns on the culms of plants propagated by the late David Bisset from rhizome cuttings of the plain green form of *Phyllostachys viridis* (P.I. 77257) at the U.S.D.A. Barbour Lathrop Plant Introduction Garden, Savannah, Georgia (McClure 1957:64–65). In any case, vegetative propagation is the most reliable means available for multiplying bamboo clones to secure genetically uniform plants for critical study or exploitation.

Whether the somatic chromosome complement in bamboos is always uniform throughout the body of the plant, either in number or in constitution, is a question that still awaits systematic investigation. Discrepant records as to somatic chromosome number have been reported for both *Bambusa arundinacea* and *Dendrocalamus strictus,* and for hybrids between *B. arundinacea* and a sugarcane variety produced at Coimbatore, India (Janaki Ammal 1938:925; Richharia and Kotwal 1940:1033; Parthasarathy 1946:234).

Uniformity in size of plants is an important—really essential—specification for material to be used for replicated treatments in all experimental studies, whether carried out in the laboratory or in the field. Moreover, uniformity of size has a financial value in relation to the problem of packing and handling large numbers of plants.

Small versus large plants

The principal advantage of large, rooted propagules (such as clump divisions) over small ones, is in the economy of time required for them to reach mature stature. Personal experience confirms this in situations where the prompt establishment of a few large plants is an urgent objective. However, as the distance through which the plants must be transported becomes greater, certain disadvantages of large propagules become more noticeable. And, of course, large plants are unsuitable for most laboratory studies.

The advantages of small, rooted plants, and the importance of their potential role in experimental studies, seem not to have been discussed in print. In 1946, I requested of the Federal Experiment Station at Mayagüez 50 plants each of several species of bamboo to be shipped, by air, to each of five Agricultural Experiment

Stations in Latin America. They were to form a nucleus for experimental work on the potential utility of these species of bamboo in relation to the local economy of the respective countries. The director then in charge of the station at Mayagüez informed me that to fill this request would exhaust the station's supply of some of these bamboos, and deplete its bamboo budget as well, since one man-day of labor was required to prepare five propagules (clump divisions), and the air-freight rates on such heavy propagules as were then being produced by the station made the cost of shipment by air prohibitive.

Collaborative studies were later carried out at Mayagüez, with a view to producing rooted plants in the size and numbers required. Well-rooted little plants of several species, weighing from 1 to 4 ounces, were produced in large quantities. Packages containing 100 to 200 of such plants, isolated from each other in moist sphagnum, and with the roots protected by a coating of colloidal material, were light enough to be handled and shipped cheaply. Moreover, the thousands of plants produced made no appreciable inroads on the available supply of propagating material. We did not achieve the goal of uniform size, and we did not find wholly satisfactory methods for propagating some of the species selected for trial.

Brief progress reports on some of these results were published in the *Annual Report* of the Federal Experiment Station for the years 1950–1953, and by McClure and Kennard (1955). Selected details are presented on pp. 229ff. A standardized procedure for producing, at will, very small plants of uniform and specified size, by vegetative means of propagation, remains to be perfected. It seems likely that the method ultimately adopted for certain species will include: (1) the use of single-bud propagules with a very much reduced amount of accompanying mother tissue; (2) the use of an artificial nutrient medium with the possible addition of synthetic hormones; and (3) the provision of accurately controlled conditions of incubation, including at least temperature and moisture. As compared with "natural" conditions—that is, where relatively large propagules are subjected to only nominal or partial controls—this level of refinement of methods should reduce significantly the influence upon the end results that ordinarily is exerted by variables such as innate differences between species and between groups of species, and even differences due to topographical origin of buds and time of year when the propagating material is severed from the mother plant.

Two groups of bamboos requiring different propagation procedures

Most of the relatively well-known bamboos have been observed to fall into two large groups that contrast more or less sharply with each other in relation to details of procedure appropriate to their vegetative propagation. The Rivières (1879:460ff) were the first to call attention to the existence of this natural grouping of the bamboos, which they discovered during the course of their studies in propagation at Hamma, Algeria. The two groups are characterized by the Rivières as follows:

1. Caespitose bamboos, with autumnal growth, and rhizomatous branches in fascicles. [Examples:] *Bambusa macroculmis, vulgaris, vulgaris vittata, hookeri, spinosa, stricta, argentea, gracilis, scriptoria.*
2. Running (rarely caespitose) bamboos, with spring growth and branches paired or in fascicles. [Examples:] *Bambusa* or *Phyllostachys mitis, quilioi, nigra, viridi-glaucescens, aurea, flexuosa, violascens; Arundinaria simoni, japonica, falcata, fortunei.*

Some of the species listed in each group now bear other specific names, and are disposed in other genera than those indicated here. The synonymy is listed in the index to scientific names.

The differences between the two groups as enumerated by the Rivières relate to the particular array of bamboo species with which they were working, and to the ecological conditions prevailing at their experiment station. The enumeration of the characteristics of the two groups given in Table 1 is based on personal experience acquired through experimental studies carried out in both the Eastern and the Western Hemispheres, and under both temperate and tropical conditions.

Among the genera whose constituent species possess characteristics that prevent their being classified as belonging strictly to either of the two groups described above are *Chimonobambusa, Chusquea,* and *Sinarundinaria,* as currently defined.

Significance of the differences between the two groups in relation to vegetative propagation

In general, given material of the proper age, bamboos of both

Table 1. Two groups of bamboos, requiring different propagation procedures.

Characters	Expression of characters	
	Group I	Group II
Climatic adaptation in relation to temperature	Tender plants that thrive best under frost-free conditions; some species are known to survive temperatures a few degrees below 32°F without serious damage	Frost-hardy plants that thrive best in climates with a marked, but not extremely cold, winter; a few species can survive temperatures a little below 0°F without serious damage
Culm initiation (active growth following breaking of buds on the rhizome) under natural conditions	Takes place typically during the summer or autumn, or at the beginning of a rainy season following a relatively dry period; apparently is controlled primarily by moisture levels	Takes place typically in the spring, at the onset of favorable levels of temperature; apparently is controlled primarily by temperature levels
Rhizome Form of the constituent axes	Pachymorph, *i.e.*, short, thick; internodes asymmetrical, broader than long (Fig. 2)	Leptomorph, *i.e.*, long, slender; internodes symmetrical, longer than broad (Fig. 3)
Form of the lateral buds	Dome-shaped; the apex intramarginal	Boat-shaped; the apex distal
Growth habit of individual axes	Determinate	Indeterminate
Clump habit typically	Caespitose; the plant a single dense tuft of culms	Diffuse; plant with culms distant from each other
Culm origin	Distal to the rhizome	Normally lateral to the rhizome
Culm branches (midculm range)	Basally swollen; recapitulating the form of the rhizome; the dominant ones often bearing root primordia of spontaneous origin *in situ*	Basally not swollen; not recapitulating the form of the rhizome; not known to bear root primordia of spontaneous origin *in situ*
Transverse venation of leaf blades	Usually obscure	Clearly manifest
Typical genera	*Bambusa, Dendrocalamus, Elytrostachys, Gigantochloa, Guadua, Oxytenanthera*	*Arundinaria, Phyllostachys, Sasa, Semiarundinaria, Shibataea, Sinobambusa*

groups respond most favorably to propagation by means of clump divisions, since such propagules are, or may be, in every respect complete plants. Of the vegetative fractions that are not complete plants, rhizomes of the right age range may generally (but not always) be counted on to give the greatest assurance of favorable response to conventional procedures. Other types of vegetative fractions, which range in size from whole culm cuttings to branch cuttings, represent larger or smaller arrays of buds from the aerial parts of the plant. It is in such propagules that the differences in behavior as between bamboos of the two groups are most marked under traditional methods of propagation, where the control of the environment is nominal.

When we have learned enough about the physiology of bud dormancy and awakening, nutrition, and tissue differentiation in the bamboos to be able to provide the right cultural conditions, it may be possible to establish a new plant from any viable bud from any bamboo whatsoever.

Clump Divisions. Clump division is the traditional, and perhaps the most generally prevalent, method of propagating bamboos vegetatively.

SEASON OF PREPARATION. Active growth of young shoots from buds on the rhizome usually is initiated during the summer in bamboos of Group I, and during the spring in bamboos of Group II. The commonly recommended practice is to process vegetative propagules just before the initiation of the annual period of active bud growth in the axes involved. This applies particularly to clump divisions. Apparently there is some sentiment in Great Britain (Thomas 1957:248) in favor of dividing the clump after the buds on the rhizome have begun to push, or at any time when the plant is in active growth. There are no published data of a scientific nature to support this practice, but it should be investigated, since a fuller knowledge of the species involved, the details of procedure, and the underlying reasons for the practice may reveal useful leads. Ecological conditions peculiar to some parts of the British Isles may be a critical factor favoring the practice.

SIZE OF PROPAGULE. The simplest method is the division of a clump into two equal parts, retaining the root system, branches, and foliage of each part as fully intact as possible. Provided they are properly set out, such propagules generally give the highest degree of success in terms of survival rates and rates of subsequent development, and are least exacting in respect to care after planting. However, as the scale on which propagation is needed increases, the prodigal use of large propagules sets a limit to the scale of operations by exhausting the reservoir of material.

At the other extreme of size in clump divisions are propagules each composed of the lower part of a single culm with the rhizome axis basal to it. These are designated by Deogun (1937:121) as "offsets." Offsets have many advantages over larger propagules; namely, ease of preparation, accessibility of material of optimum age, and economy of material. Their principal disadvantages are their more exacting demands in respect to care after planting and lower survival rate. Single-culm clump divisions (Fig. 89, *1*) are the traditionally preferred method of propagation for certain bamboos: *Dendrocalamus strictus* (cf. Deogun 1937:121) and *Bambusa tuldoides* (see pp. 000ff). For some bamboos, however (for example, *Bambusa textilis*), single-culm propagules have not, in my experience, given as good survival rates or yields as offsets consisting of two culms (2-year-old mother and 1-year-old daughter) left attached together.

AGE OF MATERIAL. In small clump divisions, the age of the rhizome appears to be of critical importance.

According to Deogun (1937:121) 1- or 2-year-old offsets of *Dendrocalamus strictus* give superior results, while propagules consisting of material 3 years or more in age give progressively poorer results.

Hageman *et al.* (1949) describe the successful use of dynamite to reduce the labor of breaking up old clumps of large bamboos for propagation. The indiscriminate use of such material, without discarding the older parts, gave very poor results in terms of survival rates in afforestation projects carried out by the Tropical Forest Experiment Station in Puerto Rico. This is to be expected, since many of the propagules consisted wholly of material much more than 3 years old.

PREPARATION OF CLUMP DIVISIONS. The critical point to keep in mind in separating a clump division from the mother plant is the importance of severing the rhizome at the right place in its complex branching system. Since the rhizome system is usually completely hidden under the soil, the first step to be taken is to learn the position and orientation of the various connected axes of the rhizome that are to be included in the propagule. This involves exploratory excavation. With experience, one learns where and how to find the place where a cut is to be made.

Clump divisions taken from the periphery of the clump are apt to give superior results, because they can be selected so as to contain material of optimal age, and because they lend themselves more readily to the use of proper methods of extraction.

In the preparation of clump divisions, a distinct procedure in cutting the rhizome is required by bamboos of each of the two groups. In Group I, the rhizome should be severed only at one point—at the neck (Fig. 2) of the oldest rhizome axis in the propagule. The cut should be made at the slender neck in order to minimize the damage to the rhizome and keep the raw surface as small as possible. Moreover, the tissues at this point appear to have a greater resistance to decay. This may be due to the presence of a higher proportion of lignified tissue in relation to parenchyma, and a smaller amount of stored food, in the tissues of the neck than in the tissues of the rhizome proper.

In Group II, either one or two cuts may be needed (Fig. 87, *C*), depending upon the orientation of the propagule to the rest of the plant, and the rhizome need not be severed at a neck.

Once the composition of the propagule to be lifted has been determined, and the rhizome has been severed as described, the aerial part of the propagule may be pruned before the roots are severed—or promptly thereafter—in order to minimize loss of water through the leaves. It is advisable to retain as much foliage as may, by ample irrigation and protection of the propagule from sun and wind, be kept from wilting until the root system is reestablished.

The root system of the propagule should be preserved as nearly

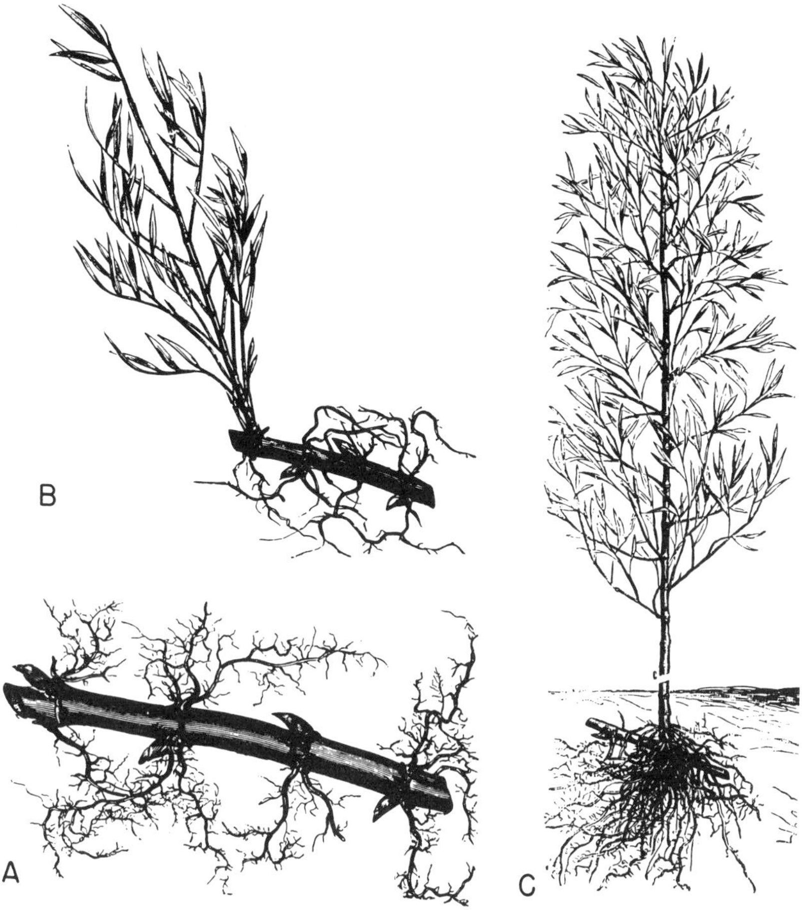

Fig. 87. Propagation of bamboos of Group II. *Phyllostachys viridi-glaucescens,* propagated (*A, B*) by means of rhizomes alone: (*A*) rhizome segment ready for the propagating bed, as illustrated by A. and C. Rivière 1879:Fig. 12; (*B*) the same propagule, represented (*ibid.,* Fig. 13) as the "result" of this method of propagation. As explained on p. 226, the propagule as illustrated had not yet produced a rooted plant. (*C*) Propagation by means of single-culm clump division. The details illustrate certain characteristics of bamboos of Group II that have significance in relation to the method of preparing clump divisions of bamboos with leptomorph rhizomes. From A. and C. Rivière 1879:Fig. 10.

intact as possible. The roots are best preserved and protected by keeping them in a ball of earth when the propagule is taken from the mother plant.

PRACTICAL EXPERIENCE—GROUP I. *Bambusa tuldoides.* Notes on the course of development, for a period of 6 years, of a small experimental planting of 1-year-old, single-culm propagules (offsets) of *Bambusa tuldoides* (as *B. breviflora* Munro, Lingnan University Bamboo Garden No. 2669) were published by McClure (1938*b*). Eighty-seven out of 90 propagules (96 percent) survived. The over-all annual increase in number, and the maximum dimensions, of new culms produced each year, are shown in Table 2.

At the end of the second season the ratio of new culms to older ones stood at a high of 4.77. The ratio fell off sharply in the following year, and continued to decline. The removal of 167 single-culm propagules in 1933, and 125 in 1934, apparently was responsible for an appreciable exaggeration of the rate of decline in the ratio of new culms to old, but even assuming that the ones that were removed would (if they had not been removed) have produced new culms at the same rate as the ones left behind, the

Table 2. Annual increment of culms from 87 single-culm clump divisions of *Bambusa tuldoides* planted in April 1931 at Canton, China; data from McClure 1938*b*.

Year	Maximum culm dimensions: Height (m)	Maximum culm dimensions: Diameter (cm)	Number of new culms	Total number of culms at year end	Number of single-culm offsets removed	Ratio of numbers of new culms to old
1931	3.0	2.2	87	87[a]	—	1.00
1932	6.0	4.0	415	502	167[b]	4.77
1933	7.5	5.3	348	683	125[b]	1.04
1934	12.0	5.8	333	891	—	.59
1935	12.0	5.8	281	1172	—	.31
1936	12.0	6.1	193	1365	—	.16
1937	12.0	6.2	258	1623[c]	—	.18

[a] This figure represents new culms only, and excludes the original propagules; the ratio of new culms to propagules happens to be 1 for the season during which the planting was made.

[b] Culms removed in the following March, that is before the season's production of new culms of that year began.

[c] Counting 292 one-culm propagules removed, the total number of culms produced is 1915.

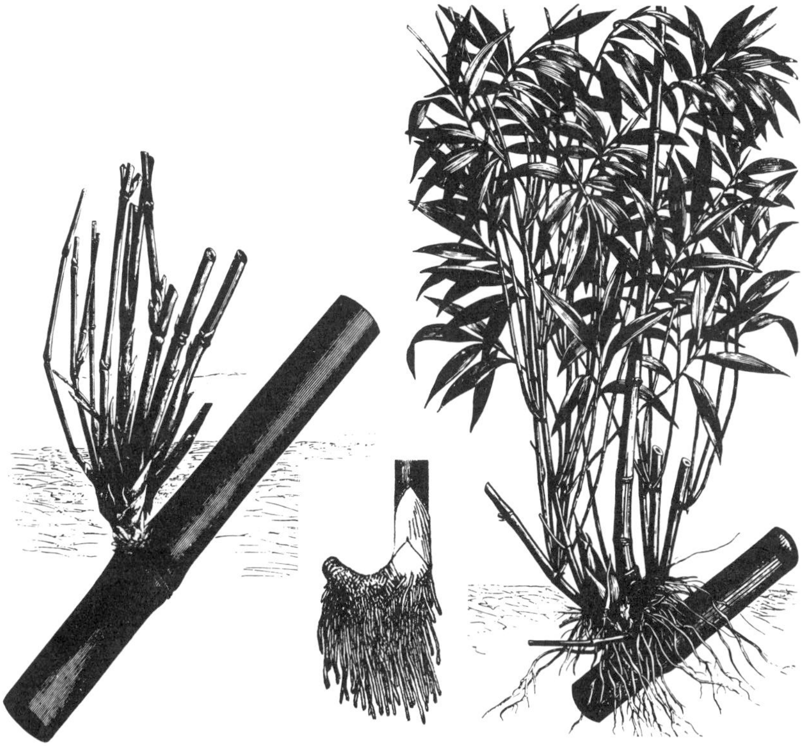

Fig. 88. Propagation of bamboos of Group I. (*Left and right*) An unidentified species of *Gigantochloa,* illustrated by A. and C. Rivière 1879:Figs. 6 and 7 (as *Bambusa macrocalamis*), propagated by means of culm segments: (*left*) culm segment bearing a young branch complement, prepared for propagation; (*right*) the principal branch has developed roots and has given rise, from one of its basal buds, to a young shoot which may become a rooted culm, thus establishing a new plant. See p. 258. (*Center*) Propagation by means of single-culm clump divisions: *Bambusa tuldoides,* cultivated at Lingnan University, Canton, China, under LUBG no. 2669. Original.

ratios given in the table for the years 1934 and 1935 would still have declined progressively.

The performance of a single propagule is shown in Table 3. Although this plant performed outstandingly in the first 2 years, it showed in later years a steady decrease in the ratio of new culms to old that is similar to that shown by the whole array.

It appears that, for this species, under the particular conditions of this experiment, (1) 1-year-old, one-culm propagules of this

Table 3. Annual increment of culms from one single-culm clump division of *Bambusa tuldoides* planted in April 1931 at Canton, China. Data from McClure, 1938*b*.

Year	Maximum culm dimensions: Height (m)	Maximum culm dimensions: Diameter (cm)	Number of new culms	Total number of culms at year end	Ratio of numbers of new culms to old
1931	3.0	3.0	4	4[a]	4
1932	5.0	4.6	12	16	3
1933	7.5	5.0	18	34	1.1
1934	12.0	5.8	18	52	0.5
1935	12.0	5.8	21	73	.4

[a]This figure represents new culms only, and excludes the original propagule.

bamboo give excellent results, and (2) the best time to break up the plants for further propagation by clump division would be in the spring, exactly 2 years after the propagules were put in. This indication should, of course, be checked by repetitions of the experiment.

Dendrocalamus strictus. Deogun (1937:121) says that offsets (clump divisions) provide the only method that has given success in the vegetative propagation of *Dendrocalamus strictus.*

> The offsets should be prepared from one-season-old culms as far as possible and, never from more than 2-season-old culms. It is necessary that some portion of the rhizome with a bud be kept and care taken, when planting, not to injure the 'eyes.' The culm may be cut down to 2′ or 3′ . . . The degree of success attained is very variable, but may be as high as 100% . . . The success of the method depends in part on the vitality of the rhizome stock used and the time of year when it is planted. If the rhizomes are taken from young healthy stock and planted immediately at the break of rain, success may be expected, but if the rhizomes are taken from old stock and planted much before the rains, complete failure may result. The weather of the year must also be an important factor.

Melocanna baccifera. The vegetative fraction of the plant that has proved most convenient for the propagation of *Melocanna baccifera* is a single-culm clump division. These should be made from the youngest culms, while the lateral buds of the rhizome are still dormant, or before they have pushed more than 2 or 3 in. Most of the culm, and the long, slender rhizome neck, may be discarded, for convenience. This leaves a propagule scarcely distinguishable from a rhizome cutting. Figure 89, *1, 2* shows a typical propagule, and the growth produced by such a propagule in 24 months. Forty-five out of 50 such propagules planted in May 1948 at Chocolá in Guatemala survived and produced vigorous plants.

Oxytenanthera abyssinica. The Nyasaland Forestry Department, Zomba (1944:11–12) gives an excellent account of the techniques developed locally for the successful propagation of *Oxytenanthera abyssinica* by single-culm clump divisions.

Rhizomes Alone. GROUP I. Published knowledge of the details of the procedures and precautions relating to the propagation of bamboos of this group by means of rhizomes alone is very meager. The following references are typical. Dabral (1950:313) says: "The best method, established by comparative experiments and statistical analysis by the Provincial Silviculturist of Madras, is by rhizome planting." The results of these experiments apparently have not been published. Dabral does not mention any species or any details of the method he recommends, without reservation, as the best.

Ahmed (1956:530) indicates that *Bambusa tulda* is successfully propagated in connection with afforestation projects at Singbhum (Bihar, India) by means of rhizomes planted *in situ.* "With slight care 80% success has been achieved; 95% survival is not uncommon." No details of procedure (for example, age of material, preparation of propagules, season of planting, or type of care) are given.

In the recorded cases where propagation of bamboos of Group I has been accomplished by means of rhizomes alone, these were planted directly in the field. For this purpose rhizomes have the advantage of greater convenience, being lighter and less bulky than clump divisions. However, systematic trials should be carried out to establish whether offsets may have certain advantages—for example, in the possible retention of some foliage, and the presence of stored food in the tissues of the culm to nourish new growth.

My own interest was directed to the use of rhizomes alone for propagating bamboos of this group when I observed the spontaneous production of small rooted shoots on fragmentary and mutilated

1

2

3

4

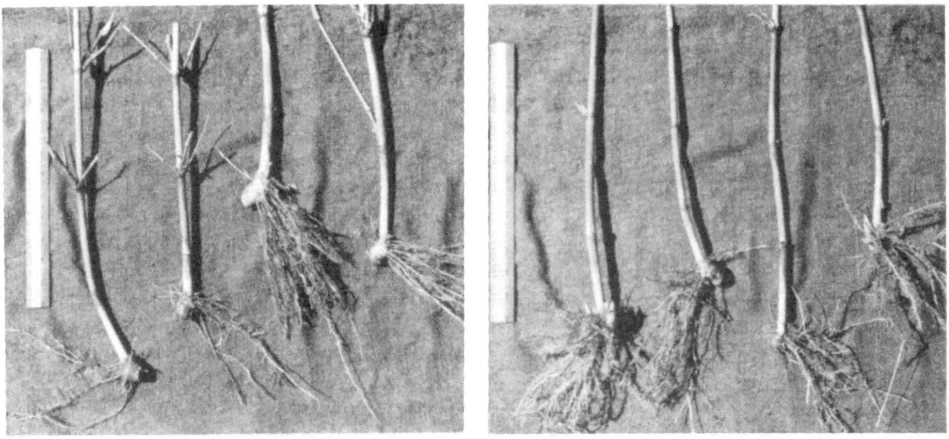

5

Fig. 89. Propagation of bamboos of Group I.

1, 2. Propagation by means of single-culm clump divisions, *Melocanna baccifera. 1.* One of 50 propagules brought from Jamaica in May 1948 for cultivation at Chocolá, Guatemala, under P.I. 164567. *2.* Plant developed from a similar propagule in 24 months; in time, the long rhizome necks will give rise to an open clump habit (see text, p. 36).

3–6. Propagation by means of rhizomes alone. *3, 4. Bambusa tulda,* cultivated at the Federal Experiment Station, Mayagüez, Puerto Rico, under P.I. 21002; two rhizome cuttings, both removed from the mother plant at approximately the same age (under 1 year) that have been in the propagating bed for 6 months. Typically, as shown here, the uppermost bud of each cutting (a bud actually borne at the base of the lowermost internode of the culm) awakens promptly and produces a plant, the first axis of which usually is poorly rooted. As a rule, the buds borne on the rhizome proper awaken more tardily, but produce plants all culms of which, including the first, are usually well rooted. The divergent behavior of buds at different levels of the propagule is construed as being related to differential levels of auxin concentration. The disparity in the incidence (and promptness) of pushing of the buds, in the rhizome proper of these two ostensibly identical propagules is not understood. Conceivably, it could be related to an unrecorded difference in the actual age of the two propagules at the time of the removal from the mother plant. *5, 6. Gigantochloa apus,* cultivated at the Federal Experiment Station, Mayagüez, Puerto Rico, under P.I. 99573. Rooted culms from the top bud of rhizome cuttings (actually the lowest bud of the culm itself) removed at 9 months from the time the rhizomes were put in the propagating bed. *5.* Plants that were topped and defoliated at 6 months, while still attached to the rhizome. *6.* "Check"—plants that were not topped or defoliated at 6 months. The defoliation of the culms was intended to break the dormancy of the buds of the rhizome proper. It did not have the desired effect. The difference in root development as between the two lots of plants is typical also of other pairs of lots from other species similarly treated, and is construed as being related to the active production of auxin by the foliage of plants in the check (untopped) lots, and the cessation of auxin production in the defoliated lots. Both lots of plants shown here were exposed to darkness for 12 days just prior to the taking of these pictures. This treatment stimulated (or at least was accompanied by) fresh root development in both lots—less in the defoliated lot (*5*). It will be noted that the superior root development in the check lot (*6*) is associated with the recent awakening of basal buds in plants 1 and 4. This is construed as possibly related to the active removal of auxin in the vicinity of basal buds during the recent active production of new roots.

rhizomes of *Bambusa longispiculata* (P.I. 93573) that had been discarded when the clump divisions of this and other large species of clump bamboos were prepared by the wholesale method formerly employed by the Federal Experiment Station in Puerto Rico. The viable buds of these fragments developed into small plants while the rhizomes were lying in the open or partly covered by dębris.

In 1949, I set up an experiment (with F. Montalvo) at the Federal Experiment Station in Puerto Rico to determine how rhizomes less than 1 year old (with all buds dormant) would respond as propagules for the production of small, rooted plants. Data on the array of material tested, dates planted, initial survival rates, and so forth, are given in Table 4. Figures 90 and 91 illustrate the preparation and handling of the rhizomes of *Bambusa longispiculata*. Figure 89, *3–6* shows some of the responses of two other species to propagation by rhizomes alone.

It seemed logical to begin with single rhizomes, taken from the periphery of the clump. Apart from the matter of convenience, there is no reason to limit the size of the propagule; in fact, units embracing two or more axes left attached to each other might well give better results in some bamboos. However, anyone who has excavated the rhizome system of a large bamboo will understand that the simplest procedure is to take away single rhizomes from the periphery of the clump. Further penetration into the mass of rhizomes greatly complicates the task, and increases the labor required to prepare each propagule. Rhizome propagules

Table 4. Initial response of five species to propagation by rhizomes alone.

Species	P.I. No.	Number of rhizomes planted	Date planted, 1949	Survival (percent)	Average height of shoots June 10, 1949 (ft)
Bambusa longispiculata	93573	150	Mar. 18, 31	98.10	6.5
B. textilis	80872	150	Apr. 1–7	85.33	2.5
B. tulda	21002	156	Apr. 1–7	94.23	4.5
B. tuldoides	21349	102	Apr. 1–7	99.00	4.0
Gigantochloa apus	99573	100	Apr. 1–7	100.00	3.5

embracing more than one axis would be cumbersome to handle and uneconomical of space in the propagating bed.

In the present study the age of the material was automatically fixed at less than 1 year, because the peripheral rhizomes must be processed before their lateral buds have begun to push—and event that takes place about 1 year after their own emergence. The rhizomes used were lifted in March and April, and were, therefore, about 9 or 10 months old.

The principal concern in this experiment was the production of small, rooted plants for shipment by air to experiment stations on the mainland of tropical America. As it turned out, the processing of the plants proved to be such a pressing task that the detailed statistical aspects of the study had finally to be abandoned to give way to the selection, processing, and shipping of the plants that met the specifications "small" and "well rooted," and the recording of notes on a small array of plants given special treatment.

The following major problems were encountered in the course of this exploratory study: (1) meager development of roots on culms arising from the distal buds of the rhizomes; (2) decay of the rhizomes; (3) slowness of the more proximal buds of the rhizomes to break dormancy.

Meager development of roots. The distal buds, located at the point at which the rhizome axis is transformed into a culm, were the first to develop. The growth of buds at more proximal positions on the rhizome was very much delayed, and most of the shoots came up singly, at long intervals. Most surviving rhizomes still had dormant buds after having been under observation for 3 years.

One of the first difficulties encountered was the weak rooting of the first shoots to appear from the rhizomes. These first shoots appeared promptly and had made a good showing by June 10, 1949 (see Table 4). However, they were too sparsely rooted to be trusted to survive shipment without first having been established in a nursery bed and allowed to produce additional, more adequately rooted culms.

As a device to get new, more fully rooted culms, a new treatment was introduced into the experiment June 10–18. A few plants in each species were topped and defoliated by cutting back each culm (while it was still attached to its rhizome) just above the

Fig. 90. Preparation of rhizome cuttings for the propagation of bamboos of Group I; *Bambusa longispiculata,* cultivated at the Federal Experiment Station, Mayagüez, Puerto Rico, under P.I. 93573. (*Above*) First step; severing the rhizome at the slender neck region. (*Below*) Second step; separating the rhizome from the culm.

Fig. 91. Preparation of rhizome cuttings for propagation of bamboos of Group I; *Bambusa longispiculata*, cultivated at the Federal Experiment Station, Mayagüez, Puerto Rico, under P.I. 93573. (*Above*) Third step; trimming the roots on the lower side of the rhizome so that it will lie flat and stable in the propagating bed. (*Below*) Fourth step; arranging the rhizomes on a layer of the propagating medium. The covering layer is still to be added.

second above-ground node and by trimming the remaining branches to a length of about 2 in. This left the plants entirely devoid of foliage. This was done on the theory that it would reduce the amount of auxin reaching the base of the plant, and thus release the buds of the rhizomes of these axes from the inhibition that auxin is known to induce in vegetative buds—in the dicots, at least. The remaining plants were left intact in all respects, as a check. The material was then rearranged in the beds so as to introduce another variable into the environment of the defoliated material: partial shade *vs.* full sun. Partial shade was provided by alternating the rows of defoliated material with rows of leafy material that overtopped and partially shaded it.

The plants were examined again 14 weeks later (mid-October 1949) and the following observations were made:

(1) No appreciable number of new shoots had emerged from the rhizome of either the intact or the defoliated culms;

(2) The branch buds on the defoliated culms in partial shade had all remained dormant, while those exposed to full sun had produced a dense canopy of leafy branches, in all of the species;

(3) There had been no appreciable change in the number of roots on either the defoliated or the intact culms.

Fourteen defoliated single-culm plants of *Gigantochloa apus* (P.I. 99573), with branch buds still dormant, and 17 intact single-culm plants of the same species, all with dormant branch buds, were removed from the rhizomes and stored in complete darkness, wrapped in burlap and plunged in moist cocopeat. After 1 week the plants were removed and examined. It was observed that the intact plants had dropped their leaf blades, and a considerable number of branch buds of both groups had begun to push. The culms of the two lots of plants were then partially uncovered and given subdued light and aeration. At this time there was only a slight intimation of the differences that showed up dramatically 1 week later, by which time the defoliated plants had made almost no additional root growth, while the intact controls had made a conspicuous development of new root branches (Fig. 89, *5, 6*).

Decay of the rhizomes. The decay that gradually developed in rhizomes of each group necessitated the discarding of some propagules each year. Since adequate provision had been made for drainage in the propagation beds, this susceptibility to decay fungi is attributed primarily to the relatively immature condition of the whole array of rhizomes. However, there are indications of interspecific differences in susceptibility.

Gigantochloa apus, whose rhizome tissues are relatively soft, proved to be the most highly susceptible to invasion by rot fungi, and many of its rhizomes were also invaded by some unidentified microorganisms that may have been responsible for the death of many of the young rooted plants that occurred while they were still attached to the mother rhizomes. *Bambusa longispiculata* was more susceptible in these respects than the other species of *Bambusa,* but somewhat less so than *Gigantochloa apus.* It is interesting that these were the two species that, by their superior vigor and the relative promptness with which their buds pushed, seemed most suited for propagation by rhizomes alone.

In later trials we found that sterilizing the freshly cut surface at each end of the rhizome with a 10-percent aqueous solution of Chlorox for 5 minutes and then, after they were dry, sealing the ends with melted paraffin reduced the losses from rotting and from disease to a negligible percentage, and also increased the period during which the rhizomes continued to produce new plants.

Slowness of rhizome buds to break dormancy. With the exception of occasional rhizomes, as exemplified in Fig. 89, *4,* the outgrowth of the buds in all of the species studied was very slow (Fig. 89, *3*), dragging out to as much as 3 years. This resulted in the expenditure of much time and labor in lifting the rhizomes repeatedly to remove single young rooted plants before they became too large.

The problem of breaking the dormancy of the buds of the rhizome is one that remains to be solved. A treatment that would induce all the buds to grow out promptly after the rhizomes are put into the propagating medium probably would solve the problem of decay as well. Moreover, causing several plants to arise from each rhizome at one time should decrease the average size of the rooted plants produced, and reduce the labor involved in the repeated lifting of the rhizomes to remove rooted plants. These are both desirable objectives where the shipment of large numbers of small, rooted plants by air, or the establishment of a very large commercial planting, is contemplated.

PRACTICAL EXPERIENCE—GROUP II. The Rivières (1879:505) were the first to describe—in a Western language, at least—the propagation of bamboos of Group II by means of rhizomes. In their very brief dissertation, they stress the following general conditions for success: (1) the use of only young material; (2) the use of pieces at least 15 to 20 cm long, with three or four nodes, each with an intact bud; (3) irrigation and cultivation as needed. Planting times suggested are April for cold climates; "winter" for warm climates. A planting depth of 10–15 cm and a spacing of 25 cm are suggested. No data relating to performance of different species are given. Figure 87, *A* shows a rhizome section of *Phyllostachys viridi-glaucescens* illustrated by the Rivières as ready to plant. Figure 87, *B* shows an allegedly rooted plant ("the result") produced by this method. However, careful scrutiny reveals that the propagule pictured here is far from having produced a truly rooted plant. The roots below the leafy shoot have emerged from the rhizome only. Not until a new rhizome axis has been developed and is giving rise to other rooted leafy shoots (culms) will a new plant have been established.

Tsuboi (1913:trans. 2–4) describes the Japanese method of processing rhizomes for the propagation of bamboos that, by inference, can belong only to our Group II, although he mentions neither genus nor species. Two sources or rhizomes are given: (1) those that are found in their normal position under the ground, and (2) those that emerge from the ground either within the grove or at the edge of a grove where the land falls away sharply at the brink of a ditch or a river bank. Those portions of an intact rhizome that have been exposed to light and air for some time are described as the easiest to get and safest to use. The use of healthy material 2–3 years old with intact buds is recommended. The rhizomes are cut into pieces 45–60 cm long. Tsuboi here stresses an important point, namely, keeping the material moist from the time it is dug up until it is put into the propagating bed. When the propagating bed is ready, the cuttings are planted in a horizontal position at intervals of 17.5 cm and covered to a depth of 7.5 cm with fine soil which is mixed with straw ash as a fertilizer, then firmly packed, and watered. Straw mulch is added to conserve soil moisture, and a shade is constructed over the bed to protect the young shoots from the full force of the sun's rays.

According to Tsuboi, gravelly sand is preferred to loamy or clay soil for the propagating bed, as the heavier soils are likely to induce decay. The moisture-holding capacity of the sandy soil is said to be improved by the use of liquid manure, which should be applied as the young shoots develop. In view of damage to the young roots that may result from the movement of the culms by the wind, the author recommends topping the culms, leaving only 5 or 6 branches. Removal of any plant from the nursery must be deferred until it has developed its new rhizome.

Tsuboi's directions are probably the most complete in print and appear sound in all respects. They need to be qualified—or rather supplemented—in only one important detail. They apply only to those bamboos of Group II that respond readily to propagation by rhizomes alone (for example, species of *Phyllostachys*). On the other hand, for *Arundinaria amabilis* (Fig. 92) and some species of *Sasa* and *Indocalamus*, rhizomes alone, without rooted culms attached, give very indifferent results at best, and often fail entirely to respond to the conventional conditions of propagation described by Tsuboi. Bamboos of Group II that do not propagate readily from rhizomes alone are generally characterized by (1) the frequent occurrence on the rhizome of nodes without buds and (2) a generally marked sparsity of roots on the rhizome. The underlying physiology of these characteristics may be related to the poor reproductive performance of the rhizomes of these bamboos.

Whole Culms. It is not now known when the first trial or demonstration of this method was made, but Kurz (1876:268) says that

> the propagation of bamboo from a practical point of view can be effected . . . by taking whole halms [culms] with their roots and burying them length-wise in the ground. By this process the alternating branch clusters send forth young branch shoots which gradually become transformed into stronger and stronger halms in the proportion as roots are formed. Thus large areas can be planted with little trouble.

Kurz adds that the propagation must be carried out at the onset of, or during, the rainy season. He gives no details. His observations were confined to bamboos of Group I.

It is not clear, from Kurz's word "with their roots," whether he meant that the root-bearing rhizome was left attached to the

Fig. 92. Propagation of bamboos of Group II by means of rhizomes alone; *Arundinaria amabilis,* cultivated at the Federal Experiment Station, Mayagüez, Puerto Rico, under P.I. 110509. (*Above*) Untreated rhizome cuttings being distributed on a 2-in. layer of cocopeat. They were later covered with a similar layer of the same material. The 95 pieces of rhizome used in this particular trial were from 18 in. to 5 ft in length; the shorter ones were unbranched and the longer ones mostly branched. This and numerous other trials with this and other propagating media and with other, more elaborate, treatments of the material left the strong impression that, under uncontrolled conditions, the propagation of *Arundinaria amabilis* by means of rhizomes alone is very wasteful of material, since very low yields were the rule. The mother plant from which these rhizome cuttings were removed is seen in the foreground. (*Below*) Owing to the crowded state of the rhizomes, and the heavy, compact nature of the "lateritic" clay soil in which they had developed, it was necessary to exercise great care in order to extricate them in an undamaged condition. See p. 156.

culm, or simply that the roots normally produced at the lower nodes of the culms of many tropical bamboos would be allowed to remain on. Presumably he meant culms without rhizomes, since culms with rhizomes attached are mentioned elsewhere by him as another category of propagating material. Since very little exact information has been published on the propagation of particular bamboos by means of whole culms without rhizomes, this method was explored for practical reasons.

GROUP I. Exploratory trials of three lots of whole-culm cuttings of *Dendrocalamus strictus* (P.I. 254923) and one lot of whole-culm cuttings of *Bambusa tulda* (P.I. 21002)[1] gave the following results:

Of three age groups of *Bambusa tulda* put in in August and lifted 260 days later, the 2-year-old culms gave the best over-all score, with an average of 7.16 rooted plants, with a total of 10 culms, per whole-culm cutting. It is of interest that, in the later experiment, 2-year-old culms of this species again made the best over-all score, which comes out at an average of 12 rooted plants, with a total of 77 culms, per whole-culm cutting (Fig. 93, *4*).

Whole-culm cuttings of *Dendrocalamus strictus* (Fig. 93, *1–3*) processed in March and lifted in less than 400 days later made the best over-all score with an average of 6.3 rooted plants, with a total of 15.9 culms, per whole-culm cutting; those put in in July and lifted 290 days later came second, with an average of 4.66 rooted plants, with a total of 12.8 culms, per whole-culm cutting; those planted in August and lifted 267 days later made the poorest score

Three-year-old culms made the best over-all score, their average for the three lots being 5.83 rooted plants, with a total of 13.16 culms, per whole-culm cutting. Four-year-old culms made the second best over-all score, their average for the three lots being 4.48 rooted plants, with a total of 11.88 culms, per whole culm cutting. Two-year-old culms were a close third, with 4.46 rooted plants, with a total of 10.16 culms, per whole culm cutting.

The results of a later experiment (McClure and Kennard 1955) give 2-year-old culms of this species the top score, with 1-year-old culms in second place (Table 5).

[1]Unpublished experiments carried out at the Federal Experiment Station, Puerto Rico, 1949–50, in collaboration with F. Montalvo.

1

2

Fig. 93. Propagation of bamboos of Group I by means of whole-culm cuttings; procedure developed, and results obtained, in trials of P.I. 76641, *Dendrocalamus strictus* (*1–3*) and P.I. 21002, *Bambusa tulda* (*4*) at the Federal Experiment Station, Mayagüez, Puerto Rico. *1.* Covering culms with earth after they have been trimmed and staked down. *2.* Plant developed from one bud in 384 days.

3

4

3. Base of same plant. *4.* Base and root system of a propagule produced from a bud on a branch in the lower third of a whole-culm cutting. After it was removed from the propagating bed, the plant was plunged for 1 week in coco-peat to promote the restoration of root growth. Each propagule bears a piece of the mother culm, sawed off to release the plant in preference to disturbing the root system to make the cut at the point of its origin.

Table 5. Plant production by eleven bamboos of Group I, propagated by whole-culm cuttings. (Adapted from McClure and Kennard 1955, Table I.)

Species	P. I. No.	Culm age (yr)	Culms put in	Total yield (plants)	Plants produced at			Plants per 10 ft of culm
					base	middle	tip	
Bambusa								
polymorpha	61373	1	8	101	2	26	73	3.4[a]
		2	9	104	5	56	43	3.5
		3<	9	36	9	23	4	1.2
textilis	80872	1	14	8	0	0	8	0.3
		2	16	0	0	0	0	0.0
		3<	274	109	0	8	101	0.2
tulda	21002	1	66	2	0	2	0	0.0
		2	73	1229	282	366	581	3.9
		3<	34	767	154	310	303	6.8
tuldoides	21349	1	36	72	0	10	62	0.6
		2	40	261	9	15	237	2.4
		3<	47	190	17	16	157	1.6
ventricosa	77013	1	7	75	14	45	16	2.5
		2	8	223	15	120	88	7.4
		3<	11	107	25	18	64	3.5
Cephalostachyum								
pergracile	64808	1	7	6	3	0	3	0.2
		2	9	0	0	0	0	0.0
		3<	13	2	2	0	0	0.1
	126493	1	8	109	20	44	45	3.6
		2	10	128	89	16	23	4.1
		3<	13	15	8	6	1	0.5
Dendrocalamus								
strictus	77061	1	12	49	6	24	19	1.0
		2	12	77	0	33	44	1.5
		3<	12	14	11	0	3	0.3
Gigantochloa								
apus	99573	1	6	36	36	0	0	1.8
		2	7	108	44	24	40	5.5
		3<	8	187	74	43	70	9.5
Guadua								
angustifolia	132895	1	7	99	19	19	61	3.9
		2	6	222	30	99	93	9.1
		3<	7	166	65	45	56	7.4
Sinocalamus								
oldhami	76496	1	9	31	3	0	28	1.1
		2	9	128	2	24	102	4.9
		3<	8	206	31	68	107	8.2

[a]Production of 2.5 plants or more per 10 ft of culm was rated as excellent; 1.5–2.4, as good.

The diverse performance observed, as between species and as between individual culms, in the several lots emphasizes the importance of using ample arrays of material in such studies.

Attention is directed particularly to (1) the last column of Table 5, which gives the score of each age group for each species in terms of the average number of plants produced per 10 ft of culm, and (2) columns 6, 7, and 8, which show the actual numbers of plants produced in the lower, middle, and upper parts, respectively, of the culms of each age group.

The number of culms used is, in most cases, too small to give statistically significant results. Moreover, a repetition of the experiment, under different ecological conditions—planting the cuttings after the initiation of the rainy season, for example—with more culms in each age group, probably would bring out a different pattern of behavior. However, the results obtained strongly suggest (1) that each species shows a distinct pattern of response to propagation by means of whole culm cuttings, and (2) that age of culm, and position within the culm, are variables of importance in relation to yield.

Bambusa textilis gave very poor results in this trial. Out of 304 culms put in, only 25 produced rooted plants. In an earlier trial under more favorable weather conditions, with adequate soil moisture, this species gave an average for all ages of 7 rooted plants per culm, or a score of about 5 rooted plants per 10 ft of culm. The contrast between this score and the very low score made in the 1952–1953 experiment is attributed to the extremely dry, hot conditions that prevailed in soil and air during the first week after the culms were put in for the second trial of this species, when irrigation facilities had not yet been installed. All of the lots of other species in the 1952–1953 experiment (Table 5) were protected by irrigation from the time they were put in the ground (1 week after *B. textilis*) until the rains started.

Bambusa ventricosa responded well to propagation by whole-culm cuttings, with an average score for all three age groups of 4.46 rooted plants per 10 ft of culm, and a high of 7.4 for 2-year-old culms. This is in sharp contrast with the almost complete failure of branch cuttings in three age groups which, in another experiment, lay in the propagating bed for 2 years, with a bud dormancy persisting at more than 95 percent.

Cephalostachyum pergracile manifested a marked difference in response as between the respective lots of material from the two clones compared (P.I. 64808 and 126493).

Gigantochloa apus (Table 5) made the highest single age-group score—an average of 9.5 rooted plants per 10 ft of 3-year-old culms.

Guadua angustifolia made the second highest single age-group score—an average of 9.1 rooted plants per 10 ft of 2-year-old culms. This species also made the highest over-all score—an average of 6.6 rooted plants per 10 ft of culm for all three age groups.

Since the outgrowth of buds on the cuttings continued serially over a long period, there was a wide divergence in the size of the rooted plants recovered. No comprehensive record was made, either of the number of culms in each plant harvested or of the range of maximum heights achieved. Many of the plants were still very small when the material was lifted, but others had reached impressive dimensions. Rooted plants from *Bambusa ventricosa,* for example, had produced culms 18 ft tall in about 400 days. It would be worth while to ascertain whether it is feasible to remove the rooted propagules individually as they reached the required height.

GROUP II. The Rivières (1879:461) indicate that their trials of whole-culm cuttings of bamboos of this group gave negative results. They do not list the species actually tried. I have not attempted to propagate any bamboos of this group by means of whole-culm cuttings.

Culm Segments. GROUP I. Culm segments of bamboos of Group I embracing one or, usually, two or more nodes bearing buds or branches, constitute a form of propagule in casual use in both hemispheres. The branches are usually pruned to a length of a few inches to a foot; no foliage is retained. Such cuttings are usually set upright or at an angle, with at least one node well covered. Although there is very little in print relating to this method[1] of propagating bamboos, personal observation indicates that it is commonly used for certain species such as *Bambusa vulgaris* which propagate readily by almost any of the conventional means.

[1]While this book was in galley proof, I came upon the experimental results of two fruitful studies of this method published by W. C. Lin (1962, 1964), Specialist and Chief, Lu-kwei Branch, Taiwan Forestry Research Institute, Lu-kwei, Kaoshiung, Taiwan.

The Rivières (1879:473) describe culm "cuttings" as pieces of culm, bearing fascicles of branches (Fig. 88). They call this the most rapid method of propagation. They deal with only a few large bamboos of Group I: *Bambusa vulgaris,* and unidentified species of *Gigantochloa* (as *Bambusa macroculmis*), and *Dendrocalamus hookeri* (as *Bambusa hookeri*), all species in which root primordia appear readily on the base of the branches. As observed by the Rivières, these "need only make contact with the soil in order to develop." The authors direct attention to the importance of leaving a part of the culm internode attached to the branch complement, the object of which they describe as "to support life in the branch cluster while waiting for the roots to develop." It is indicated that, once the branches of the cluster are rooted, "they may be separated into several individuals."

Dendrocalamus strictus. Some bamboos of Group I apparently do not respond favorably to this method of propagation in all circumstances, however. Deogun (1937:121) says of *Dendrocalamus strictus:*

> Culm cuttings have never struck at Dehra Dun in spite of the fact that cuttings from culms of all ages, planted horizontally, vertically, notched, etc., were tried. Orissa reports 20–30 percent success with such cuttings and Sunder Lal Pathak [reports] from Pinijaur, Patiala State (*Ind. For.* 1899, p. 307) a success of about 95% after one year and 50% after 2 years, the rest having been killed by drought. In this case cuttings with 2 nodes were made from culms 3–5 years old and [these] were planted out between 15th January and 15th February in nursery beds, well manured with leaf mould and stable manure and sunk below the ground level . . . The beds were flooded twice a week and the cuttings sprouted by the middle of March. Ninety-five percent give 1–3 shoots which grew 3′ during the first rains.

Dabral (1950) reports conflicting conclusions from his experience in the propagation of *Dendrocalamus strictus* by culm segments. It is stated (p. 313) that this species gave a poor response, and on the following page the author says that

> the indications are that for *Dendrocalamus strictus,* three-foot long cuttings taken from two-year old culms planted horizontally, one inch below the soil surface, offer prospects of successful regeneration of the species, under the soil and climatic conditions of Dehra Dun.

Similar cuttings of *Bambusa arundinacea* gave no response.

Dabral also reports general indications of success, without details of performance, in the propagation of *Bambusa polymorpha,*

Bambusa tulda, Dendrocalamus longispathus, and *Thyrsostachys oliveri,* from segments of 2-year-old culms, planted at Dehra Dun during the third week in June 1949. The slender tips of the culms were discarded and the remaining portion divided into 3-ft segments. The segments contained, in most cases, about three nodes each. The branches were trimmed to a length of 3 or 4 in., with care not to injure the buds. The segments were planted horizontally in trenches and covered with 1 in. of soil, well tamped.

Troup (1921:993) used culm segments of undesignated species of bamboos (all probably of Group I); he says:

> Stem cuttings without rhizomes attached are very uncertain. Vertical stem cuttings are usually taken by cutting the culm down as low as possible, where it joins the rhizome, so as to include the lowest nodes, which tend to produce rootlets. Culms one year old should be employed.

The superiority of 1-year-old material was not borne out by my own experience with culm segments of *Bambusa vulgaris* var. *vittata,* an account of which follows.

Bambusa vulgaris var. *vittata.* In connection with the establishment, in 1948, of a 20-acre experimental field planting of *Bambusa vulgaris* var. *vittata* at Teleman, in the valley of the Polochic River, Guatemala, I incorporated in the routine an experiment to test the influence of age of material on the performance of culm segments as cuttings.

One thousand nineteen of the 1696 culm segments planted were basal cuts, and the performance record presented here is confined to this array. The material was divided into the following six age groups: (1) less than 2 months; (2) about 6 months; (3) 12–18 months; (4) 24–30 months; (5) 36–40 months; (6) 48–60 months.

It is impossible to determine the age of bamboo culms accurately, unless they have been marked in the year of their emergence. However, familiarity with certain visible characteristics associated with aging makes it possible to arrange them with respect to age in a series that gives meaning to the performance records of the several groups.

The method used in planting the cuttings is shown in Fig. 94. A severe drought that followed planting resulted in a high percentage of fatalities. The influence of age on survival, however, and on the performance of the material is clearly shown in Table 6, and in Fig. 95.

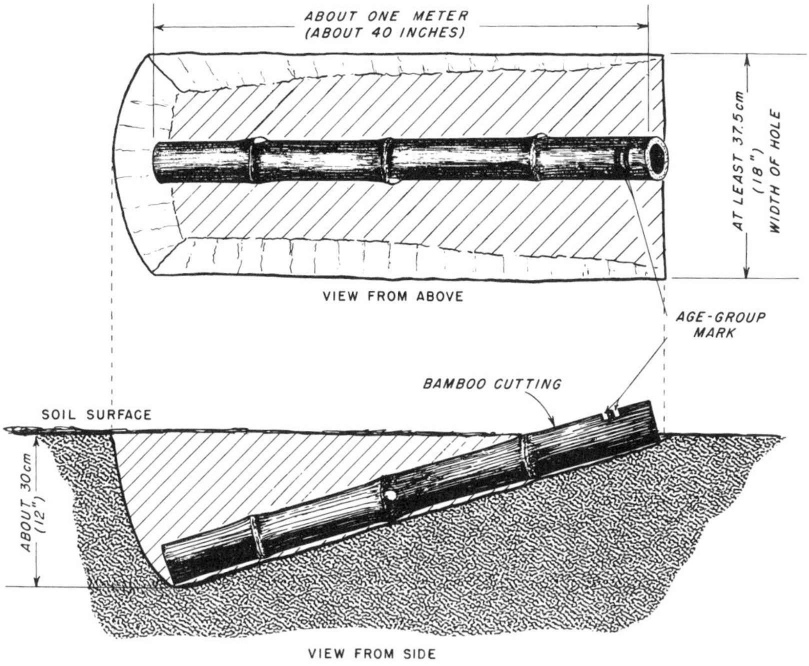

Fig. 94. Propagation of bamboos of Group I by means of culm segments with buds but no branches; *Bambusa vulgaris* var. *vittata.* Design developed for an experimental planting at Teleman, Alta Verapaz, Guatemala, showing the orientation given the cutting as it is planted. Meter-long basal cuts of culms, in six age groups, were used for determining the effect of age of culm material on yield of rooted plants. See Table 6 and Fig. 95.

Table 6. Performance of 1-m basal segments of culms of *Bambusa vulgaris* var. *vittata* put in as cuttings November–December 1948; data from field notes completed in July 1949; the calculated percentages were rounded to nearest whole number.

Age group (mo)	No. of cuttings	Produced rooted shoots (percent)	Produced unrooted shoots (percent)	Died (percent)
<2	110	20	22	58
ca. 6	256	19	18	63
12–18	283	26	19	55
24–30	228	30	19	51
36–40	98	51	29	20
48–60	44	50	34	16

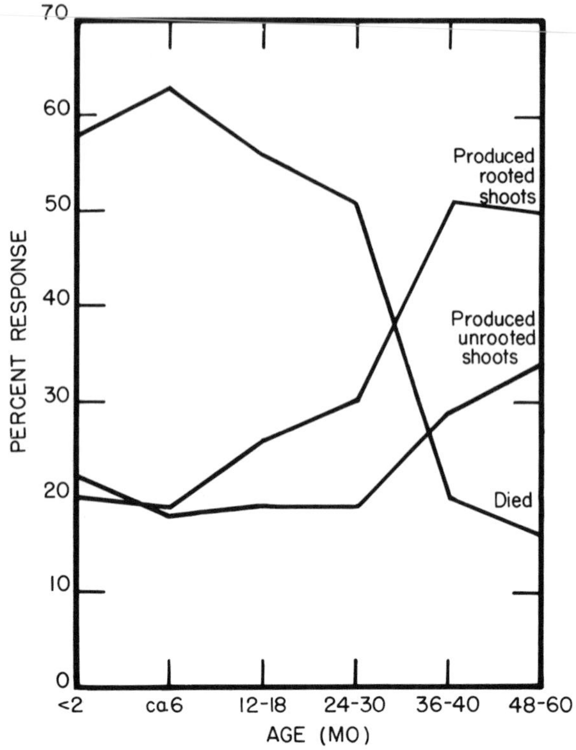

Fig. 95. Performance of 1019 culm sections (basal cuts only) by age groups; *Bambusa vulgaris* var. *vittata;* Teleman, Polochic Valley, Guatemala, December 1948-June 1949.

With the permission of Dr. J. Van Overbeek, reference is made here to unpublished results of an exploratory study he carried out at Mayagüez, Puerto Rico, in the spring of 1945, and reported to me by letter, with photographs, at that time.

Representative cuttings, including branch-bearing segments of culms, of *Bambusa vulgaris* were placed in sand in a humid propagating frame, in late April or early May, without treatment with auxin. Thirty-six days later, rooting had taken place in numerous branches, principally those above the level of the sand (Fig. 99). In exceptional cases, a few slender branches that were covered with sand produced roots.

In nature, of course, the principal mid-culm branches of *Bambusa vulgaris* form root primordia spontaneously, *in situ.* However, this apparently is the first record of the successful propagation of a bamboo by means of the slender distal branches of the culm. It is also the first record of the artificial stimulation of bamboo root growth with moist air as the propagating medium. The high humidity prevailing in the propagating frame is presumed to have been responsible for activating the growth of the existing root primordia. This demonstration suggests further experimentation. Propagules rooted out of contact with a solid rooting medium would be especially desirable for certain experimental studies. And such a convenient source of clean root tips would have important advantages for chromosome studies.

GROUP II. The Rivières intimate (1879:504) that bamboos of Group II are capable of giving rise to new plants from the basal segment of a culm, planted as a propagule. Their claim is documented by an illustration (p. 503, Fig. 11) here reproduced as Fig. 96, *left.* Although their claim is true for some bamboos of Group II, it is obvious that the conditions of success in this type of cutting are not demonstrated by the figure. The branches that have arisen from a bud just at the surface of the ground have no roots of their own. Not having produced roots while the basal nodes were still in a meristematic condition, such axes never could have roots. Neither could they in the given situation, ever produce other culms that will root themselves, since they have no buds below the surface of the ground. And, most important of all, there is no bud on any part of the original propagule itself that is covered with earth. There must be a bud in this position, and it must give rise to a rooted culm; and from a rooted culm there must emerge a rhizome—the only permanent source of new culms. Until this condition has been fulfilled, it cannot be said that a complete plant has been produced. I observed such a defective propagule of *Phyllostachys pubescens* in my garden at Lingnan University for several years. It remained throughout that period just such a "plant" as the Rivières illustrate in their Fig. 11, never producing a rhizome or a new culm.

A propagule fulfilling the conditions for success can be secured from bamboos of Group II, only (1) as the basal cut of a culm of small stature, where buds are produced at even the basal nodes, or, better, (2) by using the underground part of the culm, where

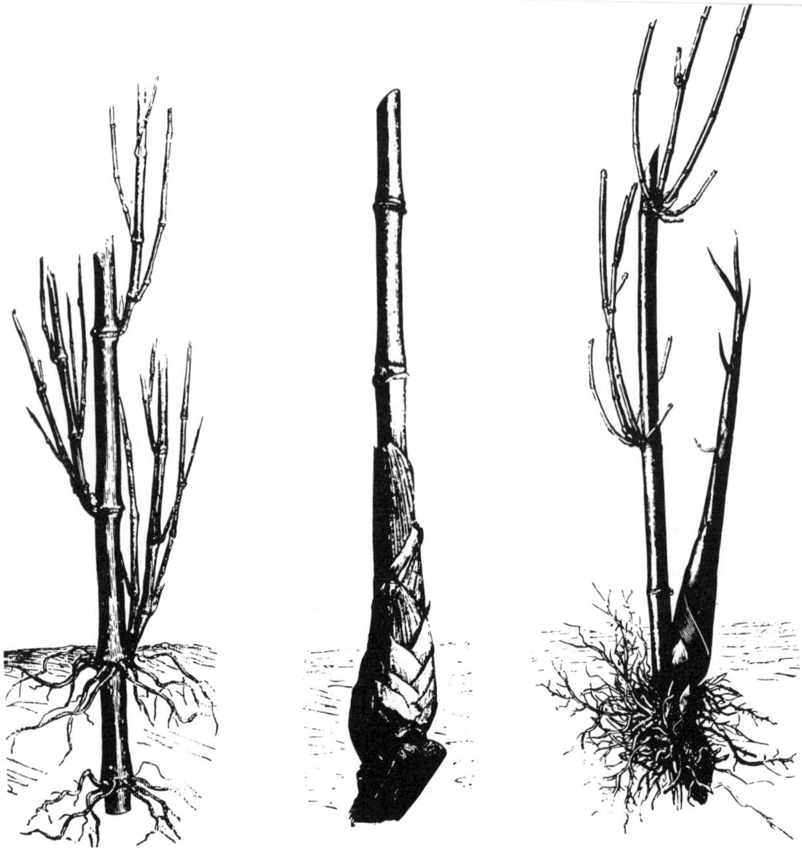

Fig. 96. *Left.* Propagation by means of a culm segment with branches. Group II: *Phyllostachys viridi-glaucescens.* A rooted propagule, consisting of the basal segments of a culm cut off at ground level, as illustrated by A. & C. Rivière 1879:Fig. 11. As explained elsewhere, this propagule has not given rise to a complete plant (See p. 239).

Center and *right.* Propagation by means of single branches. Group I. *"Bambusa macroculmis."* (*Center*) Young single branch cutting prepared for propagation. The sheaths conceal the root primordia on the swollen "rhizomatous" base. Note the "heel" of tissue from the mother culm, whose retention is stressed by the Rivières as a condition of successful propagation by single branch cuttings. (*Right*) A young shoot (still without roots) has arisen from one of the basal buds. When such a shoot becomes a rooted culm, a new plant will have been established. Figures at center and right from A. & C. Rivière 1879:Figs. 8 and 9. See p. 226.

the nodes often have dormant buds (for example, some species of *Phyllostachys*). In some bamboos of Group II, however, even this underground part of the culm is sometimes entirely devoid of buds (for example, *Arundinaria amabilis*). In view of the uncertainties that plague the search for suitable propagules of this type in bamboos of Group II, it is advisable to depend upon other methods of propagation.

The Rivières (1879:461) signalize their failure to propagate bamboos of Group II by means of culm segments bearing branch complements by the statement that "bamboos with summer growth bear only nonrhizomatous branches which cannot be used for propagation."

Since no other reference to the use of culm segments or branches of bamboos of this group for vegetative propagation could be found, and since recorded experience in bamboo propagation is generally based on procedures of a relatively primitive nature, the limitations ascribed to propagating material from certain sources is probably a function rather of the crudeness of the methods used than of the source of the material. This view is supported by the fact that *Arundinaria amabilis,* a typical member of Group II, has been propagated successfully by means of culm segments bearing full complements of branches when these are exposed to a controlled range of temperature and humidity. The conditions and results of this exploratory study by Dr. John Creech have not been published.

Single Branches. GROUP I *vs.* GROUP II. No record of the successful propagation of any bamboo of Group II by means of single branches as cuttings has come to my attention. The discussion that follows concerns only bamboos of Group I. These are characterized (p. 209) by the resemblance of the mid-culm branches to the mother culm in having a swollen basal portion that "recapitulates" the rhizome. In certain bamboos of this group (species of authentic *Cephalostachyum* and *Schizostachyum*) the branch complements in the mid-culm range are composed of numerous small, subequal branches, in which the resemblance to the mother culm and rhizome is not strikingly manifest. In other genera, however (*Bambusa, Dendrocalamus, Gigantochloa, Guadua, Sinocalamus*), the primary branch generally greatly exceeds the other members of the branch complement in size and in resemblance to the mother culm and its rhizome. In some species this resemblance includes the spontaneous generation of root primordia on the rhizomelike basal part of the primary branch, *in situ.* The Rivières (1879:476) refer to such branches as "rhizomatous." As mentioned repeatedly elsewhere, the formation of root primordia can take place only while the tissues of the branch base are still in a meristematic condition, and not after they have matured.

For propagation by conventional methods, single-branch cuttings should be made, as a rule, only from such root-bearing dominant central axes of branch complements in the mid-culm range. In the experiments reported below, branches that were not already rooted did not produce roots. The Rivières state that in their experience failure always resulted from the use of the smaller branches of the complement.

Single-branch cuttings were first illustrated by the Rivières (see Fig. 96, *center* and *right*). However, they confine their discussion (1879:469–476) chiefly to culm segments bearing whole branch complements, and give scant attention to single-branch cuttings. It is probable that the Chinese farmers were the pioneers in this field. Certain bamboos of Group I are still regularly propagated in China by means of single-branch cuttings (McClure 1938*a*:Pl. 6). Taking a cue from the Chinese farmers, I succeeded in 1925 in establishing single rooted branches of *Sinocalamus beecheyanus,* in sphagnum, for shipment to the United States.

Cobin (1947 reported the successful use of branch cuttings to propagate *Sinocalamus oldhamii, Bambusa vulgaris, B. vulgaris* var. *vittata,* and *Gigantochloa verticillata,* all species in which root primordia appear spontaneously in abundance on the swollen part of the principal branch, *in situ,* at mid-culm nodes. White (1947*b*) studied the effect of season of the year and of root-promoting substances on the rooting of branch cuttings of nine species of bamboos. Branch cuttings 12–18 in. long from 2-year-old culms were processed in lots of 50 each at four seasons: March, June, September, and December. Just before being planted, each lot of 50 cuttings was divided into five sublots of 10 each; the basal portions only of the cuttings of four of the sublots were dipped for 5 seconds into alcoholic solutions (Cooper) of, respectively, (1) 5 mg/ml indole-3-acetic acid, (2) 2 mg/ml indole-3-butyric acid, (3) 2 mg/ml α-naphthylacetamide, (4) 0.1 mg/ml 2,4-dichlorophenoxy acetic acid. The fifth sublot was used as a control. The March, June, and September lots, each totaling 50 processed cuttings, were planted in moist sand in a greenhouse; the lots similarly processed in December were planted in sand outside under partial shade.

White says (p. 393):

> No differences in rooting were found which could be associated with root-promoting treatments. Therefore, the data from [all] 50 cuttings of each species obtained at each date were summarized as units for comparison of differences among species and the time of year they were obtained.

The results, by species and by months, are shown in Table 7. The author's summary follows:

> (*a*) Treatment with root-promoting substances at the usual concentrations has no effect; (*b*) considerable variation in rooting exists among species; (*c*) rooting varies with the month [in which] cuttings are obtained; (*d*) the best month for rooting cuttings varies with the species; and (*e*) rooting may be associated with rainfall during the month previous to obtaining the cuttings.

White could have added a further observation drawn from his results, namely, that in his experiment the best score was made by those bamboos in which the most abundant production of

Table 7. Performance (percentage rooted) of single branch cuttings of nine species of bamboo put in at four different seasons of the year. (Adapted from White 1947*b*:393.)

	Season				
Species	Mar.	June	Sept.	Dec.	Ave.
Bambusa longispiculata	2	0	0	0	0.5
polymorpha	0	0	0	12	3.0
textilis	16	25	2	0	10.5
tulda	2	0	0	12	3.5
tuldoides	0	0	0	0	0.0
Cephalostachyum pergracile	0	0	0	0	0.0
Dendrocalamus asper	20	0	20	52	23.0
Gigantochloa apus	50	46	0	50	36.5
Sinocalamus oldhami	36	24	8	16	21.0
Averages, by season	14	10.5	3.3	15.8	

root primordia takes place spontaneously, *in situ,* before the branches are removed from the mother culm. Cuttings that have no root primordia when taken from the mother plant will not themselves produce roots. Rooted plants can be secured from branch cuttings only through the sprouting of buds on the underground portion.

Two hundred branch cuttings (consisting of the basal foot-long portion of principal branches from mid-culm nodes) of each of six species of bamboos of Group I were put in at Mayagüez in July 1949 (McClure and Montalvo 1950) with the objective of producing small, rooted plants for shipment by air. In terms of the uniformly small size and adequate rooting of the plants secured, branch cuttings were found to produce the best results of any of the several methods of propagation tried (Fig. 97, *1–3*). However, the slow and irregular breaking of dormancy in the critical buds rendered the yield uneven and, in some cases, entirely unsatisfactory. Of the six species, *Gigantochloa apus* gave the best yield by this method; *Bambusa ventricosa* gave the poorest, with less than 1-percent response. It is noteworthy that most of the buds of the cuttings of *B. ventricosa* remained alive—and presumably viable—after having been in the propagating medium (1:1 cocopeat and basa) for 2 full years. No effective method for breaking the dormancy of the branch buds in refractory species was found.

Spontaneous rooting, in situ, of bamboo branches. It may be of interest to discuss this characteristic briefly from the point of view of physiology. Three points may be noted as foci of attention: (1) as observed in nature, the "propensity" toward spontaneous rooting, particularly in primary branches, is more strongly developed in some bamboos of Group I than in others; (2) bamboos that normally show no root primordia, or only a few, on the principal branches may, under certain circumstances, be induced to produce them in abundance; and (3) under either normal or certain abnormal conditions in nature, complete rooted plants may be produced spontaneously in place of a part or all of a branch complement in certain bamboos of Group I.

The relative abundance, or lack, of root primordia in branches of the bamboos tested in connection with propagation studies is reflected in the readiness with which they strike root under the conditions prevailing in the conventional propagating bed. Differences between species in the natural propensity toward spontaneous production of root primordia on branches, *in situ,* are shown in the following personal observations (where available, U.S.D.A. plant introduction numbers are given): *Bambusa longispiculata,* P.I. 93573, 0-few; *B. multiplex,* 0-few; *B. multiplex* var. *riviereorum,* P.I. 77014, 0-few; *B. polymorpha,* P.I. 61373, 0-few; *B. textilis,* P.I. 80872, 0-few; *B. tulda,* P.I. 21002, 0-few; *B. tuldoides,* P.I. 21349, 0-?; *B. ventricosa,* P.I. 77013, 0-?; *B. vulgaris,* numerous; *B. vulgaris* var. *vittata,* numerous; *Cephalostachyum pergracile,* P.I. 64808, 0-?, P.I. 126493, 0-few; *Dendrocalamus asper,* P.I. 71258, numerous; *Gigantochloa apus,* P.I. 99573, numerous; *Guadua angustifolia,* P.I. 132895, 0-?; *Guadua angustifolia* (caña mansa), numerous; *Sinocalamus oldhami,* P.I. 76496, numerous.

Under certain circumstances spontaneous rooting of branches, *in situ,* is stimulated extraordinarily. Cobin (1947:183) notes that a plant of *Bambusa textilis,* whose branches normally root very sparsely, showed an abundance of roots on its branches, a circumstance which he attributes to the poorly drained condition of the soil in which it was growing at the Fairchild Gardens, Coconut Grove, Florida. I have observed the same phenomenon in the same species under circumstances that are, perhaps, more suggestive.

Spontaneous production of rooted plants in place of branches. A small, single-culm rooted plant of *Bambusa textilis* set in the bamboo nursery of the Instituto Agropecuario Nacional at Chocolá, Guatemala, failed to produce new culms, but several branches at its lowermost branched node bore long roots, most of which penetrated the soil (Fig. 97, *5*). After the branches had been removed (Fig. 97, *6*) and set in the ground as propagules, the plant itself was excavated. The rhizome was found to be devoid of viable buds. This observation supports Cobin's suggestion that it should be possible, through suitable procedures, to induce or increase development, *in situ,* of root primordia on the branches of bamboos of this group, in anticipation of the use of such branches as cuttings.

A similar occurrence was observed in a small single-culm propagule of *Bambusa tulda,* another species whose branches normally root very sparingly. However, in this case, a complete rooted plant took the place of the branch complement at the lowermost node

(Fig. 97, *4*). This was discovered in a field planting at Mayagüez. Upon the excavation of the mother propagule it was discovered that, again, there were no viable buds on the rhizome. In 1953, I observed a similar occurrence in *Bambusa arundinacea* at New Delhi, India.

It appears that the food- and growth-regulating substances elaborated in the upper part of the propagule accumulate in the lower part of it, without finding an outlet in the normal production of new growth. Apparently as a result of the presence of this accumulated material, the lowest viable branch primordium responds by producing a complete rooted plant in place of a branch complement of the conventional form.

Other examples of the occurrence of complete, rooted plants in the place of part or all of a branch complement have been observed. The first involves a bamboo in which natural rooting of the mid-culm branches is abundant. In a variant of *Guadua angustifolia* known locally in Ecuador, as "caña mansa," whole mid-culm branch complements often take on the form of a small rooted plant. If this propensity has a genetic basis, it might well be possible, and useful, to undertake to incorporate it in other, elite, bamboos through cross breeding, in order to facilitate vegetative propagation. This form is much less thorny than the typical form of the species.

The second example involves the typical form of the same species, in which the branches at mid-culm nodes rarely bear root primordia, and apparently have never been observed producing complete plants *in situ.* It was in the course of a survey of the local production of culms of this species in the Department of Caldas, Colombia, that I found, in the very top of some of the oldest culms, perfectly formed little rooted plants, in the midst of very much overcrowded and degenerated branch complements. It seemed that the production of these little rooted plants might be related to a deterioration (state of senescence?) in the phloem of very old culms—more especially, perhaps, in the anastomosing bundles at the node at which the old branch complement that produced little plants was attached. Faulty functioning of the phloem could obstruct the downward movement of elaborated food and auxins and cause them to accumulate at the base of the branch complement in whose leaves they were elaborated. Such an accumulation of materials could, conceivably, result in the spontaneous production of little plants from the adjacent branch buds. This typical form of *Guadua angustifolia* flowers with extreme rarity, and is not known to have produced viable seeds. It seems quite likely, therefore, that its natural dissemination to new areas is accomplished primarily by means of these little rooted plants which, being attached by a single very slender and brittle "neck," would easily be carried away by a stiff wind. Some of those that happen to land in a stream could, on occasion, find a lodging place in stream-bank sites.

Chaturvedi (1947:543, Pl. 33, Fig. 1) reports and figures the occurrence of rooted aerial rhizomes, *in situ,* at the lower branching nodes of old ("parent") culms in congested plants of *Dendrocalamus strictus* at Baria, Bhind District, India.

Bambusa multiplex is another bamboo in which the spontaneous production of root primordia on mid-culm branches is generally sparse or nil. In 1948 I saw, in El Salvador, an old plant of a dwarf form of this bamboo in which several little rooted plants had emerged from mid-culm branch complements of the oldest culms. The base of the branches had taken on a wholly rhizome-like form and behavior, even to the positive geotropism of the "neck," and the diageotropism of the rhizome proper.

These observations make it possible to read with comprehension the following passage in Satow (1899:75) about another small-leaved form (Japanese: Ho-o-chiku) of the same species:

> In neglected hedges, roots [rhizomes] are found hanging down in the form of a fish-hook. The upper part lengthens into a culm. From the root [rhizome] other roots [rhizomes] branch out, gradually increasing in number so as to form a bole [the rhizome "nexus" of Arber], from which fine hair-like roots grow downward, but as they cannot reach the soil, they stop growing after attaining a *sun* [Japanese inch] in length. From the bole a number of stems grow closely together, of which the inner ones bear branches. This bole attains the size of a half-bushel measure, and yet is held on to the parent stem by a single root-fibre [the "neck" of the primary branch]. Should it be hit with violence, it comes away suddenly, and if stuck in the ground will give rise to a dense growth.

In 1955 a plant of the Chinese Goddess bamboo (*Bambusa multiplex* var. *riviereorum*) was grown in nutrient solution in a pebble medium, irrigated at 8-hour intervals. After 2 months many of the branch complements on several of the culms had proliferated to

1

2

3

4

5

6

Fig. 97. *1–3.* Propagation of bamboos of Group I by means of single branch cuttings: *1,* branch cuttings being planted in the propagating bed and pertinent data being recorded, in connection with experiments carried out at the Federal Experiment Station, Mayagüez, Puerto Rico; *2, Bambusa textilis* (P.I. 80872), a well-rooted plant developed from the distal bud of the cutting; *3, Gigantochloa apus,* P.I. 99573; this array of four cuttings demonstrates a tendency that appears to be general, namely, that buds at the base of branch cuttings in which the swollen, "rhizomatous" natural base of the branch is retained are generally slower to germinate, but usually produce a better root system, while the buds at more distal nodes push more promptly and the resulting culms generally are slower to develop a good root system. See p. 245.

4–6. Spontaneous production of rooted plants in place of "normal" branches: *4, Bambusa tulda* (P.I. 21002) at Federal Experiment Station, Mayagüez, Puerto Rico, December 1948; *5, Bambusa textilis* (P.I. 80872) at Chocolá, Guatemala, May 1951; *6,* four little rooted plants removed from the mother plant shown in *5.*

produce little rooted plants. This exploratory trial was carried out through the collaboration of the late Dr. Robert Withrow, at the Smithsonian Institution.

Natural propagation by means of bulbils. Propagation by "bulbilles," a French term for bulbils or bulbules, formed in connection with the inflorescence after the manner prevailing in some other grasses (for example, *Poa bulbosa*) is reported by Dutra (1938:145, 150). The author's words, freely translated, follow:

> In *Bambusa riograndensis* Dutra [that is, *Guadua trinii* (Nees) Nees ex Rupr.] the plant usually propagates itself by bulbils . . . During the period of flowering, the plant loses all of its leaves and covers itself with a multitude of inflorescences, of which not the thousandth part produces fruits, which fact is due, I think, to the faulty disposition of the stamens in relation to the pistils. The latter, being very small, remain enclosed, while the former, each equipped with a long filament, emerge while the floret remains almost closed. The plant seeks to make up for this defect by producing, here and there, bulbils that will insure reproduction.

Layering. In contrast with propagation by means of vegetative fractions, propagation by layers delays the removal of the propagule from the mother plant until after it has established roots in a propagating medium. The following methods may be employed. (1) Either a whole culm or only the branch-bearing part of it is bent down to the ground and into a shallow trench, and fastened in place by means of hooked or crossed stakes, with or without notching it below each branch-bearing node, and covering it with earth or any other suitable propagating medium; the subsequent routine is similar to that used for whole culm cuttings. (2) The stumps of severed culms are covered with a suitable propagating medium. (3) A culm is kept erect, with or without notching it below each branch complement, and the base of each branch complement in the mid-culm range is surrounded with a suitable propagating medium, held in place by a suitable receptacle.

Method 1. The Rivières (1879:476) recommend layering of whole culms only for Group I. They call it the "easiest" method of propagation, but they do not give data on the performance of any particular species.

In preliminary trials, carried out in Guatemala, I found that 1-year-old culms of *Bambusa textilis* and *Guadua angustifolia* (clone Caña Mansa) when bent down and covered with earth while still attached to the mother plant, without notching, satisfactorily produced little rooted plants. However, this method was too cumbersome to be considered seriously except in unusual circumstances, or for very small bamboos, such as the dwarf cultivars of *Bambusa multiplex,* and *Bambusa multiplex* var. *riviereorum,* which responds favorably but does not produce abundantly.

Method 2. Stump layers may be prepared by cutting off one or more of the culms in a clump, leaving one or two nodes with a bud or branch complement, and covering the stumps so prepared with a suitable mulch.

In January 1949, exploratory trials of this method were initiated at Mayagüez with *Bambusa longispiculata, B. textilis, B. tulda, B. tuldoides,* and *Dendrocalamus strictus.* Whole clumps were prepared for treatment by severing all culms at about 2 in. above the second node above ground. A bamboo barrier was constructed around each clump to hold the propagating medium. The number of stumps in each clump was recorded, and half of the stumps in each clump were treated with indolebutyric acid, by placing a handful of cocopeat impregnated with an aqueous solution (200 ppm) of IBA in the axil of each branch complement. The branches in the other half of the stumps were left untreated, as a check. Finally, the stumps were covered with a 1:1 mixture of cocopeat and "basa" (Fig. 98, *1–2*).

One clump of each of the five species was treated in this manner in January 1949, and a second clump of each in March 1949. Both clumps of *Dendrocalamus strictus* died outright, possibly of anoxia, which may have resulted from the excessive depth of the culture medium. Of the remaining species, only *Bambusa longispiculata* produced rooted plants (Fig. 98, *3*). The groups of rooted plants produced by the two clumps of this species were removed in June 1949, with the results shown in Tables 8 and 9.

In Table 9, an analysis of the details recorded in Table 8 shows that the treated stumps gave the better rooting, but produced fewer culms than the untreated ones. Such behavior agrees with that observed elsewhere among the flowering plants, where the development of buds is inhibited, but the initiation of root primordia is stimulated, by the local application, or presence, of auxin.

This is apparently the first recorded trial showing a difference

1 

Fig. 98. Propagation of bamboos of Group I by means of stump layers. Results achieved at Federal Experiment Station, Mayagüez, Puerto Rico. *1* and *2*. Preparation of a clump of *Bambusa tulda,* P.I. 21002: *1,* propagating medium (1:1 cocopeat and "basa") being added to the level of the tops of the stumps; *2,* stumps in foreground treated with IBA; background, untreated "check." *3. Bambusa longispiculata,* P.I. 93573, examples of small, rooted plants, propagated by the method illustrated above, and selected for shipment by air. Average weight of 56 plants of this lot, 4 oz; average weight of the 94 component culms, 2.5 oz.

2

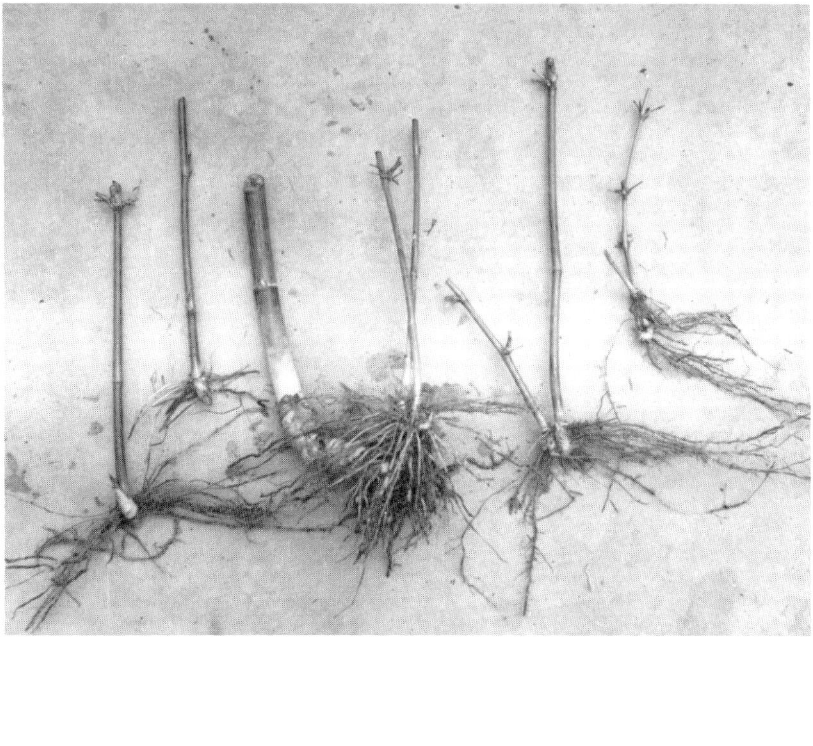

3

Table 8. Number of rooted and unrooted culms from two stump layers of *Bambusa longispiculata* treated (as shown in Fig. 99, *1, 2*) with indolebutyric acid in January and March 1949, and excavated in June 1949. (Adapted from McClure and Montalvo 1949, Table 1.)

	Date treated, 1949		
	January	March	Total
Total stumps in clump			
Treated (number)	27	25	52
Check (number)	26	35	61
Total culms from above-ground nodes	122	126	248
Treated: Rooted (number)	20	17	37
(percent)	54	27	37
Not rooted (number)	17	45	62
Check: Rooted (number)	21	11	32
(percent)	25	17	21
Not rooted (number)	64	53	117
Total culms from upper nodes of rhizomes	72	52	124
Treated: Rooted (number)	10	15	25
(percent)	67	80	74
Not rooted (number)	5	4	9
Check: Rooted (number)	29	25	54
(percent)	51	76	60
Not rooted (number)	28	8	36

between untreated propagating material of bamboo and that treated with a root-promoting substance.

Method 3. This method is really a form of air-layering, and is widely used in China in the propagation of woody dicotyledonous fruit trees. No reference to its use in the propagation of bamboos has been found in the literature.

In exploratory studies, I have tried air-layering of matured

Table 9. Performance of treated and untreated stump layers of *Bambusa longispiculata*, P.I. 93673—an interpretation of date from Table 8.

Stump layer	Number of stumps	New culms produced	Ratio of new culms to stumps	Culms rooted (percent)
Treated	52	123	2.34	50
Untreated	61	239	3.91	36

branch complements of 1-year-old culms of *Bambusa tuldoides* and current year (developing) branch complements of *Semiarundinaria fastuosa,* without achieving any success with either species. In the first case, the propagating medium, cocopeat, was held in place by funnel-shaped cups made of tarred paper. In the second case, moist sphagnum was confined in a sheet of pliofilm tied about the culm. No root-promoting substance was used in either case, and the culms were not notched.

Conclusions on past work, and future objectives

It appears that most of the published observations about the vegetative propagation of bamboos have grown out of experience of an empirical or exploratory, rather than a scientific, nature. Even where a conscious effort has been made to adopt a systematic approach to the accumulation of knowledge relating to the propagation of particular species, and to particular objectives, unassessed and uncontrolled variables have marred the quality of the documentation and prevented the achievement of conclusive results. Consequently, knowledge and practice are still largely in a retarded stage of development, and satisfactorily documented data relating to the vegetative propagation of particular bamboos apparently have not yet been produced.

Published discussions of methods of vegetative propagation of bamboos are invariably concerned primarily with the particular fraction of the plant used: clump divisions of various dimensions, from half the clump down to offsets consisting of one or more culms taken from the periphery of the plant; whole rhizome axes or segments thereof; whole culms bearing a bud or a branch complement at each node; segments of a culm, each containing one or more nodes, and each node bearing either a bud or a whole branch complement; and, finally, cuttings consisting of individual branches.

In these different kinds of propagules the mass of tissue surrounding, or communicating with, a given bud or group of buds is successively reduced. This involves a reduction of the stored food, hormones, and enzymes available to a bud and to the new shoot that develops from it. In progressively smaller propagules, the chemical constituents and physiological gradients of the tissues are exposed more and more intimately to external influences, which may be favorable or unfavorable to success. This makes progressively more urgent the necessity of supplying, artificially, basic nutrition and critical physiological stimuli to the individual bud and to the shoot that emerges from it.

Reduction in the mass of the individual propagule makes for economy of propagating material, simplifies the labor of preparing it, and reduces the requirements of space and other facilities. As the bud is deprived more and more completely of the maternal tissue that supports it, the control is perfected, the number of unassessed and uncontrolled variables is reduced, and the prospects of establishing pertinent basic principles and determining the optimal conditions of vegetative progagation for each kind of bamboo improve.

Facts and factors affecting propagation procedures

In the stems of woody dicots and gymnosperms, meristem is typically distributed in a continuous layer—the cambium. In the bamboos, on the contrary, meristematic tissue is confined to certain discontinuous and more or less widely isolated regions. The principal foci of meristematic tissue in the vegetative body of the bamboo plant are: (1) the tip of every growing axis; (2) the zone of intercalary growth just above each node of all actively elongating segmented axes; and (3) dormant buds and dormant root primordia, as long as they remain viable.

Actively growing roots are repositories of meristem but, in the bamboos, it appears that this meristem does not, under natural conditions, give rise to any kind of organ other than roots. On the other hand, the meristem of the intercalary zones of actively growing segmented axes (rhizomes, culms, or branches) may give rise to root primordia or functional roots in addition to buds (rudimentary leafy axes). There is evidence for the existence of more or less persistent islands or strands of meristem in other places within the segmented axes of the bamboo plant after elongation ceases. As far as we know at present, however, these do not play any part in the regeneration of whole organs—a process that is the principal feature of vegetative propagation. For the purposes of vegetative propagation in bamboo, viable dormant buds should be made the primary focus of attention.

With a few minor exceptions, every node of every segmented axis of a bamboo plant bears a bud—or a branch which, in turn, has a bud at every node. Theoretically, at least, each one of these buds is a potential plant. Studies in vegetative propagation should include methods for transforming as many as possible of these innumerable buds into little rooted plants.

It is important to remember that the origin of each rooted plant resulting from vegetative propagation can be traced to a single bud (Figs. 97, *2* and 99, *left*). This fact is not at once obvious when one considers relatively large plants derived from clump divisions consisting originally of several rooted culms. Once this fact is grasped, however, the notion of "rooting" bamboo cuttings (often referred to in the literature) is seen in a new light, and its limitations are apparent. A branch cutting or a whole-culm cutting often bears root primordia at its lower nodes. Only in such a case

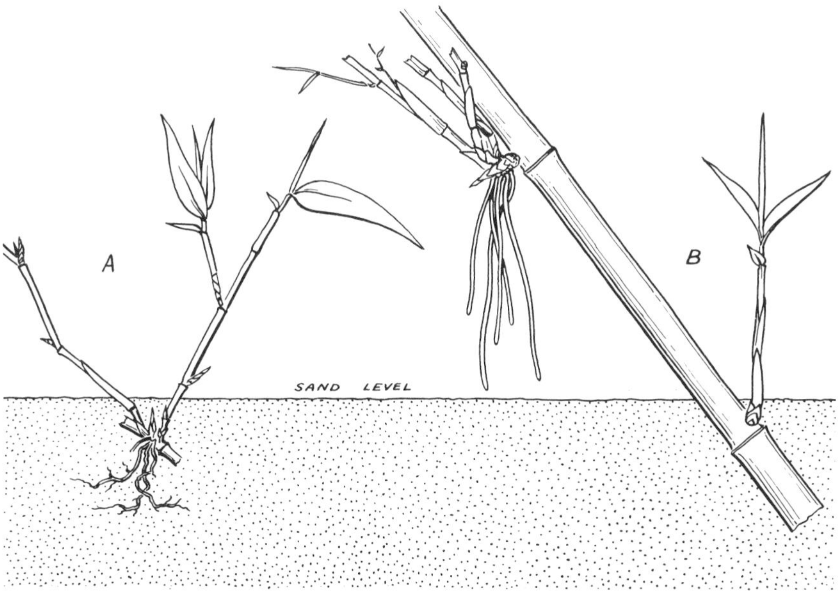

Fig. 99. Propagation of bamboos of Group I by means of culm segments with branches; *Bambusa vulgaris*. Roots developed on branches of culm segments after 36 days in a humid propagating frame: (*left*) roots developed under the sand, from a young branch on a slender culm section; (*right*) roots that developed in the humid air above the sand. Sketches based on photos by Van Overbeek.

can a bamboo cutting itself be rooted, since root primordia can be produced only from meristematic tissue, never from mature bamboo wood. However, no new plant is produced by such a rooted cutting unless and until a bud gives rise to a new axis which develops roots of its own. The above fact was demonstrated dramatically in the large-scale trial of basal culm segments of *Bambusa vulgaris* var. *vittata* referred to earlier (p. 236).

Personal observation and published experience suggest that there is a definite, optimum age range for propagules of each category (rhizomes, single branches, and so forth) from each botanical entity. The best results generally do not come from either the youngest or the oldest material. It is not clear, however, whether the basis for the differential behavior of propagules of different age resides in the maternal tissue or in the buds themselves, or both.

On the one hand, material bearing very young buds, from which the best results might naturally be expected, often gives poorer results under conventional treatment than that bearing older ones (cf. pp. 236f). This may be because the younger buds are not receiving optimum levels of nutrients in the communicating maternal tissues. Or perhaps the hormone levels in the tissues of very young propagules of certain categories are not always favorable to the awakening of the buds. On the other hand, the sudden falling off, with age, in the yield from some types of propagules—in bamboos of Group II at least, as suggested by Japanese authors (cf. Tsuboi 1913:trans., p. 2)—may be due to a loss of viability in the buds themselves. A study of this subject would be worth while from the point of view of pure science, as well as that of practical application.

Season of the year as a "variable" in procedures

As pointed out by the Rivières (1879:461), practical limitations on the propagation of bamboos must be considered in relation to the season in which the initiation of the annual period of growth takes place. Published observations and personal experience agree in showing that, once active growth has started, the tender shoots involved are apt to be interrupted in their development if the parts on which they are borne are removed from the mother plant.

Experience suggests that, under conventional conditions of propagation, processing of propagules should be limited to the period when all buds are in a dormant state. Traditional practice favors processing propagules of bamboo just prior to the initiation of its annual period of active growth. This period may be described as "late autumn (or early winter) to early spring" for the hardy bamboos (group II) and "the end of the dry season" for the tender, or tropical, bamboos (group I). However, it may prove worth while, under certain circumstances, to study the feasibility of processing certain types of propagules during the period of active growth, by providing controlled environmental conditions and suitable nutrition (cf. p. 210).

Table 7 shows that even dormant buds of propagules representing a given part of the plant may give different responses to the same routine of propagation when processed at different seasons of the year. The natural assumption is that these responses are due to seasonal differences prevailing in the physiological state of the material—differences presumably related to stored nutrients, hormonal and enzymic activity, and so forth. However, since the control of conditions surrounding the material in the propagating bed has been in all recorded cases only nominal, the observed differences in response could be in part a function of seasonal differences,[2] or differential daily fluctuations, in environmental conditions such as light, temperature, and moisture prevailing in and above the propagating bed.

Dormancy and development of buds

The natural or characteristic behavior of buds in respect to dormancy and the breaking of dormancy varies with their position in the plant, and with the species. In many species (for example, *Bambusa vulgaris*), the removal of a propagule (say a branch of a given age) from the mother plant, at a given season of the year, followed by its exposure to the conditions routinely maintained in an outdoor propagating bed, is sufficient to break the dormancy of one or more buds on the propagule. In other species (for example, *Bambusa ventricosa*), the buds of the same kind of a propa-

[2]Similar wide differences with season of the year are found in many woody cuttings. Numerous examples are listed by Thimann and Behnke-Rogers (1950).

gule of the same age, processed at the same time and exposed to the same conditions, may remain dormant for as much as 2 years. Whole-culm cuttings show a similar diversity as between species, in regard to the breaking of dormancy in their buds.

No published account of the successful use of artificial means to break the dormancy of bamboo buds at will has come to my attention. A satisfactory degree of success in the vegetative propagation of bamboos can be achieved only when routines effectively solving this problem have been established. Under certain circumstances (for example, when propagating material has to be transported) it would be useful to know also how to preserve the dormancy and the viability of the buds for some time following the severing of cuttings from the mother plant.

Propagules of any of the categories, from clump divisions to branch cuttings, sometimes fail for no obvious reason. In such cases it often turns out that, either through senescence, damage by insects, or accident, no viable buds were present on the propagule when it was put in the ground.

Rooting

New axes developed from buds on the buried part of a propagule do not always produce roots. That this may constitute a major problem has been demonstrated by data on whole-culm cuttings and culm sections of *Dendrocalamus strictus,* culm sections of *Bambusa vulgaris* var. *vittata,* and stump layers of *Bambusa longispiculata.* The same difficulty was experienced in connection with rhizome cuttings consisting of the determinate rhizomes of caespitose tropical bamboos. The primary shoots that developed from the uppermost one or two buds at the distal end of these rhizomes usually produced no roots or, at most, an inadequate number.

Therefore, one prerequisite to success in the vegetative propagation of bamboos is a knowledge of how to insure that each new axis will give rise to an ample supply of roots while the tissues in the intercalary zones of its lower nodes are still in a condition to produce meristems. With a single exception (see Table 9 and pp. 251f), applications of various hormones, such as trichloracetic, indoleacetic, and indolebutyric acids, at a wide range of concentrations have not, to date, given results differing significantly from those with untreated plants.

Because of the inexact nature and incomplete range of the controls usually maintained, the question of the best treatments and the best combination of environmental conditions for economic vegetative propagation of any bamboo, by means of any particular type of propagule, remains unanswered. Satisfactory answers can be discovered only by means of a series of experiments in which the environmental conditions surrounding the propagules are defined and precisely controlled. Furthermore, the propagules must be processed and treated in such a manner as to break the dormancy of the buds. And finally, the rooting of the resulting new axes must be assured.

Need of a new approach to the study of vegetative propagation in bamboos

The vegetative propagation of many bamboos can be made more efficient by the accumulation of further experience through systematic studies of improved design, based on certain of the conventional types of propagules. But the conquest of elite bamboos that are refractory to conventional methods of vegetative propagation requires the application of more knowledge of a basic nature, and more refined methods. Basic data relating to various unsolved problems (such as the breaking of bud dormancy, the rooting of developing axes, and getting plants from a higher percentage of available buds) are needed in order to lift the routine procedures from the empirical level to a more scientific one. It seems worth while, therefore, to make a serious attempt to develop routines and techniques designed to reduce the number of unassessed and uncontrolled variables to a minimum. The following schedule of studies is merely suggestive.

1. Selection of budwood (documented by name, age, position in the plant, season of the year);
2. Treatment to break dormancy of buds;
3. Isolation of buds, using sterile technique if it proves necessary;
4. Incubation of buds in suitable media impregnated with nutrient solution;

5. Application of auxin, if necessary, to induce rooting of the growing axis;

6. Progressive transfer of rooted propagules to potting media and cultural conditions favoring rapid subsequent development.

In any extensive investigation undertaken, the following variables should be studied individually:

1. Species or clone (begin with "elite" bamboos that are refractory to conventional methods of propagation);

2. Part of plant from which the buds are taken;

3. Age of buds (with refinements to embrace season of year at which buds are taken);

4. Methods of breaking dormancy in buds;

5. Time of application of treatment to break dormancy (including application *before* and *after* removal of budwood sticks from the plant; and *before* and *after* the removal of buds from budwood);

6. Stage of development of the buds;

7. Conditions of incubation (including temperature, moisture, and quality and quantity of irradiation);

8. Nature of culture medium (substrate);

9. Formula of the nutrient solution;

10. Constituents of atmosphere (a tentative suggestion);

11. Time, manner, and rate of application of auxin;

12. Time of removal of rooted propagules to new culture medium;

13. Conditions for subsequent development of plants;

14. State of development of rooted propagules considered optimum for various laboratory or field studies, and for nursery or field plantings.

The development, or adaptation, of the appropriate refinements of these procedures will require experience in the routines of *sterile technique,* tissue culture, the breaking of dormancy in buds, the use of hormones for stimulating root initiation, and so forth. Equipment for the automatic control of environmental conditions will be needed. Comparable studies have been made on a number of woody plants of temperate climates (see Thimann and Behnke-Rogers 1950).

Additional References. Camus 1913:190–191; Houzeau de Lehaie 1906–1908:45–51, 216–218; Ferrer Delgado 1948; Mitford, 1896: 12–18; McClure, 1938*a*, 1945*a*; McClure and Montalvo 1949, 1950; van Overbeek 1944; Pathak 1899; Piedallu 1931; Ueda 1960:41–54.

We are now entering an epoch of differential ecological, physiological
and genetic classification. It is an immense work. The ocean of knowledge is practically
untouched by biologists. It requires the joint labors of many different specialists—
physiologists, cytologists, geneticists, systematists, and biochemists.
It requires the international spirit, the cooperative work of investigators
throughout the whole world . . . it will bring us logically
to the next step: integration and synthesis.

—N. I. VAVILOV (1940)

Increased knowledge through improved methodological procedures
and a larger force of properly trained scientists working in the systematic area
of biology are the real keys to an improved classification.

—REED C. ROLLINS (1965)

Part III Bases of Classification

6 Flowering and fruiting behavior in bamboos of different genera and species

It is rarely possible to observe the whole course of the reproductive phase of the life of a bamboo plant. The recording of a full account of the significant aspects of the event is extremely difficult. Parodi's simultaneous documentation of a complete life cycle and complete reproductive cycle of *Guadua trinii* (p. 275) is unique in that it comprises first-hand observations recorded by a single individual on events from three successive generations of a given hereditary line. Raizada's report (pp. 269f) of an equally well documented case history involving events from three successive generations of *Bambusa copelandii* is second to Parodi's only in the circumstance that it depends for its completeness upon the integration of notes made independently and at different times and places by three different persons. Seifriz observed two successive flowerings of *Chusquea abietifolia* (pp. 272f), but there is only circumstantial (not documented) evidence that the same hereditary line was involved in both events.

The following limited selection of published accounts is intended to exemplify the diversity of recorded patterns of reproductive behavior manifested by different kinds of bamboo—and even by the same kind under different conditions and circumstances. Without exception, they are fragmentary and disparate in respect to details covered. In this regard the selected array of examples is representative. More elaborate perspectives on the reproductive phase in the bamboo plant are offered on pp. 82ff. It will be helpful to remember that in order to think clearly about the ontogeny of a bamboo plant, it is necessary to distinguish between its life cycle and its flowering cycle, since the two may not be coextensive.

Arundinaria amabilis McClure

Brief notes on the flowering and fruiting behavior of *Arundinaria amabilis* are recorded on p. 152.

Arundinaria auricoma Mitford

Bean (1907:229) says that flowers were noticed on plants of *Arundinaria auricoma* growing in the Bamboo Garden at Kew in 1898, and that flowers were seen every year for nine years afterward. Only a few of the culms were in flower at one time and, although these died, the vegetative vigor of the plants apparently was not adversely affected by the flowering.

Arundinaria pumila Mitford

It is recorded by Miss Rebecca Jones in a letter addressed to the U.S. Department of Agriculture, May 6, 1935, that her plant of *Arundinaria pumila* ceased all vegetative activity and flowered heavily in 1934. The flowering culms died, but the rhizomes survived and the plant began to recover its vegetative activity the following year. Plants of the same origin and lineage flowered at the same time in Golden Gate Park, San Francisco, and fruited sparingly. It is not known whether they survived.

Arundinaria pygmaea (Miq.) Aschers. & Graebner

No published record of the flowering of *Arundinaria pygmaea* since the description of the sterile plant, as *Bambusa pygmaea* by Miquel in 1866, has come to my attention. Perhaps because of having to depend wholly upon vegetative characters for its taxonomic disposition, various botanists have in the meantime placed it under the genera *Arundinaria, Sasa,* and *Pleioblastus,* with either specific or varietal status. It now appears that two different species, both persistantly sterile, have been grown in ornamental plantings and in living collections of bamboos, under the name *Arundinaria pygmaea.* The original plant was described as having solid culms. *Arundinaria vagans* Gamble is the name now applied to the one with hollow culm internodes.

Arundinaria simonii (Carr.) A. & C. Rivière

Bean (1907:231) records that odd culms of this species were flowering in the Bamboo Garden at Kew in 1892. The flowering culms eventually died, but the vegetative vigor of the plants apparently was not affected, and new culms 18 ft tall were produced simultaneously. Flowering continued in this way every year for 14 years. Then in 1904–05 every remaining culm flowered, and every plant died. Every culm of plants of the same species growing in the Temperate House at Kew flowered in 1903, produced seed abundantly, and died.

Mitford (1896:11) records in the following words his impression of the behavior of this species upon flowering and fruiting:

> *Arundinaria simoni* [*sic*] furnishes one exception of a bamboo which flowers in England without dying. It has not infrequently borne seed in this country, and has been apparently none the worse. Last year (1895) it flowered and seeded in more than one English garden. I myself gathered seed from the culm of a large clump in a garden in Surrey. The remaining culms were all in their normal condition, and there was no sign of the leafy stems being replaced by flower-bearing branchlets, or of any injury to, or exhaustion of, the plant.

Parodi (1955:134) records that, in Argentina, plants of *Arundinaria simonii* (as *Pleioblastus simonii*) flower partially, and recover their vegetative vigor afterward.

Arundinaria simonii var. *variegata* Hook. f.

Bean (1907:231) refers to the record (*Bot. Mag.* [London] t. 7146) of the flowering of *Arundinaria simonii* var. *variegata* in 1877, and states that it was flowering again at Veitch's Nursery, Coombe Wood, England, in 1907, after 30 years.

Arundinaria variegata (Sieb.) Makino

Arundinaria variegata was first described from sterile living plants by Van Houtte in 1863, under the provisional (and therefore illegitimate) name *Bambusa fortunei.* Although this Chinese plant is still popularly referred to by the trivial name "Fortunei," its legitimate name is *Arundinaria variegata,* based on *Bambusa variegata* Siebold. I have not found any published record of its having flowered during the century since it first became known to science.

Bambusa copelandii Gamble. Syn.: *Sinocalamus copelandii* (Gamble) Raizada

According to early correspondence made public by Raizada (1948), *Bambusa copelandii* flowered and fruited in the Northern Shan

States, Upper Burma, in 1896. Fruits collected by Copeland, and documented by herbarium specimens deposited in the Herbarium at Dehra Dun, were planted by Gamble in the same year. The resulting plants flowered in November 1943, and produced mature fruits in May 1944 (Fig. 55, *22*). This indicates a flowering cycle of about 48 years. From the fresh material available to him, Raizada prepared an emended description and the first illustration of this interesting bamboo to be published.

When Raizada was contemplating the transfer of this species to the genus *Sinocalamus,* he sent his manuscript and a flowering specimen to E. D. Merrill, asking for an opinion on the advisability of the proposed action. Merrill referred the inquiry to me. I wrote to Raizada indicating that the proposed transfer appeared to be reasonable in the light of the resemblance of the inflorescence of *Bambusa copelandii* to that of *B. oldhami,* which I had already placed in the genus *Sinocalamus.* Raizada then published the transfer and added, inadvertently as his own (1948:10), the last paragraph of my letter to him. These are my words: "I am, however, not at all sure that the genus *Sinocalamus* as it is known today will remain sharply set off from either *Dendrocalamus* or *Bambusa* when more of the bamboos of southeast Asia are studied in the field." Later studies have indeed improved my perspectives, and I now no longer recognize (maintain) the genus *Sinocalamus* which I set up in 1940.

Bambusa lineata Munro

According to Gamble (1896:47), *Bambusa lineata* remains in the flowering state constantly at Peradeniya, Ceylon, and at the Royal Botanic Garden, Calcutta. I saw the plants at Calcutta still in flower in 1954 and in the same year at Bogor, where they are said by the Director of the Gardens to have been in flower constantly since their introduction, which took place at least 100 years ago. Rhind (1945:14) characterizes this species as "constant-flowering" in its native habitat.

Bambusa multiplex (Lour.) Raeusch.

Plants of *Bambusa multiplex,* representing two different accessions to my bamboo garden at Canton, China, flowered and produced a heavy crop of viable seeds. Without making any further vegetative growth, the plants produced a lighter crop of flowers and fruits the second year, then died.

In the United States, the cultivar "Alphonse Karr" of this species has been observed on several occasions to remain in a flowering condition, with noticeable vegetative growth and a sparse production of fruits over long periods. Julian Nally, in a letter to the U.S. Department of Agriculture, records that his plant of this cultivar began flowering and fruiting at Gotha, Florida, in April 1937, and was still alive and in a reproductive state when the letter was written, in March 1953.

Bambusa tuldoides Munro

Plants of a given species may show diverse flowering behavior under different environmental conditions. In *Bambusa tuldoides,* as observed in southern China, death of the plant usually follows promptly upon flowering, even though very few seeds are produced. But individual plants of this species from southern China that were introduced in 1925 to the United Fruit Company's arboretum at Lancetilla, Honduras, have shown some culms in a flowering state more or less consistently ever since, with no apparent diminution of their vegetative vigor. There is no record of any of the plants having either produced fruits or died.

Chimonobambusa falcata (Nees) Nakai

Munro (1868:26) lists *Chimonobambusa* (as *Arundinaria*) *falcata* among the bamboos that flower annually. Gamble (1888:309) confirms this flowering habit. Houzeau de Lehaie (1906–1908:33) reports complete flowering, and the fruiting of a small number of 30-year-old plants in 1904, adding that the plants *almost always* die after fruiting (cf. Gamble 1921).

Bean (1907:230) writes of a doubtfully distinct form of this species, under the name *Arundinaria falcata* var. *glomerata,* that a certain number of culms of a plant growing in the Temperate House at Kew flower almost every year, and that the plant as a whole does not suffer.

Chimonobambusa hookeriana (Munro) Nakai

The first recorded observation on the events relating to the behavior of *Chimonobambusa hookeriana* at and after flowering is that of its discoverer, for whom it was named *Arundinaria hookeriana* by Munro (1868:29). Hooker writes (1854:I, 29):

> Near the top [of the pass from the Teesta to the Great Rungeet, Sikkim] I found a plant of "Praong" (a small bamboo), in full seed; this sends up many flowering branches from the root [rhizome], but few leaf-bearing ones; and after maturing its seed, and giving off suckers from the root, the parent plant dies. The fruit is a dark, long grain, like rice; it is boiled and made into cakes, or into beer, like Murwa.

Bean (1907:230) records that plants of this species (as *Arundinaria hookeriana*) growing in the Temperate House at Kew flowered in 1899, bore seed, and died, while at Glasnevin, Dublin, where it flowered at the same time, some plants died while others survived and recovered their vegetative vigor.

Chimonobambusa quadrangularis (Fenzi) Makino

Since early times the curious squarish stem and spinous culm nodes of *Chimonobambusa quadrangularis* have attracted the attention and interest of botanists and plant fanciers in China and Japan. Native of China, it was carried to Japan long ago, probably during the period when Buddhist missionaries from China were exercising their first great influence in Japan. It was introduced into Europe early in the 19th century; thus it has been known to modern botanists of both hemispheres for well over 100 years. There is no record of its having flowered during this time. Although the flowers of the square bamboo were unknown to him, Makino placed this species in the genus *Chimonobambusa* on the strength of certain vegetative morphological characters and the habit, rare in hardy bamboos, of sending up new culm shoots in winter or late autumn. This physiological peculiarity is the basis of the name *Chimonobambusa* (Gr. *cheimōn,* winter).

Chusquea abietifolia Griseb.

Two successive flowerings of *Chusquea abietifolia* in the Blue Mountains of Jamaica have been recorded. They are circumstantially related to the same hereditary line. In reference to the first (1884–85), Bean (1907:231) quoted the following statement from the writings of an eyewitness, Sir Daniel Morris (*Gard. Chron.,* Oct. 23, 1886, p. 524): "When the seed was set, the stems began to die down and apparently every plant in the island [Jamaica] died, root and all." Bean continues: "It is a curious fact that a plant [of the same species] growing in the Palm House at Kew flowered at exactly the same time and died."

Seifriz (1920) describes, from personal observation, the general flowering and death of plants of the same species in the same general locality in 1918–19. This circumstance suggests that the life cycle of *Chusquea abietifolia* is about 33 years, but the evidence that the plants involved in the two flowerings are of the same hereditary line is only circumstantial. Both flowerings were followed by an abundant production of fruits and the wholesale death of the fruiting plants.

The discovery of a few non-flowering plants of the same species was recorded by Seifriz. This is circumstantial evidence that not all plants of *Chusquea abietifolia* are monoperiodic. However, the possibility exists that these plants did not represent the same hereditary line as the one whose members flowered gregariously, or that they came from seeds produced late in the previous general flowering.

Flowering material of *Chusquea abietifolia* from living plants from Jamaica growing at Kew is described and illustrated by Hooker (1885).

Dendrocalamus hamiltonii Munro

Munro (1868:152) credits Hooker with the observation that plants of *Dendrocalamus hamiltonii* flower every year. Munro states that in the variety named *edulis* the flowers occasionally show the most capricious variations. "Some flowers contain 3 stamens tolerably perfect, the remaining 3 being converted into style-like processes; sometimes there are two distinct styles, and sometimes the style is 4-cleft; occasionally there are 7 stamens."

Dendrocalamus strictus (Roxb.) Nees

The flowering habits of *Dendrocalamus strictus* have been given much attention, and many fragmentary notes have been published (Suessenguth 1925:513; Blatter 1929:913). There is no documented record of successive flowerings of the same hereditary line, but the flowering cycle has been estimated variously as 20–65 years. Troup (1921:1006) says:

> This species commonly flowers sporadically, in isolated clumps or in small groups, almost every year; it also flowers gregariously over large tracts at long intervals, the gregarious flowering usually taking some years to complete and often progressing in a definite direction in successive years. Actually it is often difficult to distinguish between sporadic flowering on a plentiful scale and gregarious flowering, there being all stages between the two . . . The correct periodicity, however, is difficult to determine in view of the irregular manner of flowering.

Fruiting is generally abundant during gregarious flowering. The plant usually dies after heavy fruiting but may survive partial flowering (Houzeau de Lehaie 1906–1908:33). Munro (1868:148) states that the plant "does not die after flowering."

Several cases of precocious flowering in *Dendrocalamus strictus* are on record. Brandis (1899:22) quotes from the 1897 *Report* of P. Sunder Lal Pathak, Conservator of Forests, Patiala State, India, the following details. Seeds collected in June 1894 were planted early in March 1895. In February 1896, all of the seedlings were lifted and planted in baskets. In April 1896, five of the plants began to flower at the age of about 13 months. It is not stated whether any seed was produced, but the death of at least three of the plants within 3 or 4 months after flowering is recorded. Birbal records the discovery of 5-year-old seedlings in flower in a nursery where they had been abandoned. Fertile seeds were produced.

Ahmad (1937) reports that, among the 2-year-old progeny from a lot of seeds of this species collected in May 1935 and sown in a nursery at Karnpur Royal Forest, India, at the break of the monsoon in July of the same year, a single seedling was found in a flowering condition in 1937. (See also Sen Gupta 1939.) Apparently there is no record of any strain of this species in which precocious flowering is a hereditary character. See *Oxytenanthera abyssinica,* p. 276.

Guadua angustifolia Kunth

Although the vast majority of the living plants of *Guadua angustifolia* appear to flower rarely (if ever), a clone with quite different propensities has recently come to light. It was secured at Milagro, Ecuador, in 1924[1] by Johannson and was established in cultivation at Summit, Canal Zone. Although at the time of its introduction it was not suspected of differing in any way from the vast population of plants of *Guadua angustifolia* that ranges through the lowlands and up to elevations of 5000 ft in Ecuador and Colombia, the Milagro plant was discovered later by Walter R. Lindsay to be flowering very frequently, often in successive years, at Summit. Plants from the Milagro strain, established under my direction at Chocolá, Guatemala, and at Tingo Maria, Peru, have shown the same disposition to flower annually. Kennard (1955:193) records the fact that *Guadua angustifolia* flowered and set a few fruits on 1- and 2-year-old culms of young plants vegetatively propagated on the grounds of the Federal Experiment Station at Mayagüez, Puerto Rico, in 1944 and 1945. This record obviously refers to the Milagro strain, which is the only introduction of this species under cultivation at Mayagüez. In April 1948, I saw flowers on the original plant introduced there, and also on plants representing 1-year-old vegetative propagules (offsets) taken from it.

Guadua trinii (Nees) Nees

Dutra (1938:147–150, Fig. 1) was impressed by the scant production of viable seeds by abundantly flowering wild plants of *Guadua trinii* (referred to by the author as *G. riograndensis* Dutra, but the identity of the plant is well documented).

Parodi (1955) personally observed the flowering and fruiting of plants of *Guadua trinii* representing two successive generations of the same hereditary line. The elapsed time from seed to seed was exactly 30 years. Parodi's unique data are documented by preserved specimens of the flowers and fruits. After flowering and fruiting heavily one year and flowering again in the second year with little or no production of fruits, the parent plants died.

Melocanna baccifera (Roxb.) Kurz

See pp. 195f.

[1]Not 1929, as published by Higgins and Lindsay (1939:51), according to Walter R. Lindsay (personal communication).

Merostachys fistulosa Doell

Under the vernacular name Tacuara lixa, *Merostachys fistulosa,* a bamboo native to Brazil, is given a reproductive cycle of about 32 years by Pereira (1941:193). The plants are said to die completely after fruiting heavily. The pertinent observations were made in connection with the study of the plagues of rats that develop when bamboo seeds became available in quantity, a phenomenon recurrent in southern Brazil.

Nastus elegantissimum (Hassk.) Holttum

Kurz (1876:257) writes of *Nastus* (as *Schizostachyum*) *elegantissimum,* "Remarkable is bamboo ul-ul of Bandöng [Java], which flowers and dies off every third year."

Ochlandra scriptoria (Dennst.) C. E. C. Fischer

Gamble (1896:122) quotes Bourdillon as saying of *Ochlandra scriptoria* (as *O. rheedii*) that it flowers annually, not dying down after flowering.

Ochlandra stridula Thwaites

Gamble (1896:123) says, on the authority of Dr. Trimen, that plants of *Ochlandra stridula* flower regularly every year in Ceylon.

Ochlandra travancorica Benth.

Gamble (1896:126) quotes Col. Beddome, the discoverer of *Ochlandra travancorica,* as saying that "it flowers almost every 7 years and dies down" (see p. 199).

Oxytenanthera abyssinica (A. Rich.) Munro

According to an anonymous report issued by the Forestry Department at Zomba (Nyasaland Agr. Quart. Jour. 4: 10. 1944), *Oxytenanthera abyssinica* is generally regarded by the inhabitants of Nyasaland as having a flowering cycle of 30 years. A case of precocious flowering and fruiting (paedogenesis) in this species is recorded in the same report in the following words:

> A curious seeding feature of the common bamboo has been reported from the estate of R. MacFadyen, Esq. in the Lilongwe district. Seed

obtained from the Forestry Department was sown in 1930 and produced a few seedlings. These seedlings were planted out in the field in the following rainy season and two of them bore seed for three successive years immediately following the planting. *After the third year seeding ceased but the plants did not die.* All of the seed so produced was sown and about 1 percent proved to be viable. The resulting plants behaved in exactly the same way as their parents, that is, they bore seed for the first three years of their life. By collecting and sowing seeds regularly from each new plot, Mr. MacFadyen has carried out a progressive establishment of bamboos over several acres of land. This seeding of young plants has never been reported before and is of great interest. Whether it is due to some heritable trait in the individual plants first sown, or whether it is due to some local climatic or soil factors, is not known.

Phyllostachys spp.

Signs of unlimited longevity in bamboos are not restricted to plants that have never been observed to flower. The flowering of a number of species of the genus *Phyllostachys* has come under my personal observation. In all cases the plants remained in a flowering state, without maturing any appreciable amount of seed, for periods varying from 5 to 10 years. The onset of flowering was generally sudden, and was accompanied by a more or less complete suppression of the vegetative growth of the plant beyond the production of flowering twigs from existing buds on the various segmented axes. Following the period of exclusive flower production and the death of the flowering culms, vegetative activity was gradually revived, with the emergence of new culms from the rhizome system and a corresponding recession of flower production.

Unless recorded indications of continual flowering in any species of bamboo are accompanied by evidence as to the period during which a particular plant was under observation, the statement should be viewed with reservation. A clump of *Phyllostachys nidularia* (P.I. 66718) produced flowers at the U.S.D.A. Plant Introduction Garden at Savannah, Georgia, during the years 1943–1955. If the plant had been seen by me only during this flowering period and if I had not seen it in a vegetative condition either before or afterward, I might have assumed it to be a species that flowers continually. Porterfield (1926:257) says: "Every year for the past three years at Wushih, Soochow, and on Mokansan [China] the writer has collected flowering culms of *Phyllostachys nidularia.* In every case they were isolated culms with many normal vegetative culms about them."

There is evidence that certain environmental (ecological) complexes may be responsible for keeping plants of a given species constantly at or near the threshold of flowering, whereas in other conditions plants of the same species exhibit gregarious flowering only at long intervals. According to personal experience and a few published records, *Phyllostachys aurea* generally shows a long purely vegetative phase (15 to 30 years) followed by gregarious flowering and subsequent restoration of vegetative activity. This bamboo is widely used as a hedge plant in Guatemala City. Every time I have been there, perhaps a dozen times over a period of 10 years, I have seen in these hedges, here and there, individual culms in flower. No fruits have been found and no gregarious flowering has been observed in these hedges. No flowering was observed in the several clumps of this bamboo seen elsewhere in Guatemala.

Kawamura (1927) summarized recent and historical records pertaining to the flowering of several species of bamboo in Japan. The author concludes that the flowering cycle for *Phyllostachys bambusoides, Ph. pubescens* (as *Ph. edulis*), and *Ph. nigra* cv Henon (as *Ph. nigra* var. *henonis*) is roughly 60 years. For *Phyllostachys aurea,* he indicates that only one flowering period (1916–1921) is on record in Japan. He shows that the flowering state in a given species often extends over a number of years, and that the duration of the flowering state varies from species to species.

Sasa tessellata (Munro) Makino et Shibata

Bean (1907:232) says of *Sasa* (as *Bambusa*) *tessellata* that it had been in cultivation for "probably over sixty years" without flowering. And to this day, 115 years after its introduction into cultivation, no published record of the flowering of this species has come to light.

Thamnocalamus falconeri Munro

Bean (1907:230) records that plants of *Thamnocalamus* (as *Arundinaria*) *falconeri* from seeds sown in 1847 flowered simultaneously in 1876 in cultivation at Kew and at Edgecumbe in England; the species is on record as having flowered at the same time in the wild in Sikkim. Some plants at Trentham flowered in 1875 and others at Holland House did not flower until 1877. The cultivated plants all died after producing viable seeds freely, giving rise to a new generation. Odd culms among the progeny flowered at Kew during 1893 and 1894. Flowering of this species occurred elsewhere in England in 1903, 1904, and 1906. Bean states further (p. 233) that the flowering of the [then] present generation of this species had already (in 1907) extended over five seasons. Whether any of the plants of this generation survived is not indicated. Stapf (1904*a*:306) states that this species flowers at an age of 28 to 30 years. During a personal conversation at Kew in 1928, Stapf informed me that seeds of *Thamnocalamus falconeri* were received at Glasnevin, Dublin, in 1873. Plants propagated from them flowered and fruited in 1900 and 1901.

Thyrsostachys oliverii Gamble

According to Bor (1941), seeds and specimens of *Thyrsostachys oliverii* from the Katha District in Upper Burma were sent by J. W. Oliver to J. S. Gamble in 1891. Some of these seeds were planted at Dehra Dun and some at the Royal Botanic Garden, Calcutta. The dates are not indicated in Bor's account, but presumably the plantings took place in the same year. Toward the end of November 1938, Bor observed signs of flowering in all of the clumps at Dehra Dun. The first indication noted was the fading and falling of the leaves. On December 23, 1938, the first flowers were in anthesis, and ripe seeds were found on January 31, 1939. The length of the reproductive cycle for this species may therefore be calculated at about 48 years.

Bor took advantage of the unusual opportunity afforded by this event to record many interesting features of the flowering and fruiting of the plants. The reproductive structures were also described anew, and illustrated in great detail.

7 Bamboos from the point of view of taxonomy

The bamboos have not as yet been set off sharply, by definition, from the rest of the grass family. The search for the bases, and the line, of demarcation between the bamboos and the other members of the Gramineae still continues. It is more than 170 years since the first bamboo genus, *Bambusa* (as *Bambos* Retzius, *Obs. Bot.* 5:24, 1789) was described on the basis of a single species, now called *Bambusa arundinacea.* During this time 75 effectively published generic names and over a thousand specific names have been attached to plants generally recognized as bamboos. However, many of these names are either "illegitimate" or destined to fall into the limbo of synonymy.

The bamboos have been treated in the past by a few botanists (most recently by the late T. Nakai) as a separate plant family, under the name Bambusaceae. However, most agrostologists agree in retaining them in the Grass Family (Gramineae). Not all contemporary agrostologists agree with those who segregate the bamboos from the other grasses at the subfamily level. Some, perhaps the majority, prefer the more conservative procedure of retaining this and the other major divisions of the grass family at the tribe level. Such a treatment connotes less profound, or less sharp, lines of demarcation. It also gives less taxonomic weight to the observed differences between groups. At the same time, it accords greater recognition to the many lines of similarity that form a complex network of relationships, through which common characteristics link many genera of the Gramineae to two or more of the proposed major divisions of the family.

Most bamboos have a long and highly complicated life history, and an extensive plant body in which certain vegetative structures show notable variation in form within a given segmented axis. Moreover, the bamboos usually are very fragmentarily and unevenly represented by prepared specimens, even in the best collections. These conditions greatly complicate the task of comparing them with each other and with the other grasses, and satisfying the requirements of our accepted system of plant classification. This is a task to which the traditional methods, materials, and techniques of taxonomy are not perfectly adapted. For these reasons, seasoned agrostologists recognize the tentative nature of even the most conscientious and sophisticated treatment of bamboo classification.

The original description of the subfamily Bambusoideae (Ascherson and Graebner 1902:769) freely translated from the German text, follows:

> Usually stately to large grasses, in the Tropics often gigantic (with us shrubby or semishrubby). Rhizome usually short, creeping, the plants, because of this, making a thick ground cover. Stems usually stiffly erect, the tops as well as the branches usually nodding to pendulous. Branches usually numerous, often several to many arising from one node, the ultimate twigs solitary, in distichous, horizontal arrangement. Leaves with long-persistent sheaths, the articulated blade usually lanceolate to linear-lanceolate. Inflorescences often very large, paniculate, racemose or, more rarely, spicate, often clustered at the nodes. Spikelets 2-many-flowered. Empty glumes [Deckspelzen] 2-several, the lowest ones smallest, progressively larger acropetally, the uppermost shorter than the lowest lemma, the lowermost empty glumes sometimes bearing spikelets[1] in their axils. Lemmas many-nerved, awnless or, more rarely, with short terminal awns. Palea 2-many-nerved, rarely lacking. Lodicules usually 3, usually very large. Stamens 3, 6, or many. Styles 2-3, often fused at the base. Fruit not tightly enclosed by the lemma and palea.

The late Dr. Robert Pilger, in preparing his preliminary treatise on the Grass Family for the new edition of the classic German work of Engler and Prantl, *Die Natürlichen Pflanzenfamilien,* divided the Gramineae into 9 subfamilies, of which the bamboos constitute one, as the Bambusoideae. However, Pilger (1954:290) gave only a brief characterization of the bamboos, which he embodied in a

[1]In bamboos with the indeterminately branching type of inflorescence described on pp. 91ff, of the present work, the so-called "spikelets," which the authors say are borne in the axils of the lowest "empty glumes," are actually spikelet-like branches (the pseudospikelets of McClure 1934).

key to the 9 subfamilies. Translated from the German text, this reads as follows:

Culms persistent, woody, often very tall; leaf blades flat, mostly broad, usually narrowed at the base, petiolate, and articulated with the sheath; spikelet with two or more empty glumes, [inflorescence axis (omitted)] often branched and deliquescent; lemma many-nerved, not awned, or with a short, not geniculate awn; lodicules usually three; stamens 3–6–many. Fruit a caryopsis or drupe, to berrylike.

A fuller view of the group is presented here in a somewhat modified equivalent of the brief grammatical style that has been made conventional in plant taxonomy by the example of Linnaeus. Since many of the users of this description may not have a detailed knowledge of the family characteristics of the Gramineae, a disciplined usage of taxonomy is disregarded here, in that characters common to the bamboos and the other grasses are included.

BAMBUSOÏDEAE (GRAMINEAE)

Gramineae Subfam. Bambusoïdeae—Rehder (1945:78; 1949: 635); Pilger (1954:290); Keng (1959:1); McClure (1961*b*:321).

Gramina Subfam. Bambusoideae—Ascherson and Graebner (1902:769); Prat (1936:217).

Gramineae, Tribe Bambuseae—Nees (1829:520); Ruprecht (1840:91; 1839 preprint, p. 1); Bentham (1883:1094); Hackel (1887:89); Camus (1913:15); Hubbard (1948:300).

Gramina Class X, Bambusacea—Kunth (1815:75).

Bambusaceae—Link (1833:308); Trinius (1835:613); Munro (1868:10); Nakai (1933:11).

Perennial plants consisting of segmented aerial and subterranean axes bearing prophyllate buds (or branches) sheathing foliar organs, and branched or unbranched adventitious roots. Initiation of new vegetative growth typically seasonal, in temperate climates either vernal, aestival, or autumnal; in tropical climates, commonly sequential to the onset of a rainy season; in warm regions with precipitation at frequent intervals throughout the year, often more or less continuous. Each segmented vegetative axis generally more or less strongly lignified, increasingly so toward its surface; weakly so in part or throughout the plant in some taxa. Insertion of prophyllate buds and the sheaths that subtend them distichous (that is, incident on alternating sides at successive nodes) on both node, a ring of thorn-like adventitious roots in arrested development. Branch buds at culm nodes uniformly solitary in most genera (more than one in all species of the genus *Chusquea*), a 2-keeled amplexicaul prophyllum completely enclosing each branch primordium. Branches at culm nodes usually unarmed, in some species more or less strongly thorny, the thorns consisting of dwarfed and indurated twigs; thorny branches found only on the lower culm nodes in some species; in others appearing at higher levels as well. Branch complements most highly typical in form and composition at mid-culm nodes; in some species monoclade, in others pleioclade; component axes in pleioclade branch complements either subequal or unequal, the primordial branch in the letter case more or less strongly dominant; the branch complement at mid-culm nodes either remaining static basally after the completion of its initial grand period of growth or, more commonly, augmented by the subsequent development of additional leafy axes from buds at basal nodes of one or more of the component branches. Culm internodes in some taxa sulcate for some distance above the insertion of a bud or a branch complement (grooved all the way from one node to the next in all species of *Phyllostachys*); rarely somewhat square in cross section (as in mature plants of *Chimonobambusa quadrangularis*), otherwise cylindrical or nearly so; either efistulose with a pithy center (as in *Chusquea*) or fistulose, the lumen (the "pith cavity" of Hackel) containing pith in diverse forms and in different degrees of abundance, generally negligible; fibrovascular bundles of the culm wall set in a matrix of parenchyma, most closely crowded together toward the surface, progressively more widely spaced toward the center of the culm; successive internodes increasing in length (and sometimes also in diameter) below the middle of the culm; those above the middle decreasing in both length and diameter, acropetally. Each node of every segmented axis bearing a sheathing foliar organ (sheath); sheaths generally with the outer edge free throughout, but exceptionally (as in the culm sheaths of some species of *Chusquea*) with the outer edge adnate to the outer surface of the sheath for some distance from the base upward; sheaths proper either laminiferous, awned, or mucronate; laminiferous ones commonly with a bristle-bearing auricle or a tuft of bristles inserted at one or both lateral extremities of the insertion of the blade; most sheaths as a rule apically ligulate, the ligule regularly absent or rudimentary in the sheaths of certain axes (for example, necks), and elsewhere throughout the plant in some species. Culm sheaths and branch sheaths always inserted on their respective axes at levels proximal (not distal) to leaf sheaths. Culm sheaths persistent, abscissile or, in some species, incompletely abscissile; the lowest ones bladeless but mucronate, the upper ones laminiferous, the blades sessile, persistent or abscissile, and usually lacking a midrib. Leaf sheaths borne at the distal nodes of aerial segmented axes of all orders; generally differentiated sharply from typical culm sheaths and branch sheaths in features that include size, shape, texture, and locus of insertion of the sheath proper, and the form of the blade; leaf sheaths proper normally deeply inbricate and persistent; laminiferous. Leaf blades petiolate and usually at length abscissile at the level of insertion of the petiole on the sheath (sessile and abscissing only very late, if at all, in some species of *Neurolepis* and *Myriocladus*); during development tightly rolled individually, on an "axis" parallel to the midrib, with one margin at the center of the roll, afterward usually flat or nearly so (commonly needlelike in *Arthrostylidium capillifolium*); a midrib usually distinguishable (at least basally) and more prominent on the abaxial surface, the midrib in some species depressed on the adaxial surface (strongly so in *Neurolepis*). Parallel veins (secondary and tertiary) of all leaf blades (as far as known) connected by transverse veinlets, these weakly to strongly manifest on one or both surfaces, or entirely obscured to superficial view by overlying tissues. Foliage in plants with branched culms renewed periodically (commonly once a year) from buds at subdistal nodes, which give rise to leafy twigs replacing those lost by abscission; the renewal of foliage generally diminished or arrested upon the initiation of flowering; plants of some otherwise evergreen species completely deciduous under conditions of extreme drought (as in *Bambusa vulgaris* and *Dendrocalamus strictus*).

Sexual reproduction monoperiodic or polyperiodic. Flowering cycle, according to numerous but mostly weakly documented records, varying (as between species) from 1 to 60 years or more in length. Flower production commonly gregarious; in some populations erratic as to timing; in some cases involving only a part of the plants, or only a part of the culms in individual plants of a given population; flowering in some polyperiodic species (as in

Arundinaria amabilis and *Phyllostachys nidularia* under personal observation) sometimes continuous for 5–15 years of a given cycle, full vegetative vigor being restored gradually as flower production subsides. Anthesis progressing acropetally in each spikelet. Fruit production abundant to sparse or nil—partial to apparently complete self-sterility common in many species.

Inflorescences terminal or lateral to vegetative axes; consistently of either determinate or indeterminate branching within a given genus. Flowering branches leafy or leafless. Determinate inflorescences typically pedunculate (exceptionally reduced to a single spikelet) the branching habit paniculate, racemose, or spicate-racemose, usually symmetrical, rarely secund. Indeterminate inflorescences typically consisting of sessile pseudospikelets, these often forming more or less dense tufts or heads. Bracts subtending primary branches of determinate inflorescences generally weakly developed, or (excepting the lowest) lacking entirely, those subtending the lowest primary branch typically more strongly developed than those of the more distal ones, but in some genera reduced to a transverse line; bracts subtending the branches of all orders in indeterminate inflorescences always well developed, in some species all, or the lower ones only, laminiferous. Spikelets sessile or pedicellate, one-to-many flowered, terminating in either a functional floret, a sterile floret, or a slender prolongation of the rachilla, this usually bearing a discernible rudiment at its tip; the lowermost floret in the spikelet usually preceded by one-to-several (commonly 2) empty glumes, these unequal, increasing in size acropetally, one or more sometimes lacking or rudimentary. Rachilla segments in several-flowered spikelets either short and not disarticulating between florets or, more commonly, somewhat elongate, and in most genera abscissile immediately below the locus of insertion of the lemma of each functional floret (the disarticulated florets stipitate in *Guaduella*). Lemmas usually mucronate or cuspidate, in some species muticous; fertile lemmas either embracing the palea fully only basally in functional florets (the florets then either open or closed at anthesis) or convolute and tightly enclosing the palea except for an aperture at its very apex in functional florets (the florets then not opening at anthesis). Palea in functional florets usually with a longitudinal dorsal sulcus extending from base to tip, the sulcus in some species (for example, the type species of the genera *Schizostachyum* and *Elytrostachys*) tubelike, a sulcus usually lacking entirely in the palea of a functional floret terminating a spikelet, exceptionally (as in some species of *Chusquea*) a shallow groove appearing only near the apex of the palea; the palea either more or less inflated and gaping antically (as in most genera) or enclosing the floral parts entirely; in some genera (*Oxytenanthera,* for example) tightly convolute, with a small aperture at the apex only. Flowers generally hermaphrodite; in some species regularly, and in some species only under special conditions, either the androecium or the gynoecium, or both, rudimentary or lacking entirely in part or all of the flowers of a given spikelet. Lodicules in some species (all of *Gigantochloa sensu str.*) lacking entirely, in others variable in number, in most species 3; all alike in some, unequal and unlike in others, the anterior two commonly asymmetrical and paired (these in some species functioning by turgidity to open the florets at anthesis, in others not), the posterior one symmetrical, in some species reduced in size or lacking; all as a rule showing vascular traces; all thin and translucent in some species, somewhat thickened and opaque in others; glabrous on both surfaces or pubescent abaxially, the margin usually ciliate above, rarely glabrous. Stamens usually 3 or 6 (varying between 3 and 6 in some species, as in *Pseudosasa* and *Sasaella;* 6–120 in *Ochlandra*) generally in whorls of 3; the anthers usually either yellow or greenish yellow, rarely maroon when fresh and afterward brown (as in *Bambusa vulgaris*); the filaments either threadlike and free, or flat and either triadelphous or diadelphous, or all lightly connate—in some species monadelphous (inserted at the mouth of a delicate tube, as in *Gigantochloa*); the anthers at first 4-locular, the two locules of each pair uniting upon the maturation of the pollen, each cavity or sac thus formed then opening by an apical pore or by a slit that follows a lateral suture; each half of each anther extending as a lobe beyond the connective proper above and below; the connective often exserted above, in some species prominently, between the two halves of the anther, and terminating in a glabrous or penicillate point. Ovary normally solitary (sometimes duplicated in teratic flowers), 1-celled (with but a single cavity) with 1 ovule, and terminating above in 1–3 styles, these (when more than one) united at the base and upward for a distance varying with the species in some genera, in others uniform; each style terminating in 1, 2, or 3 (rarely 4 or 5) stigmas.[2] Proliferations of the stigmatic surface of diverse character and arrangement: usually capillary in form; smooth or papillate; simple or branched; of bottlebrush, pinnate, or secund insertion when simple; dendroid when branched; predominantly white, in some species purple or red (exceptionally, white with purple ones intermixed, according to Kurz 1976:266). Fruit a caryopsis, or caryopsislike, normally one-seeded and indehiscent, sessile (rarely quasi-pedunculate as in *Cephalostachyum capitatum*); usually falling at maturity with the floral envelope still attached basally; differing widely in size, shape, and anatomy; the attachment of the seed clearly shown by a longitudinal groove or sulcus (the "hilum" of Hackel) in some species, invisible or only weakly discernible externally in others; the embryo relatively short-lived among the bamboos as compared to that in other grasses generally—a dormant stage apparently lacking in some species, as in *Melocanna baccifera,* where germination often takes place before the fruit falls to the ground.

Type genus: *Bambusa* Schreber (nom. conserv.); type species: *Bambusa* (as *Bambos*) *arundinacea* Retzius.

Pilger (1854:291) made a tentative suggestion that the controversial genus *Streptochaeta,* comprising the tribe Streptochaetae of Hubbard, might be included in the subfamily Bambusoideae. Parodi (1961) has proposed the addition of the tribes Olyreae, Phareae, and Parianeae to this subfamily. However, these tribes were excluded from consideration in the preparation of the foregoing description, which embraces only the "true" bamboos in the traditional sense: the tribe Bambuseae *sensu lato.* In the taxonomic literature of this group there are 76 effectively published generic names and over 1000 specific names. Of these names, many are destined to fall into synonymy.

[2]Some agrostologists may prefer the following form of description for this part of the bambusoid pistil: Ovary . . . contracted above into a stylar column; the stylar column in some species very short (even subobsolete), in others (some species of *Arthrostylidium*) bulbous, more commonly slender and more or less elongate; rarely unbranched and terminating in a single stigma; usually divided into 2 or 3 (rarely 4 or 5) stigmatic branches.

Appendix I[1] Generic Key to Bamboos Under Cultivation in the United States and Puerto Rico

As currently classified in taxonomic literature, the bamboos known to be under cultivation in the United States represent 21 genera. The following key indicates the characteristic features by which these genera may be distinguished. The classification of many of these genera still rests, however, on tenuous and uneven criteria. Some of the generic concepts were originally based wholly or largely on characters of their rarely available flowers and fruits. Other genera were segregated wholly on the basis of vegetative characters.

The following key is based on vegetative characters as far as is possible at present. However, it has been necessary to rely upon characters of the reproductive organs for differentiating certain genera. This necessity reflects the uneven characterization of genera that still prevails. With the present lack of popular knowledge of the bamboos, the identification of many kinds must still be left to the specialist.

Owing to the still tenuous bases of their segregation as genera, the names of *Sasamorpha* (from *Sasa*) and *Pleioblastus* (from *Arundinaria*) have not yet been accepted for general use in the United States.

1a. Rhizomes pachymorph, i.e., the rhizome proper short, thicker than the culm that emerges from its tip, and not "running."

2a. Rhizome neck much longer than the rhizome proper; clump habit diffuse, i.e., with widely spaced culms; midculm branches slender, subequal. MELOCANNA.

2b. Rhizome neck shorter than the rhizome proper; clump habit caespitose, i.e., with culms rather close together.

3a. Inflorescence of indeterminate growth, i.e., the spikelet-like ultimate branches (typical prophyllate pseudospikelets) clothed basally with small, overlapping amplectant bracts, the bracts subtending buds that continue

[1]This appendix is unrevised from the original edition (1966).

for some time to produce other branches (pseudospikelets); stamens 6; transverse veinlets of leaf blades not visible or only weakly manifest.

4a. Branches at midculm nodes all slender, subequal.

Culm sheath proper and sheath blade conspicuously thickened, hard, and brittle; the auricles of the culm sheath prominently developed; style bearing 2 stigmas. CEPHALOSTACHYUM.

Culm sheath proper and sheath blade not conspicuously thickened or indurated; the auricles of the culm sheath not prominently developed; style bearing 3 stigmas. SCHIZOSTACHYUM.

4b. Branches at midculm nodes very unequal, the primary (middle) one in each complement more or less strongly dominant.

5a. Culms weak, clambering, not self-supporting, the midculm internodes very long and very thin-walled, fragile and easily crushed; spikelets typically with a single perfect flower.

Spikelets pedicellate, the lemmas and paleas loosely convolute; fruit sulcate, the pericarp moderately thickened at the very apex only, the apex crowned by the pubescent, rather fragile, base of the style. ELYTROSTACHYS.

Spikelets not pedicellate, the lemmas and paleas tightly convolute; fruit not sulcate, the pericarp greatly thickened and indurescent throughout, the apex narrowed and prolonged in a tapered persistent, glabrous beak. OCHLANDRA.

5b. Culms self-supporting, not clambering, the midculm internodes not conspicuously elongate, the walls not fragile or easily crushed.

6a. Rachilla segments very short, not disarticulating.

Spikelets few-flowered, the terminal floret of each spikelet usually well developed, perfect; stamen filaments free. DENDROCALAMUS

Spikelets several-flowered, the terminal floret of each spikelet rudimentary; stamen filaments fused or connected by a membrane. GIGANTOCHLOA.

6b. Rachilla segments elongate—up to half the length of the lemma—promptly or tardily disarticulating.

Keels of palea winged. GUADUA.

Keels of palea not winged. BAMBUSA.

3b. Inflorescence of determinate growth, i.e., emerging complete, the ultimate branches not typical pseudospikelets, and not continuing to branch; spikelets pedicellate; stamens 3; transverse veinlets of leaf blades clearly manifest (obscure in *Thamnocalamus falconeri*).

Lower branches of the inflorescence prophyllate, each subtended by a well-developed bract. THAMNOCALAMUS.

Lower branches of inflorescence not prophyllate, not subtended by a well-developed bract. SINARUNDINARIA.

1b. Rhizomes leptomorph, i.e., the rhizome proper long, "running," generally more slender than the culms that arise from its lateral buds.

7a. Inflorescence of indeterminate growth, the ultimate branches prophyllate pseudospikelets terminating in nonpedicellate spikelets.

8a. Internodes of rhizomes, culms, and branches each strongly flattened or channeled throughout its length on one side—above the point of insertion of a bud or branch complement.

Branches at midculm nodes typically in pairs, robust, strongly unequal; occasionally with a small, third one developed between them. PHYLLOSTACHYS.

Branches at midculm nodes typically 3 to 5 in each complement, slender, very short, subequal. SHIBATAEA.

8b. Internodes of rhizomes, culms, and branches cylindrical or nearly so.

Culm sheaths fugacious, abscissing completely; auricles and oral setae typically very prominently developed (but sometimes lacking entirely) on culm sheaths; bracts at the base of pseudospikelets smaller than the lemmas, without leaflike apical appendage. SINOBAMBUSA.

Culm sheaths abscissing promptly but incompletely, remaining lightly attached at the middle of the base for a short time after they dry; auricles and oral setae lacking on culm sheaths; bracts at the base of pseudospikelets larger than the lemmas, often bearing a small leaflike apical appendage. SEMIARUNDINARIA.

7b. Inflorescence of determinate growth, the ultimate branches terminating in pedicellate spikelets.

9a. Branches typically solitary at midculm nodes.

10a. Culm nodes more or less prominently inflated above the sheath scar; culms commonly tillering, usually ascending, not strictly erect; oral setae of leaf sheaths scabrous; stamens 6 in each flower. SASA.

10b. Culm nodes not prominently swollen above the sheath scar; culms normally not tillering, more or less strictly erect.

Oral setae wholly lacking in leaf sheaths; stamens 6 in each flower. SASAMORPHA.

Oral setae of leaf sheaths few or none, glabrous; stamens 3 (rarely 4 or 5) in each flower. PSEUDOSASA.

9b. Branches typically more than one at each midculm node; stamens 3 in each flower.

11a. Culm sheath blade minute; rapid growth of shoots (new culms) initiated in autumn or late summer. CHIMONOBAMBUSA.

11b. Culm sheath blade well developed; rapid growth of shoots (new culms) initiated in spring to early summer.

Oral setae of leaf sheath scabrous. ARUNDINARIA.

Oral setae of leaf sheath glabrous. PLEIOBLASTUS.

Glossary

". . . . There's a glory for you!"
"I don't know what you mean by 'glory,' " Alice said.
Humpty Dumpty smiled contemptuously. "Of course you don't—till I tell you. I meant 'there's a nice knock-down argument for you!' "
"But 'glory' doesn't mean 'a nice knock-down argument,' " Alice objected.
"When *I* use a word," Humpty Dumpty said, in rather scornful tone, "it means just what I choose it to mean—neither more nor less."
"The question is," said Alice, "whether you *can* make words mean so many different things."
"The question is," said Humpty Dumpty, "which is to be master—that's all."

LEWIS CARROLL

The following terms are defined with special reference to the usages I have adopted for describing various characteristics of the bamboo plant. Several of these terms (indicated by an asterisk*) are newly coined; some others (indicated by a dagger†) are given new or extended meanings to describe conditions not satisfactorily covered by conventional usage, or by botanical glossaries. Published sources give additional (alternative) meanings for some of these same terms—meanings that are in more general (or special) use elsewhere.

It will be obvious to any perceptive person who undertakes to prepare a glossary that his path is beset with many pitfalls. Some of these traps have been laid by uncritical conventional interpretations, or by the inadvertent misuse, of established terminology in technical papers, glossaries, and dictionaries. Other dangers lurk in the labyrinths of semantics and syntax. —The improved light cast by the growing body of recorded observations may have changed, progressively, the mental image of what is to be described. The diversity of the products of morphogenesis, as they appear in the different botanical groups, demands alertness to the need for critical discrimination in the use, or the modification, of the traditional sense of conventional terms. Ambiguities and solecisms may subvert the most determined effort to produce definitions that are both accurate and clear. The end product of this effort may never be pushed beyond the reach of constructive criticism and revision. In any case, unremitting scholarship will not go unrewarded in the long run, for a clear definition of terms is an aid to clear thinking.

Lengthy annotations are appended to some of the definitions in this glossary. These are designed to promote the reader's understanding of the special pertinence of the terms to the morphology of the bamboo plant. They should also bring the bamboos into clearer perspective in relation to the other Gramineae. However, only a first-hand study of the bamboos themselves can adequately illuminate some of the features found exclusively in this group of plants.

Abaxial (L. *ab,* away from; *axis*). Away from an axis;—designates that surface of an appendage (such as a culm sheath or a sheath blade) which faces, in a structural sense, away from the axis on which it is inserted or relies for support. The terms abaxial and adaxial (q.v.) should be employed instead of upper and lower, or dorsal and ventral, whose use—with reference to the two faces of a foliage leaf or a reflexed sheath blade, for example—is apt to perpetrate ambiguity, and so cause confusion. This is because the orientation of such a structure may vary—in time, and from one botanical entity to another—especially among the bamboos. Therefore, neither upper and lower, nor dorsal and ventral, will always be correlated reliably with anatomical dorsiventrality, to which important structural basis all pertinent observations must be oriented unmistakably in order to satisfy the requirements of scientific discipline, dependability, and clear thinking. Cf. *Adaxial.*

Abscissile (L. *abscissilis,* fr. *abscissus,* cut off). Susceptible of being cut off or disarticulated by the formation of an abscission layer;—applies to the petiole of a deciduous leaf or, as in many bamboos, to the segments of a rachilla. See *Deciduous.*

Acropetal (Gr. *akron,* the highest point; L. *peto,* I go toward). Progressing toward an apex, as when the development or emergence of appendages on any axis takes place serially in the direction of the apex of the axis. Cf. *Basipetal.*

Acropetally (adv.). Toward an apex.

Adaxial. Toward an axis;—designates that surface of an appendage which faces, in a structural sense, toward the axis on which it is inserted, or relies for support. Cf. *Abaxial.*

Adventitious (L. *adventicius*). Occurring in a location other than the usual one. In the bamboos, and in other grasses as well, the principal complement of roots is adventitious, arising at the nodes of culms and rhizomes, and not from the primordial root. In the genus *Chusquea* (apparently uniquely among the bamboos) adventitious branch buds arise independently, both at the right and at the left of the one at the usual site, which is median to the base of a subtending culm sheath. See *Constellate.*

Anastomosis (Gr. *anastomōsis,* a new outlet). "Union of one vein with another, the connection forming a reticulation" (Jackson).

Androecium (Gr. *andros,* male; *oikos,* house). The stamen complement of a flower. I have not followed Arber (1934:152 *et passim*) in adopting the spelling (andraeceum) advocated by Kraus (1908). Brown (1954), Jackson, and Nybakken all give androecium. Cf. *Gynoecium.*

†*Antically;* a.w. *anticously* (L., forward, in front). Toward the front; introrsely or adaxially. Jackson records diverse and even antonymous definitions of antical, mentioning the synonymous use of introrse as "occasional." The bamboo palea often gapes antically.

**Apsidate* (L. *apsis,* arch). In convex arcuate array;—indicates the pattern of insertion of the components of a pleioclade branch complement, where the individual axes of the primary array emerge in an arch, having developed from buds that arise distichously on the periphery of a dilated primordium that remains adnate to the surface of the culm—as in all known species of the genus *Merostachys,* and some species of *Arthrostylidium.* See pp. 53ff and Figs. 31 and 32. Cf. *Constellate; Gremial; Restricted.*

Assimilation tissue. Mesophyll;—the interior ground tissue of a leaf blade.

Awned. Bearing a bristlelike apical appendage called an awn (OE.). The empty glumes of the spikelet in the type and some other species of *Neurolepis* are awned. In *Streptochaeta,* the lemma terminates above in a very long, apically coiled structure called an awn (L. *arista*) by Nees, the author of the genus; Trinius called it a tail (L. *cauda*). The distinctions between the ways in which the different types of sheathing structures or their foliar appendages terminate apically are not sharp. Technical terms (such as apiculate, awned, cuspidate, and mucronate) and their conventional definitions do not clearly distinguish the diverse expressions that are a part of the morphological intergradation that baffles efforts at hard-and-fast categorization here, as in many other facets of plant structure.

Bamboo (*bambu,* a vernacular word of undetermined Oriental origin). The wood of bamboo culms; a plant so classified.

Bamboos. A taxonomic group of plants comprising the tribe Bambuseae of the Bambusoideae, a subfamily of the Gramineae; living plants, or culms (stems) severed from plants of this group.

Basipetal (L. *basis,* base; *peto,* I go toward). Progressing toward a base, as when the development or emergence of appendages on any axis takes place serially in the direction of the base of the axis. Cf. *Acropetal.*

Basipetally (adv.). Toward a base.

Blade (AS. *blæd,* leaf). "The limb or expanded portion of a leaf" (Jackson). In the bamboos, a thin, expanded, chlorophyll-bearing, sessile or petiolate, apical appendage (lamina) of a sheath proper. See *Leaf.*

†*Bottlebrush* (bottle brush, a brush of cylindrical shape, with the bristles uniformly distributed on the axis and oriented approximately at right angles to it). A borrowed term, given a slightly modified form, and used here (in the adjectival sense apparently not recorded in dictionaries) to describe the approximate pattern of distribution and orientation of the proliferations of the stigmatic surface in certain bamboos, a pattern loosely called plumose or aspergilliform by some authors. Cf. *Dendroid.*

Bract (L. *bractea,* a thin plate of metal). A sheathing organ, bladeless or laminiferous, found within or immediately below a bamboo inflorescence. A bract may be empty or it may subtend a branch bud or a branch. In bamboo inflorescences of determinate branching, bracts are usually rudimentary or lacking. Prophylla are by some authors referred to as bracts, or bracteoles.

†*Branch complement.* The array of branches that develop at a single culm node, including the primordial one and any that arise by proliferation from buds at its proximal nodes. See Figs. 26, *4,* 27, 29, 30, and 32; see also *Monoclade; Pleioclade.*

†*Branch sheaths.* The sheathing organs borne singly at each node of an aerial vegetative branch of any order (except the culm itself), excluding the neck sheaths and the leaf sheaths.

†*Bud* (ME. *budde*). "The nascent state of a flower or branch" (Jackson). In the usage here adopted, the term bud is applied only to those primordial vegetative or reproductive branches that (1) are enclosed in a prophyllum, and (2) have a resting stage (Figs. 28 and 42). Those (such as branch primordia in determinate inflorescences generally; and root primordia, for example) that lack either one or both of these features are referred to simply as primordia. Branch primordia in the determinate inflorescences of *Glaziophyton, Greslania,* and some other genera are prophyllate, but they apparently have no resting stage.

Caespitose (L. *caespitosus,* fr. *caespes,* a sod). "Growing in tufts, like grass" (Jackson);—describes the normal clump habit of bamboos with pachymorph rhizomes (Fig. 16, *1*), except where the rhizome neck is very much elongated, as in *Melocanna baccifera.* Cf. *Diffuse.*

†*Caryopsis* (Gr. *karyon,* a nut; *opsis,* likeness). "A one-celled, one-seeded, superior fruit, with pericarp united to the seed" (Jackson, under Cariopsis). A caryopsis is indehiscent, that is, its pericarp does not open to liberate the seed. In some bamboos (*Melocanna baccifera,* for example)

the seed is free from the pericarp at maturity. See Figs. 55 and 62.

Centrifugally (L. *centrum,* center; *fugo,* I flee from). Outward from a center.

Centripetally (L. *centrum,* center; *peto,* I go toward). Toward a center.

Chartaceous (L. *charta,* paper; *-aceus,* made of, or belonging to). Paperlike; papery.

† *Circumaxial* (L. *circum,* around; *axis*). Completely around an axis;—describes the typical reach of the locus of insertion of a sheathing appendage, or an assemblage of adventitious roots. Jackson's definition, under Circumaxile, is limited to the special sense, "surrounding a central axis which separates when the fruit splits open."

Circumnutation (L. *circum,* around; *nutans,* wavering). The circular movement that commonly takes place in the distal portion of a growing axis, causing the apex to follow a spiral path as elongation proceeds.

Congruent (L. *congruens,* conformable). In agreement; consistent; harmonious.

Connate (L. *connatus,* born at the same time). "United, congenitally or subsequently" (Jackson), in reference to the union of members of a set of homologous parts (such as the filaments in a stamen complement) that takes place in some bamboos. See *Monadelphous; Diadelphous; Triadelphous.*

Connective (L. *connectivum*). A very narrow strip of tissue upon the approximately opposite sides of which the two parts of an anther (that is, the two pairs of locules) are inserted longitudinally. Jackson describes the connective as "distinct from the filament," of which it appears to be an apical extension.

* *Constellate* (L. *constellatio,* a cluster or group of fixed stars; a constellation). Arranged in a constellationlike cluster. As defined here (in an adjectival sense not found in dictionaries), constellate characterizes the pattern of insertion of branch buds (or branches) where more than one primary bud emerges independently at a midculm node—a condition thus far found among the bamboos only in the genus *Chusquea.* Cf. *Apsidate; Gremial; Restricted.*

Corniform (L. *cornu,* horn). Shaped like a horn. See Fig. 26, *12.*

Culm (L. *culmus,* stalk, stem). A segmented aerial axis that emerges from a rhizome, and forms a part of a gramineous plant;—the term is used most commonly with special reference to bamboos (Fig. 1). Syn.: halm, haulm, haum.

† *Culm node* (L. *nodus,* knot). "The 'knot' of a grass stem" (Jackson). The term node is generally applied in a loose, comprehensive sense (*sensu lato*) to that complex locus (see text, p. 12)—the junction of adjacent internodes in a segmented axis of a gramineous plant. Hackel (Lamson-Scribner and Southworth 1890:2, 3 and Fig. 1) stressed the impor-

tance of the "constantly overlooked" difference between the culm nodes and the sheath nodes in nonbambusoid grasses. However, neither Hackel's description nor his illustration does justice to the distinctive features of these structures as they appear in the bamboos. Culm node is here defined with special reference to that level within the node (*sensu lato*) where secondary elongation (intercalary growth) takes place, and where branch buds and adventitious roots are inserted. Each culm node (*sensu stricto*) is located just above a sheath node (q.v.), from which it is usually distinguishable by the transverse thickening or ridge (the "supranodal ridge" of some authors) that appears at the level of insertion of a bud or a branch complement. Branch node and rhizome node should be similarly differentiated, respectively, from the corresponding sheath nodes. Cf. *Sheath node.*

Culm sheath. One of the sheathing organs borne singly at each node of the culm proper, below the level at which the sheaths of foliage leaves take their place. See Figs. 36 and 37. Kurz (1876:268) anticipates this usage with "halm-sheaths."

Culm shoot. Bamboo shoot; a young culm in any stage of its development short of maturity in height. See *Shoot.*

Cuspidate (L. *cuspis,* a point). Terminating apically in "a sharp, rigid point" (Jackson) in place of a blade (as in some sheathing organs). Cf. *Mucronate.*

Deciduous (L. *deciduus*). Falling off; "falling in season" (Jackson). Cf. *Persistent.*

Declined (L. *declinatus,* turned aside). "directed obliquely" (Jackson);—describes the habit or orientation of the culms in some species of bamboos.

Deliquescent (L. *deliquescens,* melting away, disappearing). Becoming dispersed or expended, as when the main axis of any structure (the rachis of an inflorescence of *Aulonemia quexo,* for example) loses its identity among its own branches, or "loses itself by repeated branchings" (Jackson). Cf. *Excurrent.*

Dendroid (Gr. *dendron,* tree; *-oid,* having the form of). Treelike; bearing branches of more than one order;—describes the pattern of proliferation of the stigmatic surface in certain bamboos.

Depauperate (L. *depauperatus,* impoverished). Reduced in size or functional efficiency or both;—characterizes pseudospikelets, florets, individual organs, or even whole plants whose development is impeded or restricted by adverse conditions, external or internal.

Dermogramme (Gr. *derma,* skin; *gramma,* something drawn). A French term used by Prat (1936:178) to designate any example of his diagrammatic illustrations of the cellular details of the structure of the epidermis of any grass or bamboo. See *Spodogram.*

†*Determinate* (L. *determinatus,* having fixed limits). Of limited growth;—characterizes an axis, or a system of immediately related axes (a determinate inflorescence, for example), whose development or potential for development is confined to a single, definitely limited "grand period of growth," after which no meristem persists, as a rule. The occasional persistence of meristem at the tip of individual rachillas sometimes prolongs the grand period of growth in a determinate inflorescence atypically. A rachilla may then continue to produce new florets until the spikelet is greatly elongated—to as much as a foot, while its normal length would be 3 to 4 inches. Again, this meristem may revert to the vegetative state, prolificating to produce a leafy axis. Both of these teratic phenomena have been observed in *Arundinaria dolichantha* (see p. 103). See *Proliferation* (where prolification is defined), and *Semelauctant;* cf. *Indeterminate.*

Diadelphous (Gr. *dis,* twice; *adelphos,* brother). Having "two groups of stamens";—cognate with Diadelphia, the name of "a Linnean class having the stamens in two bundles of brotherhoods" (Jackson). Diadelphous describes the condition of a stamen complement of a flower when its members have their filaments connate to form two distinct groups. In the bamboos, the relative numbers of stamens in the respective groups may vary from one genus to another. Cf. *Monadelphous; Triadelphous.*

Diageotropic. Adjectival form derived from *diageotropism,* which is defined under *Geotropism* (q.v.).

Diaphragm (Gr. *diaphragma,* a partition wall). The transverse internal layer of parenchyma found at the level of every sheath node. It is reinforced by the crossing over and anastomosis of vascular bundles, and forms a rigid structural element that lends strength (mechanical resilience) to the segmented vegetative axes of bamboos. "A dividing membrane or partition" (Jackson).

Diffuse (L. *diffusus,* spread out). Growing in open array;—characterizes the normal mature clump habit typical of most bamboos with leptomorph rhizomes, and those whose pachymorph rhizomes have a greatly elongated neck, as in *Melocanna baccifera,* for example. This clump habit was referred to earlier (McClure 1945*b*:278) as dumetose. Cf. *Caespitose.*

Distal (L. *disto,* I stand apart). Remote; far out;—designates loci of insertion, or structures, situated at or near the tip of an axis. Cf. *Proximal.*

Distichous (L. *distichus,* consisting of two rows). "Disposed in two vertical ranks, as the florets in many grasses" (Jackson). Distichous also describes the pattern of insertion of buds and branches, and the sheathing appendages on all segmented axes, both vegetative and reproductive, of the bamboo plant. As tacitly recognized by agrostologists, but generally not mentioned in published definitions, the two ranks are normally inserted on opposite sides of an axis.

Distichy (a substantive apparently not yet found in any dictionary). Arber's word for the two-ranked arrangement of buds, branches, and sheathing appendages generally characteristic of segmented axes in the Gramineae. See *Distichous.*

Embryotegium (Gr. *embryon,* a foetus; *tega,* a covering). Literally, embryo cover;—the "little shield" (Sp., *escudete*) of Parodi (1961). "A callosity in the seed coat of some seeds near the hilum, and detached by the protrusion of the radicle on germination" (Jackson). Since Jackson's definition is based on the nature of the embryotegium in deciduous seeds, it is not strictly applicable to its manifestation in a typical gramineous caryopsis. Hackel (Lamson-Scribner and Southworth 1890:20, Fig. 6) describes and illustrates it (without, however, using a technical name) as "a place where the embryo lies covered only by the pericarp and plainly visible on the outside." The position of the embryotegium (as described by Hackel) is clearly visible in caryopses with a thin pericarp, as in the common cereal grains, and in bamboos of certain genera, such as *Arundinaria* and *Bambusa.* In other genera, including *Melocanna* (Stapf 1904:Figs. 3 and 40) and *Ochlandra,* where the pericarp is thickened, the embryotegium is ordinarily not externally recognizable. See Fig. 55.

Empty glumes. Sheathing structures, normally present in every bamboo spikelet (usually 2, sometimes 1 or 3, exceptionally more; rarely lacking entirely) inserted on the rachilla, immediately below the lowest lemma. Empty glumes do not subtend either branch buds or flowers. In some bamboos, the distinction between empty glumes and sterile lemmas is not clear. See *Glume;* cf. *Lemma.*

Endosperm (Gr. *endo,* within; *sperma,* seed). "The nutritive tissue formed within the embryo sac" of a seed (Webster)—referred to by earlier writers as "albumen" (Jackson). In some bamboos, as in *Melocanna baccifera,* the embryo consumes the endosperm as fast as it is formed (Stapf 1904).

Excurrent (L. *excurrens,* running out). Extending through, as when the main axis of any structure (the rachis of the paniculate inflorescence in *Indocalamus sinicus,* for example) maintains its identity among its own branches; "where the stem remains central, the other parts being regularly disposed round it" (Jackson). Cf. *Deliquescent.*

†*Fibrous* (L. *fibra,* thread, filament). Numerous, slender, and not conspicuously tapered;—the meaning of the term as conventionally used to describe the gross aspect of the roots of bamboos and other grasses, a definition apparently not recorded in dictionaries or glossaries. Jackson's definition of the term is not applicable here.

Fistulose (L. *fistula,* a pipe). Hollow; having a lumen;—characterizes the internodes of culms and branches in most bamboos.

†*Floret* (dim. of OF. *flor,* flower). One of the units into which a spikelet breaks up when the rachilla segments disarticulate. Regardless of whether the rachilla segments are abscissile or not, a floret consists of (1) a segment of the rachilla, (2) the lemma that is inserted upon it, (3) a branch (the axis of a flower) subtended by the lemma, (4) a prophyllum (the palea) of the axis of the flower, and (5) the parts of the flower that are inserted on its axis. Attention is directed especially to the fact that the gramineous floret includes structures from axes of two orders, while the flower is confined to a single axis. The distinction between floret and flower is not always observed in published works on agrostology. The loose use of the two terms interchangeably is to be avoided. Cf. *Flower.*

†*Flower* (OF. *flor*). That portion of a branch of the rachilla that is distal to its own prophyllum (the palea), together with the reproductive organs (androecium or gynoecium or both) borne by it (Fig. 53, *B, C*). The lodicules, when present, are included in this concept of the flower, but the palea is not. Cf. *Floret.*

Flowering branch. A leafy or leafless segmented axis that bears one or more inflorescences. A flowering branch is distinguishable from an inflorescence proper in that it retains all of the morphological characteristics of the vegetative state, with the sole exception that in some cases it does not produce foliage leaves. The inflorescence proper, on the other hand, always has morphological features peculiar to it that are not found on the flowering branch proper. In describing and illustrating bamboos, some authors fail to distinguish the two. Arber (1934:108) characterizes a leafless flowering bamboo as a "truly gargantuan inflorescence." Cf. *Inflorescence.*

Foliar organ. See *Sheathing organ.*

Fugaceous (L. *fugax,* fleeting). Promptly deciduous; "soon perishing" (Jackson).

Fusiform (L. *fusus,* spindle). Spindle-shaped; circular in cross section, thickest in the middle and tapering toward each end. See *Subfusiform.*

Geniculate (L. *geniculatus,* with bended knee). "Bent abruptly at an angle, like the bent knee" (Webster). When a growing culm adjusts the direction of its orientation (as in situations where negative geotropism operates to effect the restoration of a fallen or deflected culm to an upright posture), the adjustment takes place through differential elongation in the upper and lower portions, respectively, of the zone of intercalary growth immediately above a node, and the node involved becomes geniculate.

†*Geotropism* (Gr. *gē,* the earth; plus *tropism,* defined by Webster as "the

innate tendency of an organism to react in a definite manner to external stimuli"). A physiological potential, and the consequent act, involved in the assumption and maintenance of a particular direction of growth by an axis or organ with reference to the force of gravity, or to the centrifugal force generated by rapid circular motion. As commonly used, geotropism ambiguously embraces (and confuses) two component phenomena: the visible physical response, and the more fundamental, invisible response, which is physiological. Jackson's definition: "the force of gravity as shown by curvature in nascent organs of plants" is incomplete. Three distinct classes of geotropic responses are recognized: positive, negative, and transverse. The last-named is called *diageotropism.*

† *Girdle* (AS. *gyrdel,* belt). A conspicuous horizontal band of tissue inserted circumaxially at the sheath nodes of some bamboos (*Melocanna compactiflorus* and some species of *Chusquea,* for example; see Fig. 38). The girdle is a visible and anatomically distinct basal part of each sheath proper on any vegetative axis of the plant wherein it occurs; it is particularly conspicuous at the nodes of the culm. As long as the corresponding intercalary zone of the embraced internode is in active growth, a girdle is expansile in all of the dimensions that correspond to its surface, and it takes on an asymmetrical form at geniculate nodes. Upon the abscission of the sheath, the girdle remains attached at the sheath node. The numerous definitions of girdle drawn by Jackson from botanical literature cover a wide diversity of applications given the term, but no one of them corresponds to the one added here. Cf. *Sheath callus.*

Glume (L. *gluma,* a hull or husk). "The chaffy, two-ranked members of the inflorescence of grasses and similar plants" (Jackson). See *Empty glumes.*

Gregarious (L. *gregarius,* of or pertaining to a flock or herd). "Growing in company" (Jackson). Gamble (1896:viii) makes gregarious mean simultaneous, to describe the flowering behavior manifested when all of the members of a given generation of bamboo plants (from seeds of a common origin) enter the reproductive state at approximately the same time. Cf. *Sporadic.*

† *Gremial* (L. *gremialis,* of a lap or bosom). "Growing in a pollard-like cluster" (Brown 1954). The botanical application of the term is extended here to characterize the insertion of a pleioclade branch complement in certain bamboos (*Guadua spinosa,* for example) where a tuft of several slender, subequal branches arises from a basal extension of the primary branch. This apparently unsegmented body of ramiferous tissue remains low and adnate to the surface of the culm. Anatomical studies

of its early stages are needed to clarify the fundamental details of the mature stage of this peculiar structure, but the distinct tuft of branches appears to be independent of those arising distichously from the divergent, clearly segmented, proximal portion of the primary branch proper. Cf. *Apsidate; Constellate; Restricted.*

Gynoecium (Gr. *gynē,* female; *oikos,* house). The complete pistil, consisting of the ovary (the ovulary of Schaffner 1934), the ovule(s), style(s), and the stigma(s), of a single flower (Fig. 53). I have not followed Arber (1934:120 *et passim*) in adopting the spelling (gynaeceum) advocated by Kraus (1908). Jackson and Nybakken both give gynoecium. Cf. *Androecium.*

Halm, haulm, haum (terms used in British agrostology). See *Culm.*

Hermaphrodite. Having both androecium and gynoecium;—designates a flower, or flowers, so equipped: commonly symbolized by ⚥ or ⚦.

Imbricate (L. *imbricatus,* covered with tiles). "Overlapping like roofing tiles and shingles" (Brown 1954). Sheathing organs at successive nodes of an axis are imbricate when each one exceeds, in length, the internode it embraces.

†*Indeterminate* (L. *indeterminatus,* unlimited). "Not terminated absolutely" (Jackson). Continuing apical growth characterizes an individual axis as indeterminate. The term is here given another connotation (namely, iterauctant) with particular reference to the continuing ramification of a system of immediately related axes, as exemplified by the indeterminate inflorescences of certain bamboos. In such inflorescences, the buds at the proximal nodes of each new order of branches bear fresh bodies of meristem. The development of new branches from these buds may be continuous (or intermittent and reactivated on a seasonal basis) for a period lasting in some cases for several successive years. Each branch of an indeterminate inflorescence (an axis with its appendages) has the appearance of a spikelet. Because of this, it is called a pseudospikelet. Each pseudospikelet has its own independent grand period of growth. The active period of the meristem at the apex of the rachilla that terminates each pseudospikelet is limited. Thus, while the branching of indeterminate bamboo inflorescences is of a continuing nature, each individual branch is apically determinate in its growth. See *Iterauctant; Pseudospikelet;* cf. *Determinate.*

Inflorescence (NL. *inflorescentia,* fr. L. *inflorescens,* flowering). A discrete aggregation of spikelets associated with a common primary rachis or a common peduncle. Exceptionally, an inflorescence may comprise but a single spikelet. The inflorescence in the bamboos of most genera is usually either strictly determinate or indeterminate as to its branching habit. The occurrence of individual features of an intermediate nature is rare. Cf. *Flowering branch;* see also *Determinate; Indeterminate; Pseudospikelet.*

Insertion (L. *insertio,* a putting in). "The mode or place of attachment of an organ or a part, as the parts of a flower" (Jackson). A locus of insertion is a point, line, or area that is coextensive with the connection between a given axis or organ and any appendage or structure to which it gives rise. See *Locus.*

**Iterauctant* (L. *itero,* I repeat; *auctans,* increasing, growing). Embracing more than one grand period of growth. The branching system, or the branching habit, of an indeterminate bamboo inflorescence may be said to be iterauctant. Cf. *Semelauctant;* see *Indeterminate; Pseudospikelet.*

†*Leaf* (AS.). Foliage leaf; bamboo leaf; the chlorophyll-bearing, usually petiolate, blade of a leaf sheath proper. This concept of leaf, in the sense bamboo leaf, includes the petiole when it is present, but excludes the leaf sheath proper. Leaf is here defined in this circumscribed sense in order to avoid a commonly encountered ambiguity occasioned by its indiscriminate use in the literature. The word is commonly used by agrostologists, without modification or qualification, to refer to any one of the morphologically, anatomically, and functionally divergent forms of sheathing organs borne on the several kinds of vegetative axes of a bamboo plant. This imprecise use of the term, in disregard of the importance of the manifest differences thus loosely covered by it, apparently is due to the generalized and unqualified nature of the definition of "leaf" given by Jackson: "the principal appendage or lateral organ borne by the stem or axis." Cf. *Sheath; leaf sheath;* see also *Sheathing organ.*

Leaf sheath. One of the leaf-bearing sheaths inserted at the distal nodes of each aerial vegetative axis of a bamboo, whether culm, branch, or twig. The conspicuous part of a leaf sheath is the usually petiolate blade inserted at the apex of the leaf sheath proper. Cf. *Leaf; Sheath.*

Lemma (Gr. a husk). The glume that subtends a flower. A lemma that subtends a functional (or fully formed) flower is called a *fertile lemma* in order to distinguish it from a *sterile lemma,* whose subtended flower is either depauperate, rudimentary, or lacking. Cf. *Empty glumes.*

**Leptomorph* (Gr. *leptos,* thin; *morphē,* form). A term coined especially to designate a slender, elongate type of rhizome proper described on p. 25 (see Fig. 3). Cf. *Pachymorph.*

Ligulate. Having a ligule.

Ligule (L. *ligula,* a little tongue). A thin, apical extension of a sheath proper, adaxial to the locus of insertion of the sheath blade (or leaf petiole). The outer rim of the little cup revealed by the abscission of the petiole of a foliage leaf is sometimes referred to as the "outer ligule," especially when it is conspicuously developed, as it is in some variants of *Arundinaria tecta,* for example. The term "inner ligule" is then applied to the adaxial structure conventionally known simply as the ligule (see Fig. 35). The ligule is reduced to a mere line, or is even lacking entirely, in the sheaths of certain species of bamboos.

Locus (L. place). The locale in which some event takes place, or where something is to be found. Locus is here given the connotation, point, or line, or area, principally in the phrase "locus of insertion." See *Insertion.*

Lodicule (L. *lodicula,* a small coverlet). One of the small, usually thin, delicate and transparent structures (by some authors referred to as "scales") inserted usually in a single whorl of 3, immediately below the stamens in the bamboo flower (Fig. 53). The lodicules are relatively large, opaque, and parchmentlike in the flowers of known species of *Streptochaeta.* Their number is variable in some bamboos (up to 12 or more in *Ochlandra stridula*); in known species of *Gigantochloa,* typical lodicules are lacking entirely.

Lumen (L. opening). "The space which is bounded by the walls of an organ, as the central cavity of a cell" (Jackson); the central cavity of a hollow internode of any segmented axis of a bamboo plant (Fig. 20).

Meristem (Gr. *meristos,* divisible). A body of tissue in which cell division and differentiation are active or potential.

**Metamorph* (Gr. *meta,* implying change; *morphē,* form). Of intermediate form;—a term proposed here to designate certain underground portions of segmented axes, whose transitional character cannot be clearly indicated by the use of existing conventional terms. See *Metamorph I; Metamorph II.*

**Metamorph I.* A transitional axis that occupies a position between the culm neck and the base of the culm proper, where no clearly defined rhizome intervenes;—in some cases doubtfully distinct from an elongated culm neck (Fig. 14).

**Metamorph II.* A transitional axis intermediate in form and position between the apex of a rhizome (either pachymorph or leptomorph) and the culm into which the rhizome is transformed apically. It appears where the transformation of the apex of a rhizome into a culm takes place gradually and not abruptly (Figs. 12 and 16, *2*).

**Metamorphological* (Gr. *meta,* beyond, plus *morphological*). Pertaining to the techniques, the areas of investigation, or the data of disciplines other than morphology.

Monadelphous (Gr. *monos,* one; *adelphos,* brother). United in a single brotherhood;—an adjective cognate with Monadelphia, the name of "a

Linnean class in which the anthers are united by their filaments into a single brotherhood" (Jackson), and conventionally used to describe the stamen complement of a flower when all of its members are united by their connate filaments. Cf. *Diadelphous; Triadelphous.*

Monocarpic. See *Monoperiodic.*

**Monoclade* (Gr. *monos,* single; *klados,* branch). Comprising but a single branch (Fig. 29, *A*);—a term proposed here to designate branch complements so constituted. The monoclade state is facultative when buds are present at the proximal nodes of the primordial branch; it is obligate when buds are lacking at this level. The monoclade state of branch complements is obligate in most known species of the genus *Sasa;* it is facultative in some species of the genus *Arundinaria.* Cf. *Pleioclade;* see also *Branch complement.*

**Monoperiodic* (Gr. *monos,* one; *periodos,* a completed course). Having but a single reproductive cycle within the lifetime of a plant;—characterizes bamboos (such as *Bambusa arundinacea*) that flower but once, then perish. Monocyclic suggests itself for use in the sense here given the newly coined term, monoperiodic, but it already bears the disqualifying connotation annual, with reference to the life span of a plant—besides indicating sets of parts (such as sepals and petals) that comprise but a single whorl or cycle. In the interest of clarity, the term monocarpic, commonly used to indicate the monoperiodic character of certain bamboos, should be reserved (with its orthographic variants; see Jackson) for reference to a pistil comprising but a single carpel. Cf. *Polyperiodic.*

Monopodial (Gr. *monos,* one; *pous, podos,* foot). Having the form of a monopodium, which is defined by Jackson as "a stem of a single, and continuous axis." This term was used earlier (McClure 1925) to designate the type of rhizome described as leptomorph (q.v.). Cf. *Sympodial.*

†*Morphogenesis* (Gr. *morphē,* form; *genesis,* beginning or origin). "The production of morphological characters" (Jackson). As defined here, morphogenesis connotes not only its external manifestations, but also, and primarily, the nature and sequence of events in the program of genic monitoring of physiological states associated with the emergence of the structural features that determine the characteristic form of a plant.

Mucronate (L. *mucronatus,* pointed). "Possessing a short, straight point, as some leaves" (Jackson). Cf. *Cuspidate; Muticous.*

Muticous (L. *muticus,* docked). "Blunt, awnless" (Jackson);—used primarily (when applicable) to describe the apex in small sheathing appendages on the various axes of the bamboo inflorescence.

†*Neck* (AS. *hnecca*). The constricted basal part, characteristic of all, or most, of the segmented vegetative axes of a bamboo plant. See pp. 12f and Figs. 2 and 3.

**Neck sheath.* One of the reduced, bladeless sheaths that clothe the constricted proximal part, the neck, of various vegetative axes in the bamboo plant. Neck sheaths are sometimes referred to as *cataphylls* (Gr. *kata,* down; *phyllon,* leaf).

Obsolete (L. *obsoletus,* worn out). "Wanting or rudimentary; used of an organ which is scarcely apparent or has vanished" (Jackson).

Obterete. Circular in cross section, tapering progressively from one end to the other, and smallest at the proximal end (for example, a rhizome neck). Cf. *Terete.*

**Pachymorph* (Gr. *pachys,* thick; *morphē,* form). A term coined especially to designate a short, thick type of rhizome proper (Fig. 2) described in detail on p. 24. Cf. *Leptomorph.*

Palea (L., [a piece of] chaff). The prophyllum of the axis of a gramineous flower. Jackson's definition: "the inner bract or glume in grasses, called 'Palet' by North American writers" is too vague to be useful. See *Prophyllum.*

†*Panicle* (L. *panicula,* a tuft). A determinate inflorescence with branches of more than one order. Jackson's definition of panicle is not applicable here.

Paniculate (L. *paniculatus,* tufted). Having some or all of the characters of a panicle (q.v.).

Parenchyma (NL., fr. Gr. *parenchein,* to pour in beside). Fundamental tissue; ground tissue; a tissue, such as pith, composed of thin-walled, undifferentiated, isodiametric cells (adapted, from Jackson).

†*Pedicel* (L. *pedicellus,* dim. of *pes,* foot). The stalk of a spikelet, that is, the distal segment or internode of the axis immediately below the glumes that mark the base of a spikelet. The ultimate branches of a determinate inflorescence are all pedicels, since each terminates in a spikelet. In spicate racemes, and in pseudospikelets, the pedicels are usually very short. Cf. *Peduncle; Rachis.*

Pedicellate. Having a pedical.

†*Peduncle* (L. *pedunculus,* dim. of *pes,* foot). The stalk of an inflorescence, that is, the unbranched segment of the inflorescence axis that is immediately below the rachis. In determinate inflorescences, the first sheath-bearing node below the first branch of the inflorescence is here taken arbitrarily as marking the base of the peduncle, and the locus of insertion of the first branch of the inflorescence as marking the apex. In those rare cases where a determinate inflorescence includes only one spikelet, the peduncle is not distinguishable from the pedicel. In indeterminately branching inflorescences, the peduncle is usually very short. An exception appears where the primary pseudospikelet is terminal (instead of lateral) to a flowering branch. Here the peduncle is more elongate (see Fig. 45). Cf. *Pedicel; Rachis.*

Pedunculate. Having a peduncle.

Pergamineous (L. *pergamena,* parchment). "Like parchment in texture" (Jackson).

Persistent (L. *persistens,* remaining in place). Not deciduous;—applies to organs that remain in place after they have fulfilled their natural functions (for example, the culm sheaths of *Arundinaria tecta*).

†*Phototropism* (Gr. *photos,* light; *tropos,* a turning). "A tropism in which light is the orienting stimulus, as in the turning toward a light of a plant shoot or a tube worm, and in the creeping away from a light of a blowfly larva" (Webster). "The act of turning towards the sun or a source of light" (Jackson, under Heliotropism, of which he gives Phototropism as a synonym). As understood and defined here, the term phototropism encompasses a complex of phenomena (visible and invisible) involved in certain biotic responses to incident light. When the corresponding physiological potential is present in the appropriate living tissue of an axis, an organ, or an organism, it induces a movement, or the assumption of a particular orientation, with respect to the principal source of incident light. Two contrasting classes of phototropic responses are recognized: positive and negative. Most published definitions of this and other tropisms fail to make clear the distinction between the two component phenomena, namely, the visible physical response, and the more fundamental, invisible response, which is physiological.

†*Pinnate* (L. *pinnatus,* feathered). "Featherlike" (Webster): having lateral appendages distributed in two continuous series inserted, respectively, on opposite sides of an axis and antrorsely oriented;—a borrowed term used here to describe the pattern of insertion and orientation of the proliferations of the stigmatic surface in certain bamboos. Jackson and Webster both confine the botanical application of the term pinnate to the arrangement of the leaflets on the rachis of a compound leaf.

Pistil (L. *pistillum,* a pestle). See *Gynoecium.*

P.I. Abbreviation of "Plant Introduction," prefixed to a permanent identifying number assigned by the U.S. Department of Agriculture to each lot of living plant material (seeds, plants, or cuttings) accessioned in its record of plant introductions. These numbers are published, along with pertinent names and documenting notes, in a continuing series of Plant Inventories.

**Pleioclade* (Gr. *pleiōn,* more; *klados,* branch). Consisting of more than one branch (Fig. 29, *B–F*);—a term here proposed for designating branch complements so constituted. Cf. *Monoclade;* see also *Branch complement; Pleiogeny.*

Pleiogeny (Gr. *pleiōn,* more; *genos,* birth). "An increase from the parental unit, as by branching or interpolation of members" (Jackson);—a general term for the particular examples described under *Proliferation* and *Pleioclade,* as pertaining to vegetative axes, and under *Indeterminate,* as pertaining to reproductive axes.

Plicate (L. *plicatus,* folded). "Folded into plaits, usually lengthwise" (Jackson); marked by longitudinal ridges suggesting a prior condition of being folded like a collapsible Chinese fan;—an appearance shown by some bamboo leaf blades.

Polycarpic. See *Polyperiodic.*

* *Polyperiodic* (Gr. *polys,* many; *periodos,* a complete course). Having many reproductive periods, alternating with vegetative periods, within the lifetime of one plant;—characterizes bamboos (such as plants of the known species of *Phyllostachys*) that flower repeatedly (and usually periodically) during an indefinite life span. Polycyclic suggests itself for use in the sense here given polyperiodic, but is rejected because of its long-established use with reference to the occurrence of an indefinite number of whorls of parts, such as sepals or petals in a flower. Pleiocyclic is already burdened with the connotation perennial, which is incompatible as a meaning alternate to that here given polyperiodic. Moreover, monocyclic, which would pair with the pleiocyclic, already means annual, in reference to the life span of a plant. In the interest of clarity, the term polycarpic, sometimes used to indicate the polyperiodic character of certain bamboos, should be reserved (with its orthographic variants) for reference to ovaries with numerous carpels (see Jackson). Cf. *Monoperiodic.*

Primordial (L. *primordialis*). "First in order of appearance" (Jackson). See *Primordium.*

Primordium (L., the beginning). An axis or an outgrowth of an axis in its earliest recognizable condition, or in an early dormant state. In the bamboos, a branch primordium borne on a segmented vegetative axis is always enclosed in a prophyllum; an adventitious root primordium is never so enclosed. See *Prophyllum.*

Proliferation (L. *proles,* offspring; *fero,* I bear). "Bearing progeny as offshoots" (Jackson). The term is brought into focus here with reference to the rapid multiplication of members of a branch complement by the prompt awakening of buds at the proximal nodes of the component members. The proliferation of the culm itself by the same process (without the intercalation of a rhizome) is called stooling or tillering. Proliferation should not be confused with prolification, which Jackson differentiates as "The production of terminal or lateral leaf buds in a flower." An example of prolification is cited under *Determinate.* See also *Pleiogeny.*

Prophyllate. Bearing a prophyllum (q.v.).

Prophyllum (L., first leaf); a.w. *prophyll.* A sheathing organ, usually 2-keeled and inserted circumaxially at the first (proximal) node of a branch. In vegetative axes, and in inflorescences of indeterminate branching, the prophyllum at first surrounds the branch primordium to form a bud (Figs. 42 and 43). Jackson's definition of prophyllum limits it to what is now generally called a palea. Prophylla are by some authors called bracts, or bracteoles. See *Palea; Bud.*

Proximal (L. *proximus,* nearest). Basal; situated at or near the base of an axis or an organ;—designates loci of insertion, or structures, so situated. Cf. *Distal.*

* *Pseudospikelet* (L. *pseudo,* false; by extension, superfically resembling; *spicula,* spikelet). A spikeletlike branch of an indeterminate (indeterminately branching) inflorescence (see p. 91 and Fig. 50, *A, B*). See *Indeterminate.*

Pulvinus (L., cushion). A trōpical organ (L. *trōpicus,* pertaining to a turning) associated with movement or differential growth. Pulvini usually manifest themselves as dome-shaped eminences, commonly in pairs, strategically located in relation to the organ with whose orientation they are concerned—at or near the base of the leaf petiole (Fig. 35), for example, or a branch of an inflorescence. Pulvini function through one or the other of two physiological mechanisms: changes of turgidity and differential growth.

† *Raceme* (L., *racemus,* a bunch of grapes). A determinate gramineous inflorescence with a single order of (usually solitary) branches. Jackson's definition is not applicable here. A raceme in which some of the branches emerge in fascicles of 2 or 3 may be characterized (interpreted) as a paniculate raceme, if the extra branches in each fascicle are taken to be secondary ones arising subcutaneously from the base of a primary one. *Athroostachys capitata* bears paniculate racemes in which all branches are relatively short. This gives the inflorescence a capitate superficial appearance. A spicate raceme is one in which the pedicels are so short that the inflorescence resembles a spike.

† *Rachilla;* a.w. *rhachilla* (dim. of *rachis,* q.v.). The axis of a spikelet in any gramineous plant. Jackson defines rhachilla vaguely as "a secondary axis in the inflorescence of grasses."

Rachis; a.w. *rhachis* (Gr. *rhachis,* backbone). The primary axis of an inflorescence;—its position is terminal to the peduncle.

Reduced. Subnormal in size;—connotes also (in some applications) either a failure to fulfill a normal function, or a diminution in the expected number of parts in a set (of stamens, for example). Cf. *Depauperate.*

* *Restricted* (L. *restrictus*). Confined;—describes the insertion of a pleioclade branch complement, when its subsidiary components all arise directly

or indirectly from buds at the proximal nodes of the primordial branch, whether this axis is dominant or not. Cf. *Apsidate; Constellate; Gremial.*

Rhizome (Gr. *rhizōma,* a mass of roots). An individual component branch of the subterranean system of segmented axes that constitute the "chassis" (popularly referred to as the "rootstock") of a bamboo plant (see Fig. 1 and p. 19). A rhizome consists of two parts: the rhizome proper and the rhizome neck. Two distinct types of rhizome are differentiated: leptomorph and pachymorph (qq.v.).

Rhizome sheath. The husklike sheathing organ inserted at each node of a rhizome proper (as distinct from the rhizome neck); see Fig. 34.

Secund (L. *secundus,* following). Unilateral, as describing (1) the arrangement, or pattern of insertion, of appendages on an axis, and (2) the form thus given to the resulting ensemble. Thus, the unilateral arrangement of the spikelets on the rachis in *Merostachys* is described by Sprengel (1825, 1: 132) as a secund spike (*spica secunda*).

**Semelauctant* (L. *semel,* once; *auctans,* increasing, growing). Embracing but a single grand period of growth. A determinate bamboo inflorescence may be said to be semelauctant. See *Determinate; Iterauctant.*

Sensu lato; sensu stricto (L.). In a broad sense; in a restricted sense;—expressions used to indicate the intended scope of a given application of a scientific name or a scientific term.

†*Sheath* (ME. *schethe*). A sheathing organ, the basal part of which, the sheath proper, completely surrounds the vegetative axis on which it is borne, its locus of insertion being circumaxial. In its simplest form (as in typical neck sheaths and rhizome sheaths) the sheath proper terminates apically in a short, hard point. In its fully elaborated form—that characteristic of examples in the middle of any series—the sheath proper usually terminates above in a ligule, at the base of which is inserted a more or less expanded laminar appendage referred to in a comprehensive sense as the sheath blade. In addition, the sheath proper commonly bears, at or near each extremity of the locus of insertion of the blade, a tuft of bristles (oral setae). These may be borne either directly on the sheath proper or on the margin of an auricle. When there is a pair of auricles (one on each side) the two may be similar and subequal, or they may be unequal in size and dissimilar in shape. To avoid the ambiguity occasioned by a commonly encountered indiscriminate usage of the term leaf (q.v.) to refer to any one of the diverse forms of sheaths borne on the vegetative axes of the bamboo plant, two expedients are followed: (1) the adoption of the differential terms rhizome sheath, neck sheath, culm sheath, branch sheath, leaf sheath, and prophyllum (qq.v.); and (2) the differentiation of the sheath

proper from its appendages (Fig. 35). A clear precedent for this effort to avoid ambiguity is found in the now fairly common adoption of a precise terminology to differentiate the several types of sheathing structures borne on the reproductive axes of the bamboo plant, as bracts, prophylla, glumes, lemmas, and paleas. See *Sheathing organ.*

†*Sheath blade.* A distinct foliar part, the lamina, that is appended apically on the laminiferous culm sheaths proper and branch sheaths proper in any series. A sheath blade is distinguishable from a leaf (q.v.), first of all by the relatively proximal position of the sheath proper of the former on any aerial vegetative axis, while leaf sheaths are always inserted at the distal nodes of any culm or branch. Sheath blades are always sessile, while the leaf blades are petiolate (in all known bamboos except some species of *Neurolepis*). Moreover, characteristic differences in shape are usually quite marked. The strong divergence in form, as between leaf blades and sheath blades in the bamboos, is one of the commonest and most useful of the gross morphological distinctions between members of the Bambuseae and those of the other gramineous tribes. See *Leaf.*

†*Sheath callus* (L. *callus,* hard skin). A somewhat prominent ring of parenchymatous (not hard) tissue which remains at a sheath node after the abscission of a sheath (of a culm, especially) in some bamboos (*Phyllostachys nidularia,* for example). Cf. *Girdle.*

†*Sheath node.* The circumaxial locus of insertion of a sheath on any vegetative axis of a gramineous plant (elaborated after Hackel; Lamson-Scribner and Southworth 1890:Fig. 1). The sheath node is marked externally by a more or less prominent offset in the surface of its axis. At the level of each sheath node the crossing over and anastomosis of fibrovascular bundles take place through a diaphragm which marks the internal boundary between adjacent internodes. Cf. *Culm node;* see *Diaphragm.*

†*Sheath scar* (Gr. *eschara,* mark). Jackson defines Scar as "a mark left on a stem by a separation of a leaf, or a seed by its detachment; a cicatrix." In the bamboos, a sheath scar is a narrow, transverse, circumaxial trace, the locus of abscission of a sheath proper. A sheath scar marks the position of a sheath node (q.v.).

†*Sheathing organ.* Any sheathing structure inserted at a node of any vegetative or reproductive axis in a gramineous plant. Among the bamboos, distinguishable types of sheathing organs are rhizome sheaths, neck sheaths, culm sheaths, branch sheaths, leaf sheaths, prophylla, bracts, empty glumes, lemmas, and paleas. Unless used, and interpreted, with discretion, the often-encountered term foliar organ is apt to be ambiguous, since it may be construed as referring either to a sheath proper

alone, to a sheath proper with all of its appendages, to the blade only, or to a bladeless sheathing structure, such as a prophyllum. See *Leaf.*

Shoot (noun, fr. AS. *scēotan,* to move rapidly). "(1) A young growing branch or twig; (2) the ascending axis; when segmented into dissimilar members it becomes a stem" (Jackson). See *Culm shoot.*

Silica bodies; also *silica corpuscles* (G. *Kieselkörper*). Bodies of silica that are secreted and persist within the cells of various tissues of the plant, particularly any epidermal layer. In both size and shape, they range widely. At one extreme, they are small, lack characteristic shape, and occur solitarily or in small numbers within a given cell. At the other extreme, they may be large enough individually to fill a cell more or less completely (see *Silica cells*) and have a characteristic shape. Metcalfe (1960:xlii) lists 20 "types" (forms) of silica bodies, some of which are illustrated in Fig. 1 of the same work. He states (p. xx) that "the silica-bodies in silica-cells assume very characteristic forms when the grass leaf is mature, and are of considerable value for diagnostic and taxonomic purposes." Ohki's key to Japanese genera of bamboos (reproduced on pp. 73f) refers to the occurrence of "silica corpuscles" in the long epidermal cells of the leaf in *Dendrocalamus* [*D. latiflorus*] and in the articulation (bulliform) cells of the leaf in *Sinobambusa* [*S. tootsik*] and *Chimonobambusa* [*Ch. quadrangularis*].

Silica cells. Epidermal short cells each of which is more or less completely filled by a single silica body (Metcalfe 1960:xx).

†*Spicate* (L. *spicatus,* bearing spikes or ears). Having some or all of the characters of a spike (q.v.). A spicate raceme is a raceme in which the pedicels are so short that the inflorescence resembles a spike.

†*Spike* (L. *spica,* an ear of grain). In the bamboos, a spike is a *determinate* inflorescence in which the sessile or subsessile spikelets are inserted on a solitary rachis. Since the distinction between a spike and a spicate raceme is, in the terms of their definitions, only a matter of the relative length of "subobsolete" pedicels, it is sometimes difficult to choose between these terms for the description of a given inflorescence (as in some species of the genus *Merostachys,* for example). According to Jackson's definition (not applicable here) a spike is an "indeterminate inflorescence with flowers sessile on a common elongated axis." A loose interpretation of Jackson's definition of a spike has resulted in the occasional misapplication, in the literature, of the term spike to a bamboo spikelet (q.v.).

†*Spikelet* (L. *spicula*). A basic structural component of every normal gramineous inflorescence, comprising a segmented axis (the rachilla) and its appendages (Figs. 48 and 52). The appendages (beginning with the lowermost) are: empty glumes (usually 2, rarely more, sometimes only

1, rarely lacking entirely), lemmas (either variable or invariable in number, according to the taxon involved), and branches of the rachilla (one subtended by each lemma) each bearing a palea and the parts of a flower. In some taxa, one or more of either the distal or the proximal lemmas, or both, may be sterile by virtue of either being empty or subtending an incompletely developed flower. See *Floret; Flower.*

Spodogram (Gr. *spodos,* ashes; *gramma,* something drawn). A graphic illustration of the microscopic details of epidermal anatomy; or the special preparation on which it is based (Molisch 1920; Ohki 1923). Conventionalized or diagrammatic illustrations of these anatomical features are the "dermogrammes" of Prat (1936:178).

Sporadic (Gr. *sporadikos,* dispersed). "Widely dispersed or scattered" (Jackson); dispersed, or irregular, in time;—as when the individual plants of a given generation of bamboos (from seeds of a common origin) enter the reproductive phase at different times, or at irregular intervals. In either case, the flowering is said to be sporadic (Gamble 1896:viii). Cf. *Gregarious.*

†*Stoma* (Gr., mouth, opening; pl., *stomata*). A functional organ found commonly, but not everywhere, in the epidermis that covers photosynthetic tissues. A stoma consists of two guard cells and (in the bamboos, as far as known) two subsidiary cells (Porterfield 1937). Changes in the turgor of the guard cells result in the opening and closing of a slitlike aperture (the stoma of Jackson) between them.

Sub- (L.). A prefix indicating either an approximation or some reservation or limitation in the use of the term with which it is combined, as in subdistal, suberect, subfamily, subfusiform.

Subfusiform. Of a shape suggesting the concept fusiform, but not corresponding to it precisely; the approximate general shape of a pachymorph rhizome. See *Fusiform.*

†*Subsidary branches.* Branches of higher order arising from buds at the base of a dominant primary branch.

Subtend (L. *subtendo,* I stretch underneath). To precede on a common axis, as a foliar organ precedes (subtends) a bud or branch inserted immediately above it.

Sulcate (L. *sulcatus,* furrowed). "Grooved or furrowed" (Jackson);—as where the otherwise approximately cylindrical shape of the surface of the internode of a segmented axis is modified by one or more longitudinal depressions. In most bamboo genera the palea in functional florets is sulcate, with a single longitudinal dorsal depression.

Suture (F., fr. L. *sutura,* fr. *suere,* to sew). "A junction or seam of union; a line of opening or dehiscence" (Jackson).

Sympodial (Gr. *sym or syn,* together; *pous, podos,* foot). Having the form of a sympodium; for example, a system of related axes wherein successive branches assume the role or position of effective leadership or dominance;—a term used earlier (McClure 1925) to designate the branching habit of the type of rhizome described herein as pachymorph (q.v.). Cf. *Monopodial.*

Taxon (neo-Gr., from Gr. *taxis,* arrangement). "A taxonomic group or assemblage of plants or animals having certain characteristics in common, which we take as evidence of genetic relationship, and possessed of some degree of objective reality" (Rickett 1958). Rickett adds, "We can use the word wherever we can use 'Taxonomic group' in referring to the characteristics, dynamics, distribution, or uses of such an assemblage." Morton (1957) cautions that "where the words 'taxonomic group' cannot be appropriately substituted, the word taxon is misused."

Teratic (Gr. *teras, teratos,* monster, wonder). Abnormal; teratological, in the sense of conventional usage, with particular reference to marked deviations from the normal, or expected, morphological expression. See teratic example of *Prolification,* under *Determinate.*

Terete (L., *teres,* well-turned). Circular in cross section, tapered progressively from one end to the other, and smallest at the distal end or tip (for example, a small bamboo culm). Cf. *Obterete.*

Testa (L., covering, in a poetical sense of the substantive). The outer coat of a seed.

Tillering. "Throwing out stems from the base of a stem" (Jackson); proliferation of a culm from its basal (subterranean) buds, without the intercalation of a rhizome proper (Fig. 14).

Traçant (F., running, creeping). Ranging widely and freely;—a term used by the Rivières (1879:321 *et passim*) to characterize (1) the slender, elongate type of rhizome herein called leptomorph (q.v.), and (2) the bamboos that spread by this means.

Triadelphous (Gr. *treis,* three; *adelphos,* brother). Having "filaments in three brotherhoods";—an adjective cognate with Triadelphia, the name of "a Linnean order of plants with their stamens in three sets" (Jackson), and conventionally used to describe the stamen complement of a flower when its members have their filaments connate to form three distinct groups. In bamboos with triadelphous androecia, the three "brotherhoods" commonly comprise one, two, and three stamens, respectively. Cf. *Monadelphous; Diadelphous.*

Turgidity (L. *turgidus,* inflated). Turgor, turgescence; the firmness imparted by "the distention of a cell or cellular tissue by water or other liquid" (Jackson).

Unilateral (L. *unilateralis,* fr. *unus,* one; *latus,* a side). One-sided. See *Secund.*

† *Vernation* (L. *vernatio,* renewal). "The disposition or method of arrangement of foliage leaves within the bud" (Webster). Jackson's definition, "the order of unfolding of leaf buds," is not applicable here. In most known bamboos, the developing leaf blades are individually rolled up tightly along their long axis, with one edge at the center of the roll.

Zygomorphic (Gr. *zygos,* yoke; *morphē,* form). Symmetrically divisible by a single plane;—"used of flowers which are divisible into equal halves in one plane only, usually the anteroposterior" (Jackson). Normal bamboo flowers conform to this restricted criterion for zygomorphy (see p. 114). According to Jackson, "Sachs extends the meaning to such flowers as may be bisected in any one plane."

Literature Cited

Numbers within brackets that follow each entry indicate the pages and the figure legends where that particular authority is cited. Hitherto unpublished personal communications are also listed here.

Acuña Gale, Julian
1960 Personal communication. [185]

Ahmad, S.
1937 "Two years old bamboo seedling," *Indian Forester 63,* 856–857. [274]

Ahmed, M.
1956 "Note on the comparative success of some common forest species in afforestation in Bihar," *Indian Forester 82,* 528–530. [217]

Alston, A. H. G.
1931 See Trimen 1900.

Arber, A.
1926 "Studies in the Gramineae. I. The flowers of certain Bambuseae," *Ann. Botany (London) 40,* 447–469. [107, 112, 171]

1927*a* "Studies in the Gramineae. II. Abnormalities in *Cephalostachyum virgatum,* Kurz, and their bearing on the interpretation of the bamboo flower," *Ann. Botany (London) 41,* 47–74. [107, 112]

1927*b* "Studies in the Gramineae. III. Outgrowths of the reproductive shoot, and their bearing on the significance of lodicule and epiblast," *Ann. Botany (London) 41,* 479–488. [107, 112]

1928 "Studies in the Gramineae. IV. 2. Stamen-lodicules in *Schizostachyum.* 3. The terminal leaf of *Gigantochloa,*" *Ann. Botany (London) 42,* 181–187. [107, 112]

1929 "Studies in the Gramineae. VIII. On the organization of the flower in the bamboo," *Ann. Botany (London) 43,* 765–781. [107, 112]

1934 *The Gramineae, a study of cereal, bamboo and grass* (Cambridge University Press, Cambridge, England). Some illustrations and passages of text are reproduced with the publisher's permission. [13, 16, 17 (Fig. 5), 18, 48, 61, 67, 79, 85, 98 (Fig. 49), 106, 107, 112, 113 (Fig. 54), 114, 115, 116, 117, 122, 126, 128, 130 (Fig. 59), 164, 297, 303, 305]

1950 *Natural philosophy of plant form* (Cambridge University Press, Cambridge, England). [61, 103, 105, 107]

Ascherson, P. F. A., and P. Graebner
1902 *Synopsis der mitteleuropäischen Flora* (Engelmann, Leipzig), vol. 2, part 1. [281, 282]

Backer, C. A.
1928 *Handboek voor de Flora van Java* (Afl. 2, Ruygrok, Batavia), bamboos, pp. 1–6, 260–289. [179]

Barnard, C.
1957 "Floral histogenesis in the monocotyledons. I. The Gramineae," *Australian J. Bot. 5,* 1–20. [117]

Bean, W. J.
1907 "The flowering of cultivated bamboos," *Kew Bull, Misc. Inform.* (1907), 228–233. [268, 269, 271, 272, 278]

Beddome, R. H.
[1873] *Forester's manual of botany for Southern India* (Gantz Bros., Adelphi Press, Madras), bamboos, pp. ccxxix–ccxxxvi; pl. xxviii. [115, 197, 199, 200]

Bentham, G.
1883 "Gramineae," in G. Bentham and J. D. Hooker, *Genera plantarum* (Reeve, London), vol. 3, part 2. [282]

Bhargava, M. P.
1946 "Bamboo for pulp and paper manufacture. I–III," *Indian Forest Bull.* (n.s., "Utilization"), No. 129 (1945 [1946]). [171, 197]

Bibral, B.
1899 "The flowering of seedlings of *Dendrocalamus strictus,*" *Indian Forester 25,* 305–306. [274]

Blatter, E.
1929 "Indian bamboos brought up to date," *Indian Forester 55,* 541–562; 586–612. [283]
1929, 1930 "Flowering of bamboo. I–III," *J. Bombay Nat. Hist. Soc. 33,* 899–921; *34,* 135–141, 447–467. [84, 171, 196, 273]

Bor, N. L.
1941 "Thyrsostachys oliveri Gamble," Indian Forest Rec. (new series) Botany vol. 2, pt. 2, 221–225. pl. 65–66. [279]

Brandis, D.
1874 *Forest Flora of northwest and central India* (Allen, London); bamboos, pp. 560–570. [171]
1899 "Biological notes on Indian bamboos," *Indian Forester 25,* 1–25. [122, 126, 144, 274]
1906 *Indian trees* (Archbold Constable, London); bamboos, pp. 600–685, 719–720. [84, 162, 199]
1907 "Remarks on the structure of bamboo leaves," *Trans. Linn. Soc. London.* [2, Botany] *7,* 69–92. [71, 72, 75, 120]

Brown, R. W.
1954 *Composition of scientific words* (published by the author). [297, 304, 305]

Brown, W. V.
1958 "Leaf anatomy in grass systematics," *Botan. Gaz. 119,* 170–178. [75]
1960 "The morphology of the grass embryo," *Phytomorphology 10,* 215–223. [120]

Bryan, G. S.
1926 *Edison: The man and his work* (Knopf, New York). [2]

Bush, G. P., and H. H. Lowell, eds.
1953 *Teamwork in research* (American University Press, Washington, D.C.). [6]

Camus, E. G.
1913 *Les Bambusées—monographie, biologie, culture, principaux usages* (Lechevalier, Paris). [162, 263, 282]

Carroll, Lewis
1895 *Through the looking glass and what Alice found there* (Altemus, Philadelphia), chap. vi. [295]

Champion Papers, Inc.
Personal communications. [160, 164, 176, 192]

Chang, F. C.
1938 "A crystalline compound from the white powder found on *Bambusa chungii*," *Lingnan Sci. J. 17,* 617–622. [2, 46]
1941 "I. Alkylation of quinones with esters of tetravalent lead. II. Triterpenoid ketones from *Bambusa chungii*," unpublished dissertation, Harvard University [not seen]. [46]

Chaturvedi, B.
1947 "Aerial rhizomes in bamboo culms," *Indian Forester 73,* 543; Pl. 33, Fig. 1. [247]

Chaturvedi, M. D.
1928 "Influence of overwood on the development of bamboo (*Dendrocalamus strictus*)," *United Provinces Forest Department Bulletin,* No. 1, 1–8. [170]

Church, A. H.
1889 "Food grains of India," *Kew Bull. Misc. Inform.* (1889), 283–284. [171]

Clement, I. D.
1956 [Flowering of *Dendrocalamus strictus* at Atkins Garden, Soledad, Cienfuegos, Cuba,] *Science 124,* 1291. [168]

Cobin, M.
1947 "Notes on the propagation of the sympodial or clump type of bamboos," *Proc. Fla. State Hort. Soc.* (1947) *60,* 181–184. [242, 245]

Creech, J. L.
1950 Personal communication. [156, 241]

Dabral, S. N.
1950 "A preliminary note on propagation of bamboos from culm segments," *Indian Forester 76,* 313–314. [217, 235]

Darlington, C. D., and A. P. Wylie
1956 *Chromosome atlas of flowering plants* (Macmillan, New York); bamboos, pp. 457–458. [199]

Deogun, P. N.
1937 "The silviculture and management of the bamboo *Dendrocalamus strictus* Nees," *Indian Forest Rec.* (n.s., Silviculture) *2,* 75–173. [132, 164, 166 (Fig. 74), 167–171, 201, 204, 205, 211, 216, 235]

Dusén, P.
1903–
1906 "The vegetation of Western Patagonia," in J. B. Hatcher and W. B.

Scott, *Reports of the Princeton University Expeditions to Patagonia, 1896–1899,* vol. 8, pt. 1, "Botany," pp. 1–34. (Published by Princeton University, Princeton, N.J.) [3]

Dutra, J.
1938 "Bambusées de Rio Grande du Sud," *Rev. Sudamericana Botan. 5,* 145–152. [250, 275]

Ezard, C. T. B.
1946 Personal communication. [159]

Federal Experiment Station, Mayagüez, P. R.
1950–
1953 *Annual Report.* [207]

Ferrer Delgado, R.
1948 "La propagación del bambú por esquejes," *Rev. Agr. Puerto Rico 39,* 3–7. [263]

Fibres
1947 "Bamboo pulp: its possibilities for rayon manufacture," *Fibres* [*England*] *8,* 82–84. [197]

Fischer, C. E. C.
1934 *Gramineae,* part X of J. S. Gamble, *Flora of the Presidency of Madras* (Published under the authority of the Secretary of State in Council, London), pp. 1690–1864. [200]

Franchet, M. A.
1889 "Note sur deux nouveaux genres de Bambusées," *J. Botanique 3,* 277–284. [88]

Freeman-Mitford, A. B.
1896 *The bamboo garden* (Macmillan, London). [72, 263, 297]

Gamble, J. S.
1888 "Notes on the small bamboos of the genus *Arundinaria,*" *Indian Forester 14,* 306–314. [271]
1896 "Bambuseae of British India," *Ann. Roy. Botan. Garden Calcutta 7,* 1–133. Atlas, plates 1–119. [67, 83, 89, 115, 158 (Fig. 71), 164, 165 (Fig. 73), 167, 172, 173 (Fig. 75), 177, 188 (Fig. 81), 192, 196, 197, 198 (Fig. 86), 199, 200, 201, 270, 276, 304]
1921 "Flowering of *Arundinaria falcata* in the Temperate House," *Kew Bull. Misc. Inform.* (1921), 302–306. [271]
Unpublished sketch, courtesy of the Keeper of the Herbarium, Royal Botanic Gardens, Kew. [119 (Fig. 55, *13*)]

Garrison, R.
1949 "Origin and development of axillary buds: *Syringa vulgaris* L.," *Am. J. Botany 36,* 205–213; "*Betula papyrifera* and *Euptelea polyandra,*" *ibid.,* 379–389. [62]

Ghinkul, S. G.
1936 "Mass blossoming of bamboo in the light of phasic development theory" [in Russian; English summary], *Soviet Subtropics 10* (26), 24–29. [85]

Giovannoni, M., L. G. C. Vellozo, and G. V. L. Kubiak
1946 "Sobre as 'ratadas' do primeiro planalto paranaense," *Arq. Biol. Technol. 1,* 185–195. [86]

Godron, D. A.
1880 "Les bourgeons axillaires et les rameaux des Graminées," *Rev. Sci. Nat. viii* [n.s., *i*], 429–442. [67]

Gupta, M.
1952 "Gregarious flowering of *Dendrocalamus strictus*," *Indian Forester 78,* 547–550. [172]

Hackel, E.
1881 "Untersuchungen ueber die Lodiculae der Graeser," *Engler Botan. Jahrb. 1,* 336–361; bamboo, p. 357. [114]
1887 "Gramineae," in Engler and Prantl, *Die natürlichen Pflanzenfamilien* (Engelmann, Leipzig), Teil II, Abt. 2, "Bambuseae," pp. 89–97. In the present work, citations of this publication refer to the English translation by Lamson-Scribner and Southworth (q.v.). [48, 282, 299, 302, 314]
1899 "Enumeratio graminum japoniae," *Bull. de l'Herbier Boissier* [old ser.] *7,* 701–726. [70]

Hageman, R. H., R. Ferrer Delgado, and N. F. Childers
1949 "The use of dynamite in lifting bamboo clumps for propagation," *Trop. Agr. 26,* 122–123. [211]

Heslop-Harrison, J.
1959 "Growth substances and flower morphogenesis," *J. Linn. Soc. London 56,* 269–281. [6, 105]

Heyne, K.
1950 *De nuttige Planten van Indonesie* (ed. 3, van Hoeve, The Hague, 1950); bamboos, vol. 1, pp. 285–304. [176, 177]

Higgins, J. E., and W. R. Lindsay
1939 *Annual reports of the Canal Zone Experiment Gardens for the fiscal years 1935 and 1936* (Panama Canal Press, Mount Hope, C.Z.). [275]

Hildebrand, F. H.
1954 "Aantekeningen over Javaanse Bambu-Soorten. Notes on Javanese bamboo species" [in Dutch; English summary], *Laporan Balai Penjelidikan Kehutanan. Indonesia Forest Res. Inst. Rept.* 66. [82]

Hisauchi, K.
1949 "Seedlings and development of rhizomes of a bamboo (*Arundinaria nikkoensis Nakai*)" [in Japanese], *J. Japan. Botany 24,* 24–26. [136, 137 (Fig. 64)]

Holttum, R. E.
1946 "The classification of Malayan bamboos," *J. Arnold Arboretum 27,* 340–346. [83]
1955 "Growth-habits of monocotyledons—variations on a theme," *Phytomorphology 5,* 399–413. [85, 123]
1956*a* "Classification of bamboos," *Phytomorphology 6,* 73–90. [93, 107, 116, 120]
1956*b* "The type species of the genus *Gigantochloa* and some other notes on the genus," *Taxon 6,* 28–30. [179]
1958 "The bamboos of the Malay Peninsula," *Gard. Bull. (Singapore) 16,* 1–135. [70, 96, 106, 120, 148, 177, 179]
Personal communications. [134, 177]

Hooker, J. D.
1854 *Himalayan journals* (John Murray, London); bamboos, vol. 1. [272]
1885 "*Chusquea abietifolia,*" *Curtis's Botan. Mag.* [3] *41,* Tab. 6811. [273]

Horizon
1959 "The debut of the picture review," *Horizon 1,* pt. 4, pp. 68–69. Based on a selection of pictures and texts taken from Peter Pollack's *Picture History of Photography,* and featuring an interview between G. F. Tournachon (as photographer, with his son at the camera) and Marie-Eugène Chevreul, on the eve of the latter's one hundred first birthday. On this occasion, Chevreul quoted words from Malebranche as expressing his own "basic philosophical principle." Chevreul refers to Malebranche by surname only; but it seems reasonable to suppose that the French philosopher Nicolas Malebranche (1638–1715) is the author of the passage, "One must strive for infallibility without pretending to it." [8]

Houzeau de Lehaie, J., ed.
1906–1908 *Le bambou,* Nos. 1–10. [19, 20 (Fig. 7), 263, 271, 274]

Hubbard, C. E.
1948 "Gramineae," in J. Hutchinson, *British flowering plants* (Gawthorn, London). [282]

Hubbard, C. E., and R. E. Vaughan
1940 *The grasses of Mauritius and Rodriguez* (Royal Crown Agents for the Colonies, London). [159]

Hughes, R. H.
1951 "Observations of cane (*Arundinaria*) flowers, seed and seedlings in the North Carolina Coastal Plain," *Bull. Torrey Botan. Club 78,* 113–121. [86, 138, 204]

Humboldt, A., and A. Bonpland
1808–1809 *Plantae aequinoctiales* (F. Schoell, Paris); bamboos, vol. 1, pp. 63–69, pl. 20, 21. [179, 181]

Imle, E.
Personal communication. [125]

Indian Forester
1875– Notices. [85]

Jackson, B. D.
1949 *A glossary of botanical terms* (ed. 4, Duckworth, London). [70 *et passim*]

Jacques-Félix, H.
1955 "Notes sur les Graminées d'Afrique tropicale. VI. Les Graminées Africaines de type archaique," *J. Agr. Trop. Botan. Appl. 2,* 423–430. [75]
1958 "Notes sur les Graminées d'Afrique tropicale. XII. Structure foliaire, écologie et systématique," *J. Agr. Trop. Botan. Appl. 5,* 809–825. [120]
1962 "Les Graminées d'Afrique tropicale. I. Généralités, classification, description des genres," *Inst. Res. Agron. Trop. Paris Bull. Sci.,* No. 8. [75]

Janaki Ammal, E. K.
1938 "Chromosome numbers in sugercane × bamboo hybrids," *Nature 141,* 925. [5, 206]

Japanese Patent No. 175,685
1948 "Deodorization of chrysalis oil or fish oil" [in Japanese], issued to C. Oyama, February 27, 1948. [2]

Jones, Q., and I. A. Wolff
1960 "The search for new industrial crops," *Econ. Botany 14,* 56–68. [149]

Jones, Rebecca
1935 Letter addressed to U. S. Department of Agriculture. [268]

Kadambi, K.
1949. "On the ecology and silviculture of *Dendrocalamus strictus* in the bamboo forests of Bhadravati Division, Mysore State, and comparative notes on the species *Bambusa arundinacea, Ochlandra travancorica, Oxytenanthera monostigma,* and *O. stocksii,*" *Indian Forester 75,* 289–299, 334–349, 398–426. [172]

Kato, K.
1911 "Ueber fermente in Bambusschösslingen," *Hoppe-Seylers Z. Physiol. Chem. 75,* 456–474. [2]

Kawamura, S.
1927 "On the periodical flowering of the bamboo," *Japan. J. Botany 3,* 335–349. [84, 104 (Fig. 51), 278]

Kelsey, H. and W. A. Dayton
1942 *Standardized plant names* (American Joint Committee on Horticultural Nomenclature; J. Horace McFarland Co., Harrisburg, Pennsylvania, ed. 2). [292]

Keng, K. H. [A. W.: Keng Pai-chieh and Keng, f.]
1948 "Preliminary study on the Chinese bamboos" [in Chinese], *Natl. Forestry Res. Bur. China. Tech. Bull.,* No. 8. [22 (Fig. 9), 23, 31]

Keng, Y. L., ed.
1959 *Flora illustralis plantarum primarum sinicarum. Gramineae* [Agency for Science Publications, Peoples Republic of China, Peking], lxxiv + 1181 p., illus., index; text in Chinese; title and scientific names in in Latin. [282]

Kennard, W. C.
1955 "Flowering of the bamboo *Guadua amplexifolia* Presl in Puerto Rico," *Lloydia 18,* 193–196. [86, 275]

Kennedy, P. B.
1899 "The structure of the caryopsis of grasses with reference to their morphology and classification," *U. S. Department of Agriculture, Division of Agrostology, Bull. 19;* bamboos, pp. 35–37. [120]

Kety, S. S.
1960 "A biologist examines the mind and behavior," *Science 132,* 1861–1870. [6]

Koestler, A.
1964 *The act of creation* (Macmillan, New York). [7]

Komatsu, S., and Y. Sasaoka
1927 "Studies in Japanese plants. VIII. The occurrence of free pentose in bamboo shoots," *Bull. Chem. Soc. Japan 2,* 57–60. [2]

Koorders, S. H.
1908 "Contribution no. 1 to the knowledge of the flora of Java (third continuation)," *Proc. (Sec. Sci.) Roy. Acad. Amsterdam 11,* 129–132. [119 (Fig. 55, *13*)]

Kraus, G.
1908 "Gynaeceum oder Gynoecium?" *Verhandl. Physik-Med. Gesellsch. Wurzburg* (N. F.) *39,* 9–14. [From Arber 1934; not seen.] [297, 305]

Kunth, Charles
1815 "Considérations générales sur les Graminées," *Mem. Mus. d'Hist. Nat. (Paris) 2,* 62–75. [282]

Kurz, S.
1875 *Preliminary report on the forest and other vegetation of Pegu. Appendix B* (Baptist Mission Press, Calcutta), bamboos, pp. 91–95 and index. [197]
1876 "Bamboo and its use," *Indian Forester 1,* 219–269, pl. I–II; 335–362, pl. III–IV. [46, 61, 118 (Fig. 55, *1, 6*), 119 (Fig. 55, *10*), 120, 132, 134, 135 (Fig. 63), 159, 196, 227, 276, 288, 300]

Kutty Amma, P. R. Bhagavathi, and T. Ekambaram
1940 "Sugarcane × bamboo hybrids," *J. Indian Botan. Soc. 18,* 209–229. [5]

Lamson-Scribner, F., and E. A. Southworth
1890 *The true grasses,* by Eduard Hackel, translated from *Die natürlichen Pflanzenfamilien* (Holt, New York); Bambuseae, pp. 193–211. References to Hackel (1887) in the present work are given as pages in the English translation by Lamson-Scribner and Southworth. [48, 282, 299, 302, 314]

Li, Hui-lin
1942 "Bamboo and Chinese civilization," *J. New York Botan. Garden 43,* 213–223. [1]

Lin, W. C.
1962,
1964 "Studies on the propagation by level cuttings of various bamboos" [in Chinese with English titles and summary], pts. 1 and 2. *Bull. Taiwan Forestry Res. Institute Nos. 80 and 105.* [234]

Lindsay, W. R.
Personal communication. [275]

Link, H. F.
1833 *Hortus regius botanicus berolinensis* (Reimer, Berlin), vol. 2. [282]

Loh, C. S., T. H. Hu, P. T. Ma, and P. M. Tseng
1950 "A report on *Saccharum* × bamboo hybrids," *Ann. Proc. Agr. Soc. China* [Taiwan] *5,* 13–33, pl. I–IV. [5]

Macleod, A. M. and L. S. Cobley, eds.
1961 *Contemporary botanical thought* (Oliver and Boyd, London). [6]

McClure, F. A.
1925 "Some observations on the bamboos of Kwangtung [China]," *Lingnan Agr. Rev. 3,* 40–47. [21 (Fig. 8), 25, 27, 308, 317]

1928 "Chinese handmade paper," unpublished dissertation, Ohio State University, Columbus, Ohio. [1]
1931*a* "Studies of Chinese bamboos. I. A new species of *Arundinaria* from southern China. Pt. 1. Diagnosis," *Lingnan Sci. J. 10,* 5–10. [152, 153]
1931*b* "Studies of Chinese bamboos. I. A new species of *Arundinaria* from southern China. Pt. 2. Notes on culture, preparation for the market and uses," *Lingnan Sci. J. 10,* 295–305. [153]
1934 "The inflorescence in *Schizostachyum* Nees," *J. Wash. Acad. Sci. 24,* 541–548. [90, 91, 93, 100, 108, 281]
1935 "Bamboo—a taxonomic problem and an economic opportunity," *Sci. Monthly 41,* 193–204. [47 (Fig. 24), 88, 148]
1937 "Bamboo," a tribute to bamboo published in the program announcing the Lingnan University Research Lecture for 1936–37, entitled "Bamboo as a field for research." The lines reproduced herein were composed as an "invitation" to the lecture. [ii]
1938*a* "Notes on bamboo culture with special reference to South China," *Hong Kong Nat. 9,* 4–18. [242, 263]
1938*b* "Diary of a small experimental bamboo planting," *Lingnan Sci. J. 17,* 473–476. [214, 216]
1941 "On some new and imperfectly known species of Chinese bamboos," *Sunyatsenia 6,* 28–51. [70]
1942 "Bamboo genera," in H. Kelsey and W. A. Dayton, eds. *Standardized plant names* (ed. 2; J. Horace McFarland Co., Harrisburg, Pa.); bamboos, pp. 40–41. [292]
1944 *Western Hemisphere bamboos as substitutes for Oriental bamboos for the manufacture of ski pole shafts* (U. S. Natl. Res. Council. Committee on Quartermaster Problems. Final Rept. QMC-24). [150, 187]
1945*a* "Bamboo culture in the Americas," *Agr. Am. 5,* 3–7, 15–16. [25, 263]
1945*b* "The vegetative characters of the bamboo genus *Phyllostachys* and descriptions of eight new species introduced from China," *J. Wash. Acad. Sci. 35,* 276–293. [1, 23, 25, 116, 301]
1945*c* "Bamboo in Ecuador's lowlands," *Agr. Am. 5,* 190–192, 194. [187]
1946 "The genus *Bambusa* and some of its first-known species," *Blumea,* Supp. III, 90–112. [23, 25, 27]
1948 "Bamboos for farm and home," *U. S. Dept. Agr. Yearbook 1948,* 735–740. [2, 159]
1953 *Bamboo as a building material* (U. S. Department of Agriculture, Foreign Agricultural Service, Washington, D.C.). [1, 181 (Fig. 78), 183]
1955 "Bamboos," in J. R. Swallen, "Grasses of Guatemala," in P. C. Standley and J. A. Steyermark, eds., *Flora of Guatemala,* pt. II, *Fieldiana* (Botany), vol. 24, pt. II; bamboos, p. 2 (key to genera) *et passim;* the genera of bamboos and other grasses arranged in one alphabetical sequence. [164, 185]
1956*a* "Bamboo utilization in Eastern Pakistan," *Pakistan J. Forestry 6,* 182–186. Some problems relating to long-term procurement of bamboo for the Karnaphuli Paper Mill. [190]
1956*b* "Bamboo in the economy of oriental peoples," *Econ. Botany 10,* 335–361 [1]
1957 *Bamboos of the genus* Phyllostachys *under cultivation in the United States with a key for their field identification* (USDA Agr. Handbook 114). [52 (Fig. 27), 206]

1958 "Bamboo as a source of forage," *Pacific Sci. Assoc., 8th Pacific Sci. Congr. Proceedings* (Manila, R. P., 1953), vol. iv B, 609-644. [2]

1961*a* "Bamboo," *Encyclopaedia Britannica,* vol. 3, pp. 17-18. Reproduced in revised and emended form, with the publisher's permission. [1]

1961*b* "Toward a fuller description of the Bambusoideae (Gramineae)," *Kew Bull. 15,* 321-324. [282]

1963 "A new feature in bamboo rhizome anatomy," *Rhodora 65,* 63-65. [13, 23]

1964 "Bambusoideae," in B. Maguire, J. J. Wurdack, and collaborators, "The botany of the Guayana highland—pt. V," *Mem. New York Botan. Gard. 10,* 1-6. [42]

McClure, F. A., and W. C. Kennard

1955 "Propagation of bamboo by whole-culm cuttings," *Proc. Amer. Soc. Hort. Sci. 65,* 283-288. [207, 229, 232 (Table 5), 244]

McClure, F. A., and P. F. Li

1941 ["Illustrated guide to selected bamboo genera" (in Chinese)], *Science (China) 25,* 285-292. [39 (Fig. 19)]

McClure, F. A., and F. Montalvo

1949 "*Bambusa longispiculata* easily propagated from stump layers," *Puerto Rico Fed. Expt. Sta. Quart. Rept. Apr./June, 1949,* pp. 102-104, Table 1. [254 (Tables 8, 9), 263]

1950 "Propagation studies," *Puerto Rico Fed. Expt. Sta. Ann. Rept. 1950,* pp. 27-28. [263]

Makino, T., and K. Shibata

1901 "On *Sasa,* a new genus of Bambusaceae and its affinities," *Botan. Mag. Tokyo 15,* 18-31. [81, 94 (Fig. 46)]

Malebranche. See *Horizon*

Mathauda, G. S.

1952 "Flowering habits of bamboo—*Dendrocalamus strictus,*" *Indian Forester 78,* 86-88. [169, 172]

Metcalfe, C. R.

1954 "An anatomist's views on angiosperm classification," *Kew Bull.* (1954), 427-440. [6]

1956 "Some thoughts on the structure of bamboo leaves," *Botan. Mag. Tokyo 69,* 391-400. [72, 75]

1960 *Anatomy of the monocotyledons,* vol. 1, *Gramineae* (Clarendon Press, Oxford, England). Some passages from the text are reproduced with the permission of both the author and the publisher. [75, 315]

Mitford. *See* Freeman-Mitford

Miyake, I., and G. Sugiura

1950,
1951 "Research for the activitive [*sic*] charcoal of Nemagari-bamboo (*Sasa paniculata*) materials. Special research about the carbon of air electric battery. I, II," *Japan. Forestry Soc. J. 32,* 181-185; *33,* 207-211. [2]

Moebius, M.

1898 "Ueber ein eigentümliches Blühen von *Bambusa vulgaris* Wendl.," *Senckenbergische Naturforsch. Ges.* (Frankfurt a. M.), 81-89. [82, 90, 93, 96, 164]

Molisch, H.
1920 "Aschenbild und Pflanzenverwandschaft," *Sitzber. Akad. Wiss. Wein, Math.-Naturw. Kl., Abt. 1, 129,* 261-294; bamboo, p. 277 and Pl. II, Fig. 8. [73, 316]

Mooney, H. F.
1933 "The forests of the Orissa States," *Indian Forester 59,* 200-221. [172]
1938 "A synecological study of the forests of western Singhbhum," *Indian Forest Rec.* [n.s., Silviculture] *2,* 259-356. [172]

Morton, C. V.
1957 "The misuse of the term taxon," *Taxon 6,* 155. [317]

Munro, W.
1868 "A monograph of the Bambusaceae, including descriptions of all the species," *Trans. Linn. Soc. London 26,* 1-157. [84, 85, 89, 106, 114, 116, 117, 271, 272, 273]

Muroi, H.
1956 *Take to sasa* [in Japanese: "Bamboos and sasas"] (Tokyo). [118 (Fig. 55, *2*), 119 (Fig. 55, *16, 17, 19, 21*), 127 (Fig. 57), 138, 139 (Fig. 65), 274]

Nakai, T.
1925 "Two new genera of Bambusaceae with special remarks on the related species growing in eastern Asia," *J. Arnold Arboretum 6,* 145-153. [53]
1933 *Flora sylvatica koreana. Pars 20,* 1-55, pl. i-xiv (Forest Expt. Station, Keijyo, Japan [Korea]). [Text in Japanese and Latin, with some English.] [23, 282]

Nally, Julian
[1953.] Letter to U.S. Department of Agriculture. [271]

Nees, C. G.
1829 "Agrostologia brasiliensis," in C. F. P. Martius, *Flora brasiliensis* (Engelmann, Leipzig), vol. 1, part 1; bamboos, pp. 520-538. [89, 93, 282]
1841 *Florae Africae australioris* (Prausnitz, Glogau), vol. 1, *Gramineae;* bamboos, pp. 460-464. [89]

Netherlands Patent No. 53,471
1942 "Werkwijze voor het uitvoeren van katalytische reacties," issued to N. V. de Bataafsche Petroleum Maatschappij, The Hague, November 16, 1942. [2]

Nicholson, J. W.
1922 "Notes on the distribution and habit of *Dendrocalamus strictus* and *Bambusa arundinacea* in Orissa," *Indian Forester 48,* 425-428. [170]

Nozeran, René
1955 "Contribution a l'étude de quelques structures florales (essai de morphologie florale comparée)," *Ann. Sci. Nat., Botan. Biol. 16,* 1-224. [105]

Nyasaland Forestry Department, Zomba
1944 "The common local bamboo," *Nyasaland Agr. Quart. J. 4* (3), 8-13. [217, 276]

Nybakken, O. E.
1959 *Greek and Latin in scientific terminology* (Iowa State University Press, Ames, Iowa). [297, 305]

Ochse, J. J., in collaboration with R. C. Backhuizen van der Brink
1931 *Vegetables of the Dutch East Indies* (Archipel Drukkerij, Buitenzorg, Java); bamboos, pp. v–xii; 301–327. [177]

Ohki, K.
1932 "On the systematic importance of spodograms in the leaves of Japanese Bambusaceae," *J. Tokyo Imper. Univ. Facul. Sci. (III Bot.) 4,* 1–130. [73, 316]

Oliver, D.
1894 "*Phyllostachys heteroclada* Oliv., sp. nov.," in W. J. Hooker, ed., *Icones plantarum* (Dulau, London), vol. 23, pl. 2288 and one page of text. [117]

Page, V. M.
1951 "Morphology of the spikelet in *Streptochaeta,*" *Bull. Torrey Bot. Club 78,* 22–35. [108]

Parodi, L. R.
1936 "Las Bambuseas indigenas en la Mesopotamia Argentina," *Rev. Argentina Agron. 3,* 229–244. [186]
1945 "Sinopsis de las gramíneas chilenas del género *Chusquea,*" *Rev. Univ. (Univ. Catol. de Chile) 30,* 61–71. [3]
1955 "La floración de la *Tacuara brava* ('*Guadua trinii*')," *Rev. Argentina Agron. 22,* 134–136. [84, 269, 275]
1961 "La taxonomía de las Gramineae Argentinas a la luz de las investigaciones más recientes," *Recent advances in botany* (from lectures and symposia presented to the International Botanical Congress, Montreal, 1959; University of Toronto Press, Toronto), vol. 1, pp. 125–130. [7, 302]

Parthasarathy, N.
1946 "Chromosome numbers in Bambuseae," *Cur. Sci. 15,* 233–234. [167, 206]

Pathak, S. L.
1899 "The propagation of the common male bamboo by cuttings in the Pinjaur-Patiala forest nurseries," *Indian Forester 25,* 307–308. [263, 279]

Pereira, C.
1941 "Sobre as ratadas no sul Brasil e o ciclo vegetativo das taquaras," *Arq. Inst. Biol. (São Paulo) 12,* 175–196. [86, 276]

Piatti, Luigi
1947 "Flüssige Brennstoffe aus bambus," *Schweiz. Arch. angew. Wiss. Tech. 13,* 370–376. [2]

Piedallu, A.
1931 "La culture des bambous," *J. Agr. Prat.* (n.s.) *56,* 293–297. [263]

Pilger, R.
1927 "Ueber die Blütenstände und Aerchen der Bambuseen-Gattung *Guadua* Kunth," *Ber. Deut. botan. Gesellsch. 45,* 562–570. [90, 93]
1937 "Gramineae," in L. Diels, *Beiträge zur Kenntniss der Vegetation und Flora von Ecuador* (Biblioteca Botanica, Heft 116); bamboos, pp. 57–58. [70]
1945 "Additamenta agrostologica," *Engler botanische Jahrbucher 74,* 22, 24. [105]

1954 "Das System der Gramineae," *Engler botanische Jahrbucher 76,* 281–384. Published posthumously; edited by Dr. Eva Potztal; reviewed by Jacques-Felix (1955) and Parodi (1955). [281, 282, 288]

Plank, H. K.

1950 "Factors influencing attack and control of the bamboo powder-post beetle," *Bull. Puerto Rico. Fed. Expt. Sta.,* No. 48, pp. 1–39. [157, 172]

Po Chu-i (A.D. 772–846); see A. Waley

Porterfield, W. M.

1923 "A new feature in vascular anatomy as displayed by bamboo, particularly by the young sheath leaf," *China J. Sci. Arts 1,* 273–279. [69]

1926 "The morphology of the bamboo flower with special reference to *Phyllostachys nidularia,* Munro," *China J. Sci. Arts 5,* 256–260. [98, 99 (Fig. 50), 110 (Fig. 52), 111 (Fig. 53), 277]

1928 "A study of the grand period of growth in bamboo," *Bull. Torrey Botan. Club 55,* 327–405. [91, 123]

1930*a* "Morphology of the growing point in bamboo," *Bull. Yenching Univ. Dept. Biol. 1,* 7–15. [61, 62, 123]

1930*b* "The mechanism of growth in bamboo," *China J. 13,* 86–91, 146–153. [42, 123, 124 (Fig. 56)]

1933 "The periodicity of bamboo culm structures," *China J. 18,* 357–371. [6, 12, 42, 60]

1935 "The relation of shoot roots to shoot elongation in the bamboo *Phyllostachys nigra,*" *Am. J. Botany 22,* 878–888. [80 (Fig. 44, *A*)]

1937 "Histogenesis in the bamboo with special reference to the epidermis," *Torrey Botan. Club Bull. 64,* 421–432. [75, 316]

Prat, H.

1936 "La systématique des graminées," *Ann. Sci. Nat.* [X, Botan.] *18,* 165–258; bamboos, p. 170 *et passim.* [72, 75, 282, 300, 316]

1954 "Gradients histo-physiologiques et perfectionnement organique" (Travaux de l'Institute de Biologie Générale et de Zoologie de l'Université de Montréal, No. 62); extrait des *Rapports et Communications du VIII*^{e} *Congrès International de Botanique, Paris, 1954,* Sec. 8, pp. 294–296. [6]

1960 "Vers une classification naturelle des Graminées," *Bull. Soc. Botan. France 107,* 32–79. [6]

Raizada, M. B.

1948 "A little-known Burmese bamboo [*Sinocalamus copelandi*]," *Indian Forester 74,* 7–10. [119 (Fig. 55, *22*), 269, 270]

Ray, David

Translation. [136]

Rebsch, B. A.

1910 "The bamboo (*Dendrocalamus strictus*) forests of the Ganges Division, U. P.," *Indian Forester 36,* 202–221. [172]

Reeder, J. R.

1961 "The grass embryo in systematics," *Recent advances in botany* (from lectures and symposia presented to the IX International Botanical Congress, Montreal, 1959; University of Toronto Press, Toronto), vol. 1, pp. 91–96. [120]

1962 "The bambusoid embryo: a reappraisal," *Am. J. Botany 49,* 639–641. [120]

Rehder, A.
1945 "Notes on some cultivated trees and shrubs," *J. Arnold Arboretum 26,* 67–78. [282]
1949 *Bibliography of cultivated trees and shrubs* (Arnold Arboretum of Harvard University, Jamaica Plain, Mass.). [282]

Rhind, D.
1945 *The grasses of Burma* (Baptist Mission Press, Calcutta); bamboos, pp. 1–26. [270]

Richharia, R. H., and J. P. Kotwal
1940 "Chromosome number in bamboo (*Dendrocalamus strictus*)," *Indian J. Agr. Sci. 10,* 1033. [167, 206]

Rickett, H. W.
1958 "So what is a taxon?" *Taxon 7,* 37–38. [317]

Rivière, A. and C.
1879 "Les bambous," *Bull. Soc. Acclim.* [III] *5,* 221–253, 290–322, 392–421, 460–478, 501–526, 597–645, 666–721, 758–828. Preprinted (1878) as a separate, continuously paged publication; citations in the present work refer to the 1879 edition, which contains a few minor corrections and emendations. [16 (Fig. 4), 19, 27, 164, 168, 172, 204, 208, 213 (Fig. 87), 215 (Fig. 88), 226, 234, 235, 239, 240 (Fig. 96), 241, 242, 250, 258, 317]

Rollins, Reed C.
1965 "On the bases of biological classification," *Taxon 14,* 6. [264]

Roxburgh, W.
1795–1819 *Plants of the coast of Coromandel* (Nicol, London); bamboos, vol. 1 (1795), col. 56–58; vol. 3 (1819), col. 37–38. [132, 197]

Ruprecht, F. J.
1839 *Bambuseas monographice exponit* (Acad. St. Petersburg), pp. 1–71, pl. 1–18; reprinted, "Bambuseae," *Mem. Acad. St. Petersb.* (ser. VI) *Sci. Nat. 3* (pt. 1), 91–165 (1840). Citations in the present work refer to the 1839 edition. [89, 282]

Satow, E.
1899 "The cultivation of bamboos in Japan," *Trans. Asiatic Soc. Japan 27,* pt. III, 1–127, i–vii. According to Satow, the text of this paper consists mainly of a translation of a Japanese work by Nawohito Katayama, *Nihon chiku-fu* [Manual of Japanese bamboos] (1885). [247]

Schaffner, J. H.
1927 "Control of sex reversal in the tassel of Indian corn," *Botan. Gaz. 84,* 440–449. [6]
1934 "Phylogenetic taxonomy of plants," *Quart. Rev. Bot. 9,* 129–160. [305]

Schomburgk, R. H.
1841 *Reisen in Guiana und am Orinoko wahrend der Jahren 1835–1839,* herausgegeben von O. A. Schomburgk (Georg. Wigand, Leipzig). [42]

Seibert, R. J.
1947 "A study of *Hevea* (with its economic aspects) in the Republic of Peru," *Ann. Missouri Botan. Garden 34,* 261–352. [61]

Seifriz, W.
1920 "The length of the life cycle of a climbing bamboo. A striking case of sexual periodicity in *Chusquea abietifolia* Griseb.," *Am. J. Botany 7,* 83-94. [143, 273]

Sen Gupta, M. L.
1939 "Early flowering of *Dendrocalamus strictus*," *Indian Forester 65,* 583-585. [274]

Sharman, B. C.
1945 "Leaf and bud initiation in the Gramineae," *Bot. Gaz. 106,* 269-289. [62]

Shepherd, W. O.
1952 "Highlights of forest grazing research in the Southeast," *J. Forestry 50,* 280-283. [3]

Shibata, K.
1900 "Beiträge zur Wachstumsgeschichte der Bambusgewächse," *J. Tokyo Imp. Univ. Coll. Sci. 13,* 427-496. [13, 18, 23, 81, 123]

Shigematsu, Y.
1958 "Analytical investigation of the stem form of the important species of bamboo" [in Japanese; English summary], *Bull. Fac. Agr. Univ. Miyazaki 3,* 125-135. [42]

Siebold, Fr. de, and J. G. Zuccarini
1843 "Plantarum quas in japonia collegit Dr. Ph. de Siebold. Genera nova, notis characteristicis delineationibusque illustrata proponunt. Fasc. primus," *Abhandl. Bayer. Akad. Wiss. Math.-Phys. Kl. 3* (3), 745-749, pl. 5. [116]

Sprague, T. A.
1944 In C. R. Metcalfe *et al.,* "On the taxonomic value of the anatomical structure of the vegetative organs of the dicotyledons. Art. 7. A review," *Proc. Linn. Soc. London. 155th Session (1942-43),* pt. 3, 232-235. [6]

Sprengel, K.
1825 *Caroli Linnaei systema vegetabilium* (ed. xvi; Dieterich, Goettingen); bamboos, vol. 1, pp. 132, 249, vol. 2, pp. 112-113. [313]

Stapf, O.
1904*a* "On the fruit of *Melocanna bambusoides,* Trin, endospermless, viviparous genus of Bambuseae," *Trans. Linn. Soc. London* [2, Botany] *6,* 401-425. [121, 190, 197]
1904*b* "Himalayan bamboos. Arundinaria falconeri and A. falcata," *Gard. Chron. 35,* 304-307. [279]

Stebbins, G. Ledyard
1956 "Cytogenetics and evolution of the grass family," *Am. J. Botany 43,* 890-905. [5]

Stern, W. L. and K. L. Chambers
1960 "The citation of wood specimens and herbarium vouchers in anatomical research," *Taxon 9,* 7-13. [147]

Suessenguth, K.
1925 "Ueber die Blüteperiode der Bambuseen," *Flora oder Allgem. Botan. Ztg. 118,* 503-535. [84, 273]

Sung, Ying-hsing, ed.
1929 *T'ien kung k'ai wu* (Commercial Press, Shanghai); a compendium of illustrated articles (in Chinese) describing the arts and industries practiced in China during the Ming Dynasty (1368–1644); "Chinese papermaking," vol. 2, art. 13. [1]

Takenouchi, Y.
1926 "On the rhizome of Japanese bamboos" [in Japanese], *Trans. Nat. Hist. Soc. Formosa 16,* 37–46. [22 (Fig. 9)]
1931*a* "Systematisch-vergleichende Morphologie und Anatomie der Vegetationsorgane der japanischen Bambus-Arten," *Tahoku Imp. Univ. (Formosa) Fac. Sci. Mem. 3,* No. 1 (Botan., No. 2), pp. 1–60. Citations in the present work refer to the pagination in a manuscript English translation by B. Y. Morrison deposited in the library of the U. S. Department of Agriculture. [7, 15 (Fig. 3), 22, 23, 31 (Fig. 13), 32 (Fig. 14), 41 (Fig. 20), 63 (Fig. 34), 66 (Fig. 37), 70 (Fig. 39), 71 (Fig. 40), 72, 73 (Fig. 41), 78, 80 (Fig. 44)]
1931*b* "Morphologische und entwicklungsmechanische Untersuchungen bei japanischen Bambus-Arten," *Mem. Coll. Sci. Kyoto Imp. Univ.* (ser. B) *6,* 109–160. [45, 49]
1932 *Take no kenkyu* [Bamboo studies; in Japanese; with bamboo names in Latin] (Tokyo). Citations in the present work refer to the pagination in a manuscript English translation begun by F. A. McClure and Kam Hok-chau and completed by Saburo Katsura in collaboration with R. A. Young and Mary Schoff, and deposited in the Library of the U.S. Department of Agriculture. Some illustrations and passages of text (in translation) are reproduced with the publisher's permission. [7, 22 (Fig. 9), 23, 29 (Fig. 11), 43 (Fig. 21), 45, 77 (Fig. 42), 78 (Fig. 43), 127, 134]

Takhtajan, A. J. [a. w.: Takhtadzhian]
1959 *Essays on the evolutionary morphology of plants* (Amer. Inst. Biol. Sci., Washington, D.C.). An English translation of the Russian original (Leningrad University, Leningrad, 1954) by Olga Hess, ed. by G. L. Stebbins. [6, 105]

Thimann, K. V.
1954 "Correlations of growth by humoral influences" (Harvard University, Biological Laboratories, Cambridge 38, Mass.); reproduced from *Rapports et Communications du VIIIe Congrès International de Botanique, Paris, 1954,* Sec. 2, pp. 114–128. [6]

Thimann, K. V., and J. Behnke-Rogers
1950 *The use of auxins in the rooting of woody cuttings* (Maria Moors Cabot Foundation Publication, No. 1; Harvard University, Harvard Forest, Petersham, Mass.). [259, 262]

Thomas, G. S.
1957 "Bamboos," *J. Roy. Hort. Soc. 82,* 247–255. [210]

Thompson, J. McL.
1944 "Towards a modern physiological interpretation of flowering," *Proc. Linn. Soc. London. 156th Session (1943–44),* pt. 2, 46–69. [6, 105, 107]

Triana, J. V.
1950 "Algunos metodos de propagacion y medidas de crecimiento en las especies *Guadua angustifolia* y *Bambusa vulgaris,*" *Boletin informativo* (Biblio-

teca del Centro Nacional de Investigaciones de Cafe), No. 11, 26–31. [187]

Trimen, H.

1900 *Handbook of the flora of Ceylon* (Dulau, London), pt. V, *Eriocaulonacae-Gramineae*; "continued by Sir J. D. Hooker"; bamboos, pp. 308–319; pt. VI, *Supplement*, by A. H. G. Alston (1931); bamboos, pp. 341–342. [162]

Trinius, C. B.

1821 "Agrostographische Beyträge," in K. Sprengel, *Neue entdeckungen Pflanzenkunde* (Fleischer, Leipzig), vol. 2, pp. 33–94. [197]

1835 "Bambusaceas quasdam novas describit," *Mem. Acad. St. Petersburg* [ser. VI] *Sci. Nat. 1*, 613. [282]

Trotter, H.

1922 "Development of bamboos from natural seedlings (*Dendrocalamus strictus*)," *Indian Forester 48*, 531–536. [172]

Troup, R. S.

1921 *The silviculture of Indian trees* (Clarendon Press, Oxford, England); bamboos, vol. 2, p. 725; vol. 3, 997–1013. Some illustrations and passages of text are reproduced with the publisher's permission. [128, 129 (Fig. 58), 130, 131 (Fig. 60), 132, 169, 189 (Fig. 82), 190, 196, 197, 205, 236, 274]

Tsuboi, Isuke

1913 *Jikken chikurin zosei-ho* [in Japanese: Experimental method of establishing bamboo groves] (Forestry Association, Gifu Prefecture, Japan). Quotations used herein are drawn from a manuscript translation of pp. 13–23 of the Japanese text by Saburo Katsura, with the collaboration of R. A. Young (1936) under the title "Methods of bamboo propagation," deposited in the library of the U.S. Department of Agriculture. [226, 258]

Udagawa, T., T. Mizuno, and M. Seki

1958 "A field experiment on some rat-poisons" [in Japanese; English summary], *Japan. Min. Agr. and Forestry* (Forest Expt. Sta. Bull. No. 105), pp. 1–10. [86]

Ueda, K.

1960 *Studies on the physiology of bamboo, with reference to practical application* (Bull. Kyoto Univ. Forests, No. 30). [25, 263]

U.S. President's Commission on Increased Industrial Use of Agricultural Products

1957 *Interim report to The Congress, pursuant to Public Law 540* (84th Congress, Washington, D.C.), x, 6, 10, 69, and 14 p. [202]

Usui, H.

1957*a* "Study of the embryo on *Sasa nipponica* and *Pleioblastus chino*," *J. Japan. Botany 32*, 193–200. [120, 138, 140 (Fig. 66)]

1957*b* "Morphological studies on the prophyll of Japanese bamboos," *Botan. Mag. Tokyo 70*, 223–227. [53, 54 (Fig. 28)]

Valt'er, O., T. Krasnosel'skaya, N. Maksimov, and V. Mal'chevskiy

1910 "Ueber den Blausäuregehalt der Bambusschöszlinge," *Bulletin du Département de l'Agriculture aux Indes Néerlandaises 42*, 1–4. This paper is a brief, preliminary communication ("vorläufige Mitteilung"), in German, of results more extensively elaborated in the 1911 paper by the same authors, in Russian. The full name of this apparently rare publication is taken from its title page. [3]

1911 ["The hydrocyanic acid content of bamboo, and its distribution"; in Russian; scientific names in Latin], *Bull. Imp. Acad. St. Petersburg* (1911), 397–426. [3]

Van Overbeek, J.

1944 "Growth-regulating substances in plants," *Ann. Rev. Biochem. 13,* 631–666. [263]

[1945] Personal communication. [238, 257 (Fig. 99)]

Vavilov, N. I.

1940 "The new systematics of cultivated plants," in J. Huxley, *The new systematics* (Oxford University Press, London). Passage from p. 565 quoted with the publisher's permission. [6, 264]

Velasquez, G. T., and J. K. Santos

1931 "Anatomical study of five Philippine bamboos," *Bull. Nat. Appl. Sci. Univ. Philippines 1,* 281–318. [49]

Venkatraman, T. S.

1937 "Sugarcane-bamboo hybrids," *Indian J. Agr. Sci. 7,* 513–514. [5]

Venning, Frank D.

Personal communication [45]

Waley, A.

1929 *A hundred and seventy Chinese poems* (Knopf, New York), p. 214. The first ten lines of Waley's translation of a Chinese poem by Po Chu-i, entitled "Eating bamboo shoots," are reproduced with the publisher's permission. [xvi, 1]

Wardlaw, C. W.

1956 "The floral meristem as a reaction system," *Proc. Royal Soc. Edinburgh* [B, Biology] *66,* 394–408. [6, 105]

1959 "Methods in plant morphogenesis," *J. Linn. Soc. London 56,* 154–160. [6]

1961 "Morphology," in A. M. Macleod and L. S. Cobley, eds., *Contemporary botanical thought* (Oliver and Boyd, London). [105, 143]

Watson, E. V.

1943 "The dynamic approach to plant structure and its relation to modern taxonomic botany," *Biol. Rev. (Gr. Britain) 18,* 65–77. [6, 103, 105]

[Webster, N.]

1959 *Webster's new international dictionary of the English language* (ed. 2, unabridged; Merriam, Springfield, Mass.). In the 1959 printing here cited, and used, the section entitled "Addenda," expanded to incorporate definitions of many words not found in earlier printings, was copyrighted in 1954. [302 *et passim*]

Wendland, J. C.

1810 *Collectio plantarum* (Hahn, Hannover), vol. 2, p. 26, pl. 47. [82, 163]

White, D. G.

1947*a* "Longevity of bamboo seed under different storage conditions," *Trop. Agr. 24,* 51–53. Citations in the present work refer to the reprint repaged, 1–4. [203]

1947*b* "Propagation of bamboo by branch cuttings," *Proc. Am. Soc. Hort. Sci. 50,* 392–394. [242, 243 (Fig. 7)]

1948 *Bamboo culture and utilization in Puerto Rico* (Puerto Rico Fed. Expt. Sta. Cir., No. 29). [204]

White, D. G., and N. F. Childers
1945 "Bamboo for controlling soil erosion," *J. Am. Soc. Agron. 37,* 839–847. [79]

Whitehead, A. N.
1948 *Science and philosophy* (Philosophical Library, New York), p. 131. Passage reproduced with the publisher's permission. [8]

Wilson, C. L.
1942 "The telome theory and the origin of the stamen," *Am. J. Botany 28,* 759–764. [105]

Yoshida, N., and S. Ikejiri
1950 "Bamboo-sprouts extracts as growth-prompting [*sic*] substances for micro-organisms. Studies on the growth-promoting substances for micro-organisms. I" [in Japanese; English summary], *Shikoku Acta Med. 1,* 25–28. [2]

Young, R. A.
1946 "Bamboos in American Horticulture (V)," *Nat'l. Hort. Mag. 25,* 360. [186]
[1960] *American nurseries listing bamboos* (mimeographed, issued irregularly; New Crops Research Branch, Crops Research Division, Agricultural Research Service, U.S. Department of Agriculture, Beltsville, Maryland). [292]

Young, R. A. and J. R. Haun
1961 *Bamboo in the United States: description, culture and utilization,* with key to genera by F. A. McClure (Handbook No. 193; U.S. Department of Agriculture, Washington, D.C.). [289]

Zimmerman, W.
1961 "Phylogenetic shifting of organs, tissues and phases in Pteridophytes," *Canadian J. Botany 39,* 1547–53. The author discusses other groups of plants besides the Pteridophytes. [6, 106]

Index of Scientific Names

For each scientific name mentioned in the text, the authority is given in this index. Many of the bamboos listed are burdened with an extensive synonymy. The inclusion of synonyms here is restricted to those essential to the purposes of this treatise. The word "see" or "equals" follows an obsolete name (one relegated to synonymy), and directs attention to the name currently adopted in its stead. "See also" indicates cases where both the obsolete name and the accepted name for a given species appear in the text.

Index of Subjects

Owing to the abundance of figures related to certain indexed subjects, direct reference to illustrations is omitted in the interest of economy and simplicity. For this reason, some of the page references indicate illustrations only; others indicate text only, or both pertinent text and figure(s).